Mathematica for Bioinformatics

George Mias

Mathematica for Bioinformatics

A Wolfram Language Approach to Omics

George Mias
Department of Biochemistry and Molecular
 Biology, Institute for Quantitative Health
 Science and Engineering
Michigan State University
East Lansing, MI
USA

ISBN 978-3-319-72376-1   ISBN 978-3-319-72377-8 (eBook)
https://doi.org/10.1007/978-3-319-72377-8

Library of Congress Control Number: 2018931483

© Springer International Publishing AG 2018
This work is subject to copyright. All rights are reserved by the Publisher, whether the whole or part of the material is concerned, specifically the rights of translation, reprinting, reuse of illustrations, recitation, broadcasting, reproduction on microfilms or in any other physical way, and transmission or information storage and retrieval, electronic adaptation, computer software, or by similar or dissimilar methodology now known or hereafter developed.
The use of general descriptive names, registered names, trademarks, service marks, etc. in this publication does not imply, even in the absence of a specific statement, that such names are exempt from the relevant protective laws and regulations and therefore free for general use.
The publisher, the authors and the editors are safe to assume that the advice and information in this book are believed to be true and accurate at the date of publication. Neither the publisher nor the authors or the editors give a warranty, express or implied, with respect to the material contained herein or for any errors or omissions that may have been made. The publisher remains neutral with regard to jurisdictional claims in published maps and institutional affiliations.

Printed on acid-free paper

This Springer imprint is published by Springer Nature
The registered company is Springer International Publishing AG
The registered company address is: Gewerbestrasse 11, 6330 Cham, Switzerland

To my family

Preface

This monograph presents an introduction to using the Wolfram Language for Bioinformatics. The content stems from many years of using the Wolfram Language, previously as Mathematica, in different fields: initially as an undergraduate, doing differential equations homework as a freshman, nuclear structure calculations, as a graduate student in theoretical physics calculating Feynman diagrams and working out quantum dynamics in a system, as a postdoc in genetics initially parsing database information and mapped data, and expanding analysis in my own laboratory to mass spectrometry, transcriptomes, protein data, microarray analysis, and more. The intent of this material is to put together various training problems I have used, primarily to get students to begin thinking about a problem, and really look at the details of their solution.

The work builds gradually from basic concepts and introduction of the Wolfram Language and coding paradigms in Mathematica, to building explicit working examples derived from typical research applications using Wolfram Language code. The topics were driven from the daily bioinformatics needs of a broad audience that I have come to contact with during our research endeavors in genetics: the experimental user looking to understand and visualize their data, a beginner bioinformatician acquiring coding expertise in providing biological research solutions, and the practicing expert bioinformatician working on omics that wishes to expand their toolset to utilizing the Wolfram Language, particularly for prototyping solutions.

The Wolfram Language offers an alternative coding solution, particularly for physical or mathematical sciences researchers that may wish to expand their investigation repertoire to include bioinformatics. While R and lately Python may be the languages of choice in bioinformatics for many tasks, particularly because of free availability, I strongly believe that further availability of bioinformatics solutions in multiple languages facilitates thinking about a problem in multiple ways. Every programmer has their own opinion on their favorite language to code in, but I believe there should not be monopolies in this sense. Having worked extensively with R, Python, MATLAB, Swift, Objective C, C/C++, FORTRAN, UNIX, I also have language favorites for different usages and have adapted a lot of the material in this manuscript for these languages. However, I would first say the more material available in any language, the better, and I believe we need additional development in the Wolfram Language for bioinformatics.

The code assumes initially minimal knowledge of the Wolfram Language, but a strong willingness to work through the material as tutorials. The coding paradigms may not necessarily be the best or most optimal solution, but are offered as starting points that will hopefully help the reader expand their own code, and generate their own bioinformatics solutions in the language.

East Lansing, MI, USA George Mias
December 2017

Contents

1	**Prolog: Bioinformatics with the Wolfram Language**		1
	1.1 Bioinformatics in Modern Genetics		1
	1.2 Scope of This Monograph		2
	1.3 Prerequisites		3
	1.4 Presentation of Input and Output and Graphics		3
	1.5 Accompanying Notebooks		5
	References		6
2	**A Wolfram Language Primer for Bioinformaticians**		7
	2.1 Getting Started		7
		2.1.1 Hello World	7
		2.1.2 Hello to Wolfram|Alpha	10
		2.1.3 Getting Started Resources	10
	2.2 Syntax		12
	2.3 Variables		13
	2.4 Basic Mathematical Operations		14
	2.5 Type Casting		16
	2.6 Algebra		17
	2.7 Lists		18
		2.7.1 Collections of Expressions	18
		2.7.2 List Element Extraction	21
		2.7.3 Ordered Lists with Table	22
		2.7.4 Element Manipulation	22
		2.7.5 List Manipulation	23
		2.7.6 Matrices	24
	2.8 Functions		26
		2.8.1 SetDelayed	27
		2.8.2 Transformation Rules	28
		2.8.3 Defining Functions	30
		2.8.4 Options	31

	2.9	Programming Essentials	32
	2.9.1	Comments	32
	2.9.2	Boolean Operations and Flow	32
	2.9.3	Modules, Blocks and With	36
	2.9.4	Patterns	38
	2.9.5	Pure Functions and Slots	40
	2.9.6	Function Manipulation	42
	2.10	Strings	44
	2.10.1	Character Codes	46
	2.10.2	Sequences as Strings	47
	2.11	Importing and Exporting Data	48
	2.11.1	Importing Files	48
	2.11.2	Streams	51
	2.12	Associations and Datasets	54
	2.13	Exceptions	56
	2.14	Graphical Capabilities	57
	2.15	Additional Capabilitities Through Packages	61
	2.15.1	Importing Packages	61
	2.15.2	MathIOmica	62
	References		65
3	**Statistics**		**67**
	3.1	A Primer for Statistics with the Wolfram Language	67
	3.2	Descriptive Statistics	67
	3.3	A Short Probabilistic Sequence of Events	69
	3.4	Examples of Discrete Distributions	72
	3.4.1	Bernoulli Distribution	72
	3.4.2	Binomial Distribution	75
	3.4.3	Geometric Distribution	78
	3.4.4	Poisson Distribution	80
	3.4.5	Hypergeometric Distribution	80
	3.5	Examples of Continuous Distributions	85
	3.5.1	Uniform Distribution	85
	3.5.2	Exponential Distribution	87
	3.5.3	Normal Distribution	88
	3.5.4	Chi-Squared Distribution	94
	3.5.5	Student-t Distribution	95
	3.6	Other Distributions	96
	3.7	Data Sets	97
	3.7.1	ExampleData	97
	3.7.2	ResourceData	104
	3.7.3	Entity Types	106
	3.7.4	Data Example: The Golub ALL AML Data Set	108

		3.7.5	Example: Sandberg et al. Data by Pavlidis	111
		3.7.6	Example: Marcobal et al.	113
		3.7.7	Example Data: MathIOmica	114
	3.8	Hypothesis Testing		115
		3.8.1	Location Tests	116
		3.8.2	A High-Level Approach	120
	3.9	Additional Statistics		120
		3.9.1	Expectations and Correlations	120
		3.9.2	Moments of a Distribution	125
		3.9.3	Distribution Fit	127
	References			131
4	**Databases: E-Utilities and UCSC Genome Browser**			133
	4.1	Connecting to Databases		133
	4.2	NCBI Entrez Programming Utilities		134
		4.2.1	EInfo	134
		4.2.2	ESearch	137
		4.2.3	EPost	147
		4.2.4	ESummary	149
		4.2.5	EFetch	152
		4.2.6	ELink (Entrez Links)	154
		4.2.7	EGQuery	160
		4.2.8	ESpell	161
		4.2.9	ECitMatch	162
		4.2.10	API Keys and Usage Guidelines	162
	4.3	UCSC Genome Browser		163
		4.3.1	Sequence Retrieval	163
		4.3.2	Sequence Retrieval Details	163
	References			170
5	**Genomic Sequence Data and BLAST**			171
	5.1	Sequences of Genomic Information		171
	5.2	Parsing Files		174
		5.2.1	FASTA Sequences	174
		5.2.2	GenBank Records	176
	5.3	Sequence Alignment		179
	5.4	BLAST(n)		182
		5.4.1	BLAST API Parameter List	182
		5.4.2	Running Web Based BLAST Example 1	183
		5.4.3	BLAST Example 2	188
	References			192

6	**Transcriptomics Examples**		193
	6.1	Transcriptomic Analysis	193
	6.2	Golub ALL AML Training Set	193
		6.2.1 Golub Original Set Normalization	195
		6.2.2 Golub ALL AML Combined Dataset	198
		6.2.3 Golub Annotations	203
		6.2.4 Leukemia Differential Expression Analysis	205
	6.3	Analysis of Variance for Multiple Tests	213
	References		225
7	**Proteomic Data**		227
	7.1	Amino Acids	227
	7.2	Protein Information	237
	7.3	UniProt	237
		7.3.1 Individual Entry Retrieval	237
		7.3.2 Query Entry Retrieval	239
		7.3.3 Executing Queries	240
		7.3.4 Random Entry Generation	242
		7.3.5 Identifier Mapping	242
	7.4	NCBI Entrez Utils	243
		7.4.1 Sequence Retrieval	243
		7.4.2 From Sequence to Protein	244
		7.4.3 Search by Molecular Weight	246
	7.5	Proteins Sequence Alignment	246
		7.5.1 SequenceAlignment	247
		7.5.2 BLASTp	248
	References		250
8	**Metabolomics Example**		251
	8.1	Metabolomics Data	251
	8.2	Germ Free and Inoculated Mice Data	251
		8.2.1 Processing Imported Data	252
	8.3	Principal Component Analysis	257
	8.4	Differential Analysis GF Versus 5 Days After Inoculation	262
	8.5	Identifying Compounds Using ChemSpider	268
	8.6	Identifying Compounds and Pathway Analysis: KEGG	274
	References		280
9	**Machine Learning**		283
	9.1	A Taste of Clustering	283
	9.2	Dimensional Reduction	285
	9.3	Classification	286
	9.4	The Iris Data Classified Across Methods	289
	References		296

10 Graphs and Networks ... 297
- 10.1 Introduction to Graphs ... 297
  - 10.1.1 Vertices and Edges ... 297
- 10.2 Basic Graph Construction ... 300
  - 10.2.1 Entering a Graph ... 300
  - 10.2.2 Defining Weighted Graphs ... 305
- 10.3 Basic Graph Properties ... 306
  - 10.3.1 Degree ... 307
  - 10.3.2 Adjacency Matrix ... 308
  - 10.3.3 Weighted Adjacency Matrix ... 310
- 10.4 Some More Definitions ... 311
  - 10.4.1 Walks and Paths ... 311
  - 10.4.2 Graph Geometry ... 312
  - 10.4.3 Centrality ... 313
  - 10.4.4 Clustering Coefficient ... 314
- 10.5 Graph Examples ... 315
  - 10.5.1 Empty Graphs ... 315
  - 10.5.2 Complete Graphs ... 316
  - 10.5.3 Regular Graphs ... 318
  - 10.5.4 Cycle Graphs ... 319
  - 10.5.5 Trees ... 321
  - 10.5.6 Bipartite Graphs ... 321
- 10.6 Isomorphisms ... 322
- 10.7 Random Graphs ... 324
  - 10.7.1 Barabasi Albert Distribution ... 325
  - 10.7.2 Watts Strogatz ... 326
- References ... 328

11 Time Series Analysis ... 329
- 11.1 Time Series ... 329
  - 11.1.1 The TimeSeries Function ... 329
  - 11.1.2 Multiple Time Series Example ... 333
- 11.2 Vignette: FinancialData ... 338
- 11.3 Time Series Model Fitting ... 339
  - 11.3.1 USA BMI Data Modeling ... 341
- 11.4 Unevenly Sampled Time Series Classification Examples ... 349
  - 11.4.1 Lomb Scargle Classification in MathIOmica ... 349
  - 11.4.2 Classification Simulation Example ... 350
  - 11.4.3 Classification RNA Sequencing Data Example ... 353
  - 11.4.4 Heatmaps and Dendrograms ... 364
- References ... 372

12 Epilog: Bioinformatics Development with Mathematica 375
    12.1   Bioinformatics Development With the Wolfram Language 375
    12.2   Loading Packages 375
    12.3   The examplePackage 376
            12.3.1   Contexts 377
    12.4   Odds and Ends 379
            12.4.1   More Information on Packages 379
            12.4.2   Dynamic Interfaces and Manipulate 379
    12.5   The Wolfram Language Community 380
    References 380

Index 381

Acronyms

AIC	Akaike Information Criterion
ALL	Acute Lymphoblastic Leukemia
AML	Acute Myeloid Leukemia
ANOVA	Analysis of Variance
API	Application Program Interface
BLAST	Basic Local Alignment Search Tool
BLOSUM	BLOcks SUbstitution Matrix
BMI	Body Mass Index
CSV	Comma-Separated Values
DNA	Deoxyribonucleic Acid
E-utilities	Entrez Programming Utilities
FDR	False Discovery Rate
FPKM	Fragments Per Kilobase of transcript per Million mapped reads
GF	Germ-Free
InChI	International Chemical Identifier
iPOP	integrative Personal Omics Profiling
IUPAC	International Union of Pure and Applied Chemistry
JSON	JavaScript Object Notation
lhs	Left-hand side of an assignment
NB	Nota bene
NCBI	National Center for Biotechnology Information
ORA	Over-Representation Analysis
PAM	Point Accepted Mutation
PCA	Principal Component Analysis
PCR	Polymerase Chain Reaction
PDB	Protein Data Bank
PDF	Probability Density Function
pmf	Probability mass function
ppm	Parts per million
REST	Representational State Transfer

rhs	Right-hand side of an assignment
RID	Request Identity
RNA	Ribonucleic Acid
RSV	Respiratory Syncytial Virus
SMILES	Simplified Molecular Input Line Entry Specification
SQL	Structured Query Language
T2D	Type 2 Diabetes
TSNE	t-distributed Stochastic Neighbor Embedding
TSV	Tab-Separated Values
URI	Uniform Resource Identifier
XML	Extensible Markup Language

Chapter 1
Prolog: Bioinformatics with the Wolfram Language

1.1 Bioinformatics in Modern Genetics

With the advent of the 21st century and the unparalleled advancements in computational capabilities and technologies we are witnessing a new era in quantitative science, and particularly genetics. Data has become an indispensable part of the daily work of not only computational scientists but also wet lab experimentalists who produce the data necessary for computation, without which no modeling would be possible.

So-called omics technologies are here to stay. While it is possible to measure a small set of genes, a few proteins, a few small molecules, we now have the capabilities to study components at a systems levels, and obtain data on thousands of constituents at the same time. Such big data availability renders bioinformatics a necessity for the modern geneticists, who must have the skills to preprocess, statistically analyze and interpret such data. The modern bioinformatician is a Jack of all trades, dabbling in computer science, statistics, mathematics, signal processing and engineering, and genetics.

The shift to data-driven hypothesis-generating research is still a work in progress, as the more traditional hypothesis driven approaches are supplemented with new systems level approaches. The concept of bioinformatics has expanded over the last decade or so. The development of algorithms and computational biology approach is now coupled with informatics and the ever-increasing use of software tools to handle omics data, particularly *Big Data* as generated from high-throughput approaches for sequencing, mass spectrometry and imaging. Additionally, the increasing use of machine learning combined with data abundance invites the use of computational tools for classification and prediction to genetics.

Electronic supplementary material The online version of this chapter (https://doi.org/10.1007/978-3-319-72377-8_1) contains supplementary material, which is available to authorized users.

© Springer International Publishing AG 2018
G. Mias, *Mathematica for Bioinformatics*,
https://doi.org/10.1007/978-3-319-72377-8_1

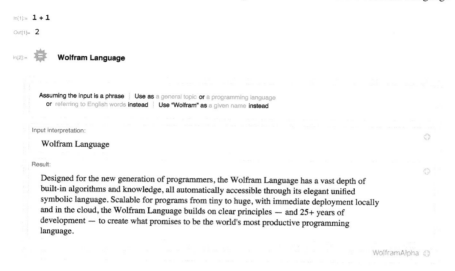

Fig. 1.1 The Wolfram Language will be used for coding purposes and approaches to bioinformatic analyses. Here the description of the language on Wolfram|Alpha [11] is displayed

1.2 Scope of This Monograph

In this work we will touch on aspects of bioinformatics and its use through the Wolfram Language (Fig. 1.1), as used through Mathematica [11, 12] to address certain problems in omics analyses. The overview and examples are biased by the author's interests, and also usage over time and present a personalized view to what is necessary to beging tackling bioinformatic analyses in the Wolfram Language. The book is not a theoretical treatise on bioinformatics or algorithm development, but rather a pragmatic hands-on introduction to what one may do using the Wolfram Language in a variety of areas in bioinformatics. As such, code is presented in detail (perhaps occasionally too much detail) including inputs and outputs throughout the book.

After an introduction to the Wolfram Language, we explore how we can carry out statistical computations using the language. Following this we introduce extensively how to access databases by the NCBI (National Center for Biotechnology Information) [6], and the University of California at Santa Cruz Genome Browser [3] in Chap. 4. Then we discuss how to carry out sequence alignments in the Wolfram Language, and use external programs such as BLAST [1, 9] through a notebook interface in Chap. 5. We discuss access to the UniProt [10] database for proteomics, as well as protein sequence analysis in Chap. 6. We analyze transcriptome [2, 7, 8], Chap. 7 and metabolomics [4] data Chap. 8, including considerations for multiple hypothesis testing and analysis of variance with multiple factors. We introduce the availability of machine learning algorithms through the Wolfram Language in Chap. 9, which is a topic of great current interest in bioinformatics. In Chap. 10 we have a basic introduction of the use of the Wolfram Language's graph and network analysis abilities, and in Chap. 11 we discuss the analysis of dynamic data using time series methods, as well as extensions to omics datasets with missing data/uneven sampling.

The exposition of material is heavily influenced by the writer's interests and remains solely within his experiences with using the software. As such, we will not discuss other packages that are available for bioinformatics in the Wolfram Language, or otherwise, and will not broach large subjects such as crystallography, image analysis, sequence mapping algorithms, and many more.

The goal is not to provide statistics or mathematical support, but encourage the user to directly work with code in the Wolfram Language and adapt the components they might find useful in their own work.

1.3 Prerequisites

There are minimal prerequisite for this book, beyond the willingness to work through the code. Some knowledge of genetics and the use of bioinformatics will be useful in understanding the background behind the examples which is not discussed in detail. No previous knowledge of the Wolfram Language is required, though programming experience would be beneficial.

1.4 Presentation of Input and Output and Graphics

The code in the book is presented in a series of input and output statements, made to resemble Mathematica notebook cells, and the numbering corresponds to the order that these would evaluate in the accompanying notebooks. The cells look like:

In[number]:= code to be executed
Out[number]= expected output

Graphical output is represented in Figures. Each Figure corresponding to an output has the corresponding **Out**[number] = label included in the caption as a reference. Please note that all graphical output in this manuscript was generated in the Wolfram Language using Mathematica. Graphics were post-processed in Adobe Illustrator and modified for size and generation of eps figures. The content of the output has not been modified.

Wolfram Language Evaluations. We will be using the Wolfram Language throughout this book. The code for each chapter is available in the *notebooks* accompanying this monograph. A basic introduction to the language is presented in the next chapter. Here, we provide just enough information to get you started and also to be able to evaluate the code in this chapter.

A *notebook* is the main working/coding document for Mathematica (extension .nb). After you start Mathematica on your computer you can open a working document for coding notebook is displayed. Alternatively a new notebook can be started using the application menus by selecting File → New → Notebook. In notebooks (extension .nb), code is put in input chunks in cells, outlined in light brackets (visible by hovering the mouse at the right edge of a notebook). Output is produced directly below the input, also outlined with a cell bracket, and output and input pairs are outlined with an overall external cell bracket. To create a new cell in a notebook move the cursor within the notebook until it appears as horizontal and click. A horizontal bar with then appear and a new cell will be created in that location where we can now type new code.

To evaluate a cell with input code, as long as your cursor is within the cell containing the code, or the cell is selected, you can type **Shift+Enter** simultaneously and the code will be evaluated. Evaluated code will be prepended with **In**[i]:= input code, where i denotes the ith evaluation. The corresponding ith output will begin with **Out**[i]= evaluation results. The tag i will be incremented automatically for the same coding session. For example,

In[1]:= 1 + 1
Out[1]= 2

In addition to entering standard Wolfram Language commands to be evaluated in the system, other forms of input allow Mathematica to interface with Wolfram|Alpha technology by prepending code input with equal signs (Fig. 1.2):

- A single equal sign followed by code, = code. The equal sign becomes an orange square equal sign and evaluation returns the result and Wolfram Language syntax.
- A double equal sign followed by code, == code: returns the Wolfram|Alpha results.
- Control and an equal sign followed by code, Ctrl+= code : inserts a free-form input Wolfram Language expression that can be used further.

We can get information regarding DNA bases using Wolfram|Alpha (Fig. 1.3).

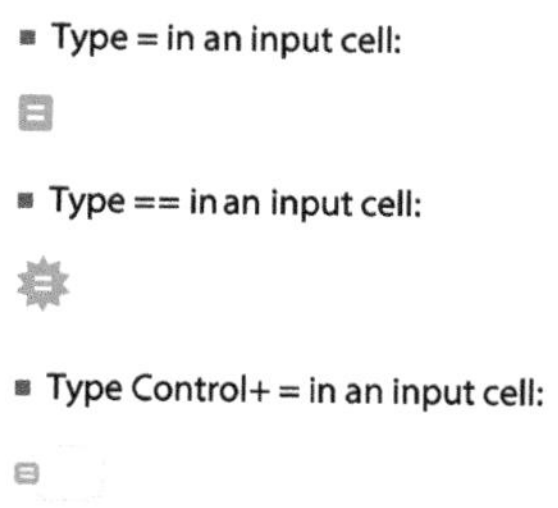

Fig. 1.2 Multiple input methods can be used to allow Mathematica to interface with Wolfram|Alpha technology. This can be achieved by entering equal signs in front of code input as shown to the right

- Type = in an input cell:

- Type == in an input cell:

- Type Control+ = in an input cell:

Fig. 1.3 We can use Wolfram|Alpha to get useful information about deoxyribonucleic acid (DNA). For example, we can look at the structure fo the DNA bases, adenine, guaninine thymine and cytosine by searching for the terms, first typing == and the keywords "dna bases structure"

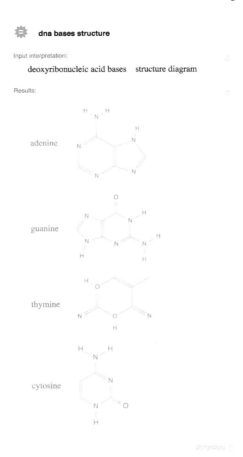

1.5 Accompanying Notebooks

To facilitate the approach taken, a set of notebooks are provided to be used to accompany the manuscript. These are currently available on github at:

"https://github.com/gmiaslab/MathematicaBioinformatics/releases"

The notebooks contain early drafts of the text, the entirety of the code and output from the manuscript, as well as all the files that are used in the manuscript.

Additionally, the manuscript uses the package MathIOmica [5], beginning with the end of Chap. 2 and for various examples. This is also available on github at:

"https://github.com/gmiaslab/mathiomica/releases"

The notebooks' code when evaluated in order correspond to all inputs and outputs in the text. On occasion a lot of output is shown, and code explained in detail, highlighting points that students in particular might have difficulty in figuring out the commands used. The choice to provide all the output was made to avoid leaving implementations to the reader. As such, we are taking a very pragmatic and practical approach in the book, that will hopefully provide insights into the use of Wolfram Language code that underlies some bioinformatics analysis.

References

1. Altschul, S.F., Gish, W., Miller, W., Myers, E.W., Lipman, D.J.: Basic local alignment search tool. J. Mol. Biol. **215**(3), 403–410 (1990)
2. Chen, R., Mias, G.I., Li-Pook-Than, J., Jiang, L., Lam, H.Y., Chen, R., Miriami, E., Karczewski, K.J., Hariharan, M., Dewey, F.E., Cheng, Y., Clark, M.J., Im, H., Habegger, L., Balasubramanian, S., O'Huallachain, M., Dudley, J.T., Hillenmeyer, S., Haraksingh, R., Sharon, D., Euskirchen, G., Lacroute, P., Bettinger, K., Boyle, A.P., Kasowski, M., Grubert, F., Seki, S., Garcia, M., Whirl-Carrillo, M., Gallardo, M., Blasco, M.A., Greenberg, P.L., Snyder, P., Klein, T.E., Altman, R.B., Butte, A.J., Ashley, E.A., Gerstein, M., Nadeau, K.C., Tang, H., Snyder, M.: Personal omics profiling reveals dynamic molecular and medical phenotypes. Cell **148**(6), 1293–307 (2012)
3. Kuhn, R.M., Haussler, D., Kent, W.J.: The ucsc genome browser and associated tools. Br. Bioinform **14**(2), 144–61 (2013)
4. Marcobal, A., Yusufaly, T., Higginbottom, S., Snyder, M., Sonnenburg, J.L., Mias, G.I.: Metabolome progression during early gut microbial colonization of gnotobiotic mice. Sci. Rep. **5**, 11,589 (2015)
5. Mias, G.I., Yusufaly, T., Roushangar, R., Brooks, L.R., Singh, V.V., Christou, C.: Mathiomica: an integrative platform for dynamic omics. Sci. Rep. **6**, 37,237 (2016)
6. NCBI Resource Coordinators: Database resources of the national center for biotechnology information. Nucleic Acids Res. **45**(D1), D12–D17 (2017)
7. Pavlidis, P.: Using anova for gene selection from microarray studies of the nervous system. Methods **31**(4), 282–289 (2003) (Candidate Genes from DNA Array Screens: application to neuroscience)
8. Sandberg, R., Yasuda, R., Pankratz, D.G., Carter, T.A., Del Rio, J.A., Wodicka, L., Mayford, M., Lockhart, D.J., Barlow, C.: Regional and strain-specific gene expression mapping in the adult mouse brain. Proc. Natl. Acad. Sci. **97**(20), 11038–11043 (2000)
9. Schäffer, A.A., Aravind, L., Madden, T.L., Shavirin, S., Spouge, J.L., Wolf, Y.I., Koonin, E.V., Altschul, S.F.: Improving the accuracy of psi-blast protein database searches with composition-based statistics and other refinements. Nucleic Acids Res. **29**(14), 2994–3005 (2001)
10. UniProt, C.: Uniprot: a hub for protein information. Nucleic Acids Res. **43**(Database issue), D204–12 (2015)
11. Wolfram Alpha LLC: Wolfram|Alpha (Access November 2017) (2017)
12. Wolfram Research, Inc.: Mathematica, Version 11.2. Champaign, IL (2017)

Chapter 2
A Wolfram Language Primer for Bioinformaticians

2.1 Getting Started

Mathematica [9] was released in 1988 by Wolfram Research (founded in 1987 by Steven Wolfram). The core of Mathematica is built on the Wolfram Language [7] which can be used to code programs. The Wolfram Language has broad applicability in multiple areas. Through using Mathematica for development and computation in the Wolfram Language, we can easily combine input and output as well as interactive information.

The Wolfram Language is a high level language, and can be used for both procedural and function programming styles. The easiest way to learn more about the Wolfram Language is through the Documentation in Mathematica. This can be found by selecting from the menu Help → Wolfram Documentation. The documentation is highly interactive and provides examples that can be executed in place, Fig. 2.1.

2.1.1 Hello World

As briefly introduced in the previous chapter, the main working/coding document for Mathematica is a *notebook*, with extension .nb. After you start Mathematica on your computer you can create a new notebook by using the application menus to select File → New → Notebook. Code is written in input cells, outlined in light brackets (visible by hovering the mouse at the right edge of a notebook). Output is produced directly below the input, also outlined with a cell bracket. Furthermore, input - output cell pairs are grouped by having an overall external cell bracket. Input and Output are actually cell styles, corresponding to input code and output results. In addition, other

---

Electronic supplementary material   The online version of this chapter (https://doi.org/10.1007/978-3-319-72377-8_2) contains supplementary material, which is available to authorized users.

© Springer International Publishing AG 2018
G. Mias, *Mathematica for Bioinformatics*,
https://doi.org/10.1007/978-3-319-72377-8_2

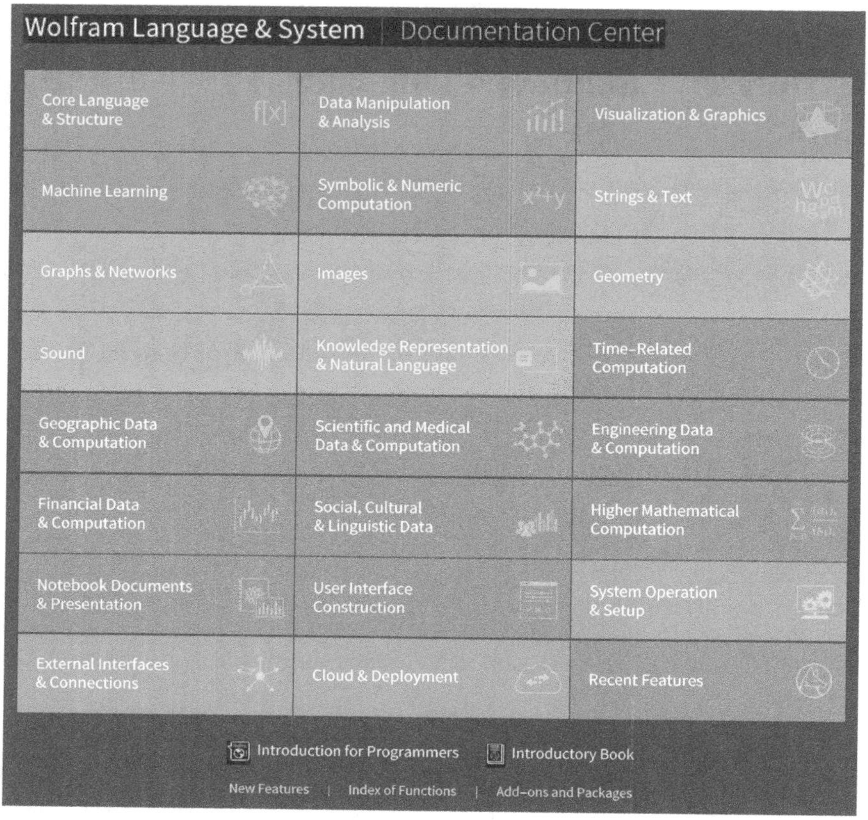

Fig. 2.1 The *Wolfram Language and System* Documentation Center offers an excellent reference for function, examples and usage for the Wolfram Language. The links to the Introduction for Programmers and Introductory Book at the bottom are particularly useful starting points for both experienced as well as new programmers working with Mathematica and the Wolfram Language

styles may be selected for cells to create a formatted document, useful for reports or presentations, or writing. The styles, such as Titles, Subtitles, Chapters, Section, etc, can be selected by first highlighting the cell to be formatted, and selecting from the application menus Format→ Style → style. A set of possible styles are displayed in Fig. 2.2, with text indicating the name of the style. Note the hierarchical structure of cell brackets corresponds to the hierarchy of styles (e.g. Chapter cells are contained within a parent Title cell, or subsections are contained within section cells).

A code notebook can have multiple brackets layered corresponding to input and output. Double clicking on the brackets can contract or expand a cell, which is indicated by arrows appearing in the brackets indicating the direction of contracted code. To create a new cell in a notebook move the cursor within the notebook until it appears as horizontal and click. A horizontal bar will then appear and a new cell will be created in that location where you can now type new code. You can evaluate

2.1 Getting Started

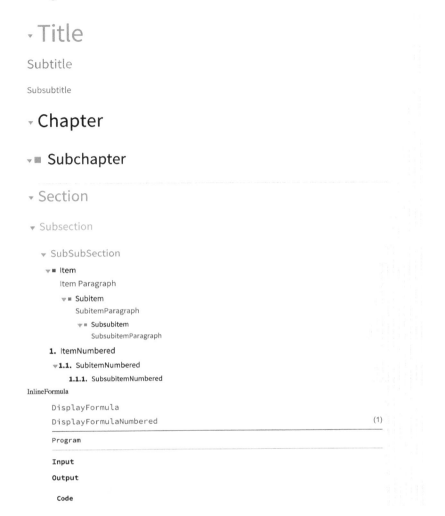

Fig. 2.2 *Mathematica* Notebook files can contain a variety of hierarchical styles

an Input cell by typing simultaneously (indicated by a "+" when we are referring to buttons) **Shift + Enter** as long as your cursor is within the cell itself, or you have already selected the cell by clicking once on its bracket. The evaluating code will be prepended with an Input cell label, **In**[i]:=, where i enumerates the evaluation in the current Mathematica session. A corresponding ith Output cell containing the result of the evaluation is then created directly below the Input cell. The Output cell is labeled **Out**[i]=, where the tag i is incremented sequentially every time new code is evaluated. For example, as a prototypical coding example, we type as input `"Hello World"` and hit **Shift + Enter** (or **Shift + Return** on a Macintosh/Apple computer):

In[1]:= "Hello World"

This will give as output the string, with the correct **Out** [1]= label:

Out[1]= Hello World

2.1.2 Hello To Wolfram|Alpha

Regular Wolfram Language code can also be supplemented through the use of the Wolfram Language's interface with Wolfram|Alpha [8] technology as mentioned in Chap. 1, Fig. 1.2. The input for using the interface needs to begin with one of the following:

- A single equal sign, = code, for returning the result and Wolfram Language syntax.
- A double equal sign, == code, for returning an interpretation and results from Wolfram|Alpha.
- Ctrl+= code, for inserting a free-form input Wolfram Language expression.

For the interface to Wolfram|Alpha to work an active internet connection is necessary. For example Fig. 2.3 shows the results of typing "= plot x^2" followed by **Shift+Return** to get a plot, or "==plot sine of 2x" and press **Shift+Enter**. *N.B. From now on we will omit the Shift+Enter, or Shift+Return step to get output results, as this is an implied necessary step in all evaluations.* The prototypical == hello world program is also available.

2.1.3 Getting Started Resources

There are many online resources, in addition to the Wolfram Documentation Center (which should be your first point of reference for functionality). For beginners, you may want to try Stephen Wolfram's An elementary introduction to the Wolfram Language [7], as of this writing available online):

In[5]:= **SystemOpen**["http://www.wolfram.com/language/elementary-introduction/2nd-ed/"]

Experienced programmers can access the fast introduction (which includes notes for Python and Java developers) that is available at:

In[6]:= **SystemOpen**["http://www.wolfram.com/language/fast-introduction-for-programmers/en/"]

Additionally, two useful and very active forums for Wolfram Language and Mathematica questions and answers are:

2.1 Getting Started   11

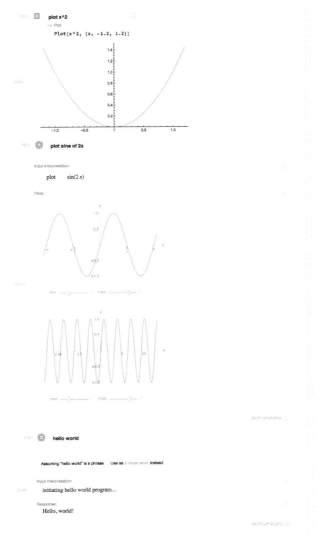

Fig. 2.3 *Wolfram|Alpha examples of* : **Out** [2]= plot x^2, **Out** [3]= plot sine of 2x, **Out** [4]= "hello world"

1. The Wolfram Community (hosted by Wolfram):

   In[7]:= **SystemOpen**["http://community.wolfram.com"]

2. The Mathematica Stack Exchange (hosted on Stack Exchange):

   In[8]:= **SystemOpen**["https://mathematica.stackexchange.com"]

In[9]:= ? Plot

Plot[*f*, {*x*, *x*$_{min}$, *x*$_{max}$}] generates a plot of *f* as a function of *x* from *x*$_{min}$ to *x*$_{max}$.
Plot[{*f*$_1$, *f*$_2$, ...}, {*x*, *x*$_{min}$, *x*$_{max}$}] plots several functions *f*$_i$.
Plot[{..., *w*[*f*$_i$], ...}, ...] plots *f*$_i$ with features defined by the symbolic wrapper *w*.
Plot[..., {*x*} ∈ *reg*] takes the variable *x* to be in the geometric region *reg*. ≫

Fig. 2.4 Out [9]= Using the inbuilt information lookup by typing ? followed by the name of the function, **Plot**. By clicking the >> on the bottom of the returned information you can access the help documentation, with additional details and examples that can be evaluated in place

2.2 Syntax

The Wolfram Language takes a different approach to defining constructs for coding. At the center of everything is the expression. Input such as numbers or strings can be though of as expressions. Functions can be defined that can assist in programming, that take as inputs different expressions. The Wolfram Language has an immense number of built-in functions that can be used to facilitate operations. A naming convention is used, where all built-in functions have full names with the first letter capitalized. For example **Plot**, **Integrate**, **Solve**, **Mean**. If multiple words are used in the naming of a function each word is capitalized, for example **ListLinePlot**, **Plot3D** - also called CamelCase. The Wolfram Language is case sensitive so selecting a consistent naming conventions can be helpful for clarifying your own code. Expressions look like:

Head[*argument*] , **HeadMultipleArguments**[*argument*$_1$, *argument*$_2$] .

This includes built-in functions, and we point out three the common characteristics: (i) Each expression has a **Head**, which is its name, (ii) followed by square brackets [], (iii) enclosing one or more, *argument*$_i$, inputs.

As you type in Mathematica, suggestions will appear to help complete the function name (autocomplete). Additionally, you can always get information about a function using a single question mark ?. For example, if you type ?**Plot** and hit **Shift+Enter** you can get information regarding the usage of the plot function. As you can see in Fig. 2.4 the **Head** of the function is **Plot** and in its primary usage, listed first, it can take as inputs a function *f* and a list {*x*, *x*$_{min}$, *x*$_{max}$, }.

Other usages are also listed. By clicking the >> on the bottom of the returned information you can access the help documentation for the function that includes multiple examples. We will use the ?**Function** method to display the returned information from the Wolfram Language throughout this book. The documentation contains many examples if you need additional usage information.

2.3 Variables

In the Wolfram Language types are usually interpreted on entry and you do not need to define them explicitly. To find the type of a expression you can check the **Head** of the input using the built-in **Head** function:

In[10]:= Head[1]
Out[10]= Integer

In[11]:= Head[1.]
Out[11]= Real

In[12]:= Head[1/2]
Out[12]= Rational

In[13]:= Head[1 + I]
Out[13]= Complex

In[14]:= Head["c"]
Out[14]= String

The expressions listed above are examples of the smallest bits of expression that can be constructed in the Wolfram Language and are called *Atomic* expressions. These expressions cannot be divided into subexpressions, and are used to build more complicated expressions. Atomic expressions include numbers, strings and symbols.

Variables and constants can be easily assigned using the = sign:

In[15]:= constant1 = 3
Out[15]= 3

We can check whether an expression, including our defined variable above, is an Atomic expression by using the function **AtomQ** which will return **True** or **False** as an output:

In[16]:= AtomQ[constant1]
Out[16]= True

All expressions that are not an Atomic expression are *Normal* expressions, and have more complexity and can take arguments. We can use variables once assigned in different computations, where they take their assigned values:

In[17]:= constant1*3
Out[17]= 9

We can also combine several variables in computation:

In[18]:= a = 2
Out[18]= 2

In[19]:= b = 3
Out[19]= 3

In[20]:= c = a + b
Out[20]= 5

You will notice that Mathematica colors the code to facilitate coding. Variables that are defined have a black color. Undefined variables are colored blue. Variables can be cleared using the **Clear** function. For example we can see that variable a has the value 2, but after clearance it has no value (notice also that in a notebook the variable a will now be colored blue).

In[21]:= a
Out[21]= 2

In[22]:= **Clear**[a]

In[23]:= a
Out[23]= a

Alternatively we can use the **Remove** function or manually clear it using =.:

In[24]:= **Remove**[b]
In[25]:= c =.

In[26]:= c
Out[26]= c

Finally, the labels of evaluation **Input** or **Output** cells can be used as variables for the same coding session. For example we evaluate again the **Out**[11] result, **Real**, above to get:

In[27]:= **Out**[11]
Out[27]= **Real**

Any **Out**[n] output can be used for the nth output. A shorthand for this is also %n. A single percentage sign % corresponds to the last output generated, and a double to the one before last, %%. This can be extended to use as many % as you need, say i to get the ith previous result. We could have used then instead of the above to get **Out**[11]:

In[28]:= %11
Out[28]= **Real**

We recommend that you use lowercase names for your variables, or at least variables where the first letter is lowercase to avoid interfering with the Wolfram Language built-in functions and variables. Note also that the Wolfram Language reserves certain names for constants, such as Pi, E, and I for π, the exponential constant e, and the imaginary unit $\sqrt{-1} = i$ respectively.

2.4 Basic Mathematical Operations

The Wolfram Language can be used to carry out basic math, similar to your calculator, although it is much more powerful than that. For example, we can add, subtract, multiply, divide and take powers:

2.4 Basic Mathematical Operations

```
In[29]:= x = 5
Out[29]= 5

In[30]:= y = 2
Out[30]= 2

In[31]:= x + y
Out[31]= 7

In[32]:= x - y
Out[32]= 3

In[33]:= x*y
Out[33]= 10

In[34]:= x/y
Out[34]= 5/2

In[35]:= x^y
Out[35]= 25
```

As we mentioned before everything in the Wolfram Language is an expression. These operations above are actually all short convenient forms of expressions. We could alternatively use **Set** to assign the variables, and use the operations **Plus**, **Subtract**, **Times**, **Divide**, **Power** for addition, subtraction, multiplication, division and power evaluations respectively:

```
In[36]:= Set[x, 5]
Out[36]= 5

In[37]:= Set[y, 2]
Out[37]= 2

In[38]:= Plus[x, y]
Out[38]= 7

In[39]:= Subtract[x, y]
Out[39]= 3

In[40]:= Times[x, y]
Out[40]= 10

In[41]:= Divide[x, y]
Out[41]= 5/2

In[42]:= Power[x, y]
Out[42]= 25
```

Multiplication is also special, as it can be entered in different ways, using the * operator between entries, using a space between entries, or having a number in front of a symbol or defined variable as you would do in an algebra class:

In[43]:= x y
Out[43]= 10

In[44]:= 2 x
Out[44]= 10

In[45]:= 2 x
Out[45]= 10

In[46]:= 2 x
Out[46]= 10

2.5 Type Casting

N[x,n] will generate a numerical approximation for number x to n significant digits. You may also use it to convert an **Integer** to a **Real**:

In[47]:= **N**[2, 2]
Out[47]= 2.0

You can use **Round**[x] can be to obtain the integer closest to number x. If you want the greatest integer less than x you can use **Floor**[x], and if you want the smallest integer greater than x you can use **Ceiling**[x]:

In[48]:= **Round**[1.01]
        Floor[1.01]
        Ceiling[1.01]
Out[48]= 1
Out[49]= 1
Out[50]= 2

Note in the above example that we have put together multiple evaluations in the same cell. You can do this as long as the separate evaluations are separated by a new line by pressing **Return** or **Enter**. You can also suppress output in the Wolfram Language by adding ; to the end of a code to be evaluated. Note that the evaluation still takes place, and you can check the value of the new variable to see that this has worked:

In[51]:= roundX = **Round**[1.01];
         floorX = **Floor**[1.01];
         ceilingX = **Ceiling**[1.01];

In[54]:= ceilingX
Out[54]= 2

You can use **Chop**[] to replace numbers less than 10^{-10} with the exact integer 0:

In[55]:= **Chop**[0.00000000000000000001]
Out[55]= 0

2.5 Type Casting

You can also convert **Real** numbers to **Rational** by using **Rationalize** [x], which will produce the nearest exact **Rational**, or within a tolerance delta if it is provided using **Rationalize** [x, delta].

In[56]:= **Rationalize**[1.012]
Out[56]= $\frac{253}{250}$

In[57]:= **Rationalize**[1.012, 0.01]
Out[57]= $\frac{84}{83}$

You can also convert expressions to **String**

In[58]:= testString = hello
Out[58]= hello

In[59]:= **Head**[testString]
Out[59]= **Symbol**

In[60]:= testStringConverted = **ToString**[testString]
Out[60]= "hello"

In[61]:= **Head**[testStringConverted]
Out[61]= **String**

Equivalently, you can convert **String** variables to evaluations:

In[62]:= testStringEvaluation = "{Sin[Pi/2],1+1,2^2}"
Out[62]= "{Sin[Pi/2],1+1,2^2}"

In[63]:= **ToExpression**[testStringEvaluation]
Out[63]= {1, 2, 4}

2.6 Algebra

The Wolfram Language can be used extensively to handle algebraic symbolic manipulations. This can be of great help when handling complicated calculations. First clear the variables x and y from any previous calculations and then use them for algebraic manipulations:

In[64]:= **Clear**[x, y]

In[65]:= (x + y)^2
Out[65]= $(x+y)^2$

In[66]:= **Expand**[(x + y)^2]
Out[66]= $x^2 + 2 x y + y^2$

In[67]:= **Expand**[(x + y)^4]
Out[67]= $x^4 + 4 x^3 y + 6 x^2 y^2 + 4 x y^3 + y^4$

In[68]:= **Factor**[32 + 80 x + 80 x^2 + 40 x^3 + 10 x^4 + x^5]
Out[68]= $(2 + x)^5$

If you want to put in symbolic math that looks like textbook math you can use the palettes for input from the menu Palettes → Basic Math Assistant. The Wolfram Language will still evaluate the code correctly and return a simple expression:

In[69]:= **FullSimplify**[x y^2 + y x^3 + x y + x y^3 + y^2]
Out[69]= y (x^3 + y + x (1 + y + y^2))

2.7 Lists

In this section we consider how we can put together and manipulate lists of objects in the Wolfram Language.

2.7.1 Collections of Expressions

Collections of expressions can be created in the Wolfram Language by putting together any kind of object in a list, which is denoted by a set of curly brackets, with members separated by commas, $\{object_1, object_2, ..., object_n\}$. For example, a set of numbers, a mixed set of strings and number, a set of symbols and strings combined:

In[70]:= listExample1 = {1, 2, 3, 4, 5};

In[71]:= listExample2 = {"a", 1, "b", 2};

In[72]:= listExample3 = {x, x^2, **Sin**[x + 2], "stringObject"};

As you can see the list is a very general object which can have entries of multiple types. Furthermore, lists can be nested:

In[73]:= nestedList = {x, {y, z}, {p, {q, r}}}
Out[73]= {x, {y, z}, {p, {q, r}}}

The expressions in the list can be more complicated:

In[74]:= nestedList2 = {x,{y,**Exp**[z] + x y},{(p+2)^(3 **Cos**[**Pi**/3x]), {q,r}}}
Out[74]= {x, {y, e^z + x y}, $(p+2)^{3\,Cos[\frac{\pi x}{3}]}$, {q, r}}}

The expressions can be visualized in terms of **TreeForm** to see the levels and structure of both the nested lists and also the functions involved in Fig. 2.5. Both of our lists have dimensions 3:

2.7 Lists

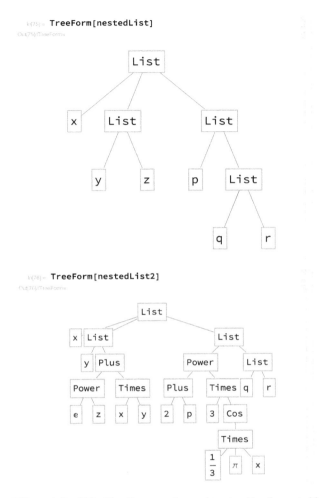

Fig. 2.5 **Out**[75]= and **Out**[76]= **TreeForm** can be used to visualize the nested levels in expressions and lists

In[77]:= **Dimensions**[nestedList]
Out[77]= {3}

In[78]:= **Dimensions**[nestedList2]
Out[78]= {3}

We can get the depth of each list:

In[79]:= **Depth**[nestedList]
Out[79]= 4

In[80]:= **Depth**[nestedList2]
Out[80]= 7

The **Head** of the list is of course **List**:

In[81]:= **Head**[nestedList]
Out[81]= **List**

We can extract the different levels corresponding to the **TreeForm** by using **Level**, and specifying the specific level with curly brackets as the second argument:

In[82]:= **Level**[nestedList,{1}]
Out[82]= {x,{y,z},{p,{q,r}}}

In[83]:= **Level**[nestedList,{2}]
Out[83]= {y,z,p,{q,r}}

In[84]:= **Level**[nestedList,{3}]
Out[84]= {q,r}

We can also get parts of a list by using double square brackets [[i]]. Let us get the first element and then the second element of our example list:

In[85]:= nestedList[[1]]
Out[85]= x

In[86]:= nestedList[[2]]
Out[86]= {y,z}

We see that the second element is also a list with parts. We can get the next level of list components by specifying an additional index in the double square brackets. Additional indices are separated by a bracket. So, to get the second list first element we have:

In[87]:= nestedList[[2,1]]
Out[87]= y

To get the second element of the second member of the list we have:

In[88]:= nestedList[[2,2]]
Out[88]= z

As you can see from the **TreeForm** results we can get the numbering of the levels simply by traversing the trees in the diagram, Fig. 2.5. Each branch moving downwards corresponds to a new level in the expression. The overall **Head** of the expression is at the very top, and in this case is **List**. We can traverse the levels to get the desired elements. Using the figure you can check the following :

In[89]:= nestedList2[[3,1,2,2,1,2]]
Out[89]= π

In[90]:= nestedList2[[2,2,2,2]]
Out[90]= y

Hence, we can get parts of expressions in addition to getting parts of a list.

2.7 Lists

2.7.2 List Element Extraction

Let us get back to list element extraction. If we consider the following list:

In[91]:= testList={a,{b,c,d,{e,f}},{g,h},i,j,k,{l,m},z}
Out[91]= {a,{b,c,d,{e,f}},{g,h},i,j,k,{l,m},z}

We can get the second list element:

In[92]:= testList[[2]]
Out[92]= {b,c,d,{e,f}}

We might want the second element of the second list:

In[93]:= testList[[2,2]]
Out[93]= c

If we use the notation **List**[[j ;; n]] we can get the jth to nth elements of a list. In our example we get the second to third elements:

In[94]:= testList[[2;;3]] (* element 2 to 3*)
Out[94]= {{b,c,d,{e,f}},{g,h}}

In[95]:= testList[[1;;5;;2]] (* elements 1 to 5 in steps of 2*)
Out[95]= {a,{g,h},j}

In real applications, we may have multi-nested lists where we want to extract all the same information across all elements of a list. For example, we consider the following list:

In[96]:= testList2={{1,2,3,4},{a,b,c,d},{"i","ii","iii","iv"}};

We now get from all members the second element:

In[97]:= testList2[[All,2]]
Out[97]= {2,b,ii}

We can also dive further into the list. We can add additional external indices to specify which elements from the previous extraction we want. So, first we get all the second elements from each member, and then construct a new list of only elements 2 and 3. Notice the curly brackets in this construction within the square brackets. These allow us to select specific elements from the same list.

In[98]:= testList2[[All,2]][[{2,3}]]
Out[98]= {b,ii}

We can traverse lists starting from the back using negative indices. For example we can get the last element:

In[99]:= testList2[[-1]]
Out[99]= {i,ii,iii,iv}

We can also get any specific element, in any way we want the order to be. For our example lists let us get the second element from the end, and place it before the first element. Again to extract specific elements in a new order we use curly brackets within the square brackets:

```
In[100]:= testList[[{-2,1}]]
Out[100]= {{1,m},a}

In[101]:= testList2[[{-2,1}]]
Out[101]= {{a,b,c,d},{1,2,3,4}}
```

2.7.3 Ordered Lists with Table

Ordered lists can be easily constructed using **Table**. The basic form of the expression is:

```
In[102]:= integerList=Table[i,{i,1,10}]
Out[102]= {1,2,3,4,5,6,7,8,9,10}
```

where the iterator runs from the minimum to maximum values {iterator, min, max}, in our case 1 through 10.

We can also create more restricted elements, by starting for example at 3 this time as a minimum for iterator i:

```
In[103]:= integerListRestricted=Table[i,{i,3,10}]
Out[103]= {3,4,5,6,7,8,9,10}
```

Additionally, we can add more nesting, by adding additional iterators, with their own maxima and minima. For example we create a set of paired integers by using the expression i,j and allowing i to run from 1 to 7 and j from 1 to 5. Note that the list associated with index i is outermost:

```
In[104]:= integerListNested=Table[{i,j},{i,1,7},{j,1,5}]
Out[104]= {{{1,1},{1,2},{1,3},{1,4},{1,5}},
           {{2,1},{2,2},{2,3},{2,4},{2,5}},
           {{3,1},{3,2},{3,3},{3,4},{3,5}},
           {{4,1},{4,2},{4,3},{4,4},{4,5}},
           {{5,1},{5,2},{5,3},{5,4},{5,5}},
           {{6,1},{6,2},{6,3},{6,4},{6,5}},
           {{7,1},{7,2},{7,3},{7,4},{7,5}}}
```

2.7.4 Element Manipulation

We can use several functions to manipulate lists. For example we can append elements to the end of a list:

```
In[105]:= Append[integerList, 11]
Out[105]= {1,2,3,4,5,6,7,8,9,10,11}
```

We can also prepend to a list:

```
In[106]:= Prepend[integerList, 0]
Out[106]= {0,1,2,3,4,5,6,7,8,9,10}
```

2.7 Lists

We can insert elements at a particular position. For example we can insert 1.5 at position 2 in our integerList:

In[107]:= **Insert**[integerList,1.5, 2]
Out[107]= {1,1.5,2,3,4,5,6,7,8,9,10}

We can also remove elements, for example we can delete the fourth part of the list:

In[108]:= **Delete**[integerList, 4]
Out[108]= {1,2,3,5,6,7,8,9,10}

We can replace specific parts of a list. For example we can replace the fourth part with 101:

In[109]:= **ReplacePart**[integerList, 4 ->101]
Out[109]= {1,2,3,101,5,6,7,8,9,10}

Note here that the second argument in 4 -> 101, uses a construct with a two character right arrow ->, rendered as →, called a rule.

Additionally, we can replace specific parts in nested lists. For example here we replace part 2 of the fourth list with the element 22, 44:

In[110]:= **ReplacePart**[integerListNested, {2,4}→{22,44}]
Out[110]= {{{1,1},{1,2},{1,3},{1,4},{1,5}},
 {{2,1},{2,2},{2,3},{22,44},{2,5}},
 {{3,1},{3,2},{3,3},{3,4},{3,5}},
 {{4,1},{4,2},{4,3},{4,4},{4,5}},
 {{5,1},{5,2},{5,3},{5,4},{5,5}},
 {{6,1},{6,2},{6,3},{6,4},{6,5}},
 {{7,1},{7,2},{7,3},{7,4},{7,5}}}

2.7.5 List Manipulation

Now let us look at operations to combine and compare lists. First we define three short lists:

In[111]:= list1={a,b,c};
 list2={d,e,f};
 list3={g,h,i};

We can use **Join** to join two or more lists:

In[114]:= list4=**Join**[list1, list2]
Out[114]= {a,b,c,d,e,f}

In[115]:= list5=**Join**[list2, list3]
Out[115]= {d,e,f,g,h,i}

If we think of lists as sets, we can carry out several set operations as well. For example we can take the **Union** of lists:

In[116]:= **Union**[list4,list5]
Out[116]= {a,b,c,d,e,f,g,h,i}

We can also look at the **Intersection** and **Complement** of lists:

In[117]:= **Intersection**[list4,list5]
Out[117]= {d,e,f}

In[118]:= **Complement**[list4,list5]
Out[118]= {a,b,c}

These set operations can take multiple inputs:

In[119]:= **Union**[list1,list2,list3,list4,list5]
Out[119]= {a,b,c,d,e,f,g,h,i}

Also, we can join lists with different levels of nesting:

In[120]:= list6={{1,2},{3,4}};
In[121]:= list7=**Join**[list1,list6]
Out[121]= {a,b,c,{1,2},{3,4}}

We can actually remove some of the nesting with **Flatten**:

In[122]:= **Flatten**[list7]
Out[122]= {a,b,c,1,2,3,4}

There are many other operations that can be done with lists, and the functions we have discussed can also take additional inputs and options. The above is only meant to be your getting started guide. As lists are fundamental in dealing with data in the Wolfram Language and very useful in programming you should get more familiar with them by consulting the in application tutorial which you can access by typing `"tutorial/ListsOverview"` in the documentation search bar.

2.7.6 Matrices

Vectors are essentially lists of expressions, and we can think of matrices as made of vectors. In the Wolfram Language we can form lists of lists:

In[123]:= row1={v11,v12,v13};
row2={v21,v22,v23};
row3={v31,v32,v33};
m1={row1,row2,row3}
Out[126]= {{v11,v12,v13},{v21,v22,v23},{v31,v32,v33}}

We can visualize these lists as matrices, using **MatrixForm**:

In[127]:= **MatrixForm**[m123]
Out[127]//**MatrixForm**= m123

This is often used in a postfix form: expression // f, which is equivalent to printing out f[expression]:

2.7 Lists

In[128]:= m1//**MatrixForm**
Out[128]= $\begin{pmatrix} v11 & v12 & v13 \\ v21 & v22 & v23 \\ v31 & v32 & v33 \end{pmatrix}$

We can also use **TableForm** for visualization:

In[129]:= m1//**TableForm**
Out[129]= v11 v12 v13
v21 v22 v23
v31 v32 v33

We can think of nested lists as matrices, where if we are indexing the components the first index is a row, and the second a column. We can retrieve the 1st row 3rd column element:

In[130]:= m1[[1,3]]
Out[130]= v13

We can retrieve the first row:

In[131]:= m1[[1,**All**]]
Out[131]= {v11,v12,v13}

We can retrieve the second column:

In[132]:= m1[[**All**,2]]
Out[132]= {v12,v22,v32}

Returning to matrices, single lists are represented as column vectors using **MatrixForm**:

In[133]:= row1//**MatrixForm**
Out[133]//**MatrixForm**= $\begin{pmatrix} v11 \\ v21 \\ v31 \end{pmatrix}$

Let us define a couple more matrices:

In[134]:= m2={{u11,u12,u13},{u21,u22,u23},{u31,u33,u33}};
m3 ={{a11,a12},{a21,a22}};
m2//**MatrixForm**
m3//**MatrixForm**

Out[136]//**MatrixForm**= $\begin{pmatrix} u11 & u12 & u13 \\ u21 & u22 & u23 \\ u31 & u32 & u33 \end{pmatrix}$

Out[137]//**MatrixForm**= $\begin{pmatrix} a11 & a12 \\ a21 & a22 \end{pmatrix}$

We can add scalars element-wise:

In[138]:= v1+m2//**MatrixForm**
Out[138]//**MatrixForm**= $\begin{pmatrix} u11+v1 & u12+v1 & u13+v1 \\ u21+v1 & u22+v1 & u23+v1 \\ u31+v1 & u33+v1 & u33+v1 \end{pmatrix}$

We can do matrix element-wise multiplication:

In[139]:= v1*m2//**MatrixForm**

Out[139]//**MatrixForm**= $\begin{pmatrix} u11v1 & u12v1 & u13v1 \\ u21v1 & u22v1 & u23v1 \\ u31v1 & u33v1 & u33v1 \end{pmatrix}$

We can carry out matrix multiplication using a period . :

In[140]:= m1.m2//**MatrixForm**
Out[140]//**MatrixForm**=
$\begin{pmatrix} u11v11+u21v12+u31v13 & u12v11+u22v12+u33v13 & u13v11+u23v12+u33v13 \\ u11v21+u21v22+u31v23 & u12v21+u22v22+u33v23 & u13v21+u23v22+u33v23 \\ u11v31+u21v32+u31v33 & u12v31+u22v32+u33v33 & u13v31+u23v32+u33v33 \end{pmatrix}$

We can also compute trace (**Tr**) and determinants (**Det**);

In[141]:= **Tr**[m2]
Out[141]= u11+u22+u33

In[142]:= **Det**[m3]
Out[142]= −a12a21+a11a22

Also available are **Eigenvalues Eigenvectors** and **Eigensystem** (which returns both eigenvalues and eigenvectors).

In[143]:= **Eigenvalues**[m3]
Out[143]= $\left\{ \frac{1}{2}\left(-\sqrt{a11^2 - 2a11a22 + 4a12a21 + a22^2} + a11 + a22\right), \right.$
$\left. \frac{1}{2}\left(\sqrt{a11^2 - 2a11a22 + 4a12a21 + a22^2} + a11 + a22\right) \right\}$

In[144]:= **Eigenvectors**[m3]
Out[144]= $\left\{ \left\{ -\frac{-a11+a22+\sqrt{a11^2-2a22a11+a22^2+4a12a21}}{2a21}, 1 \right\}, \right.$
$\left. \left\{ -\frac{-a11+a22-\sqrt{a11^2-2a22a11+a22^2+4a12a21}}{2a21}, 1 \right\} \right\}$

In[145]:= eigensystemM3=**Eigensystem**[m3]
Out[145]= $\left\{ \left\{ \frac{1}{2}\left(-\sqrt{a11^2 - 2a11a22 + 4a12a21 + a22^2} + a11 + a22\right), \right.\right.$
$\left. \frac{1}{2}\left(\sqrt{a11^2 - 2a11a22 + 4a12a21 + a22^2} + a11 + a22\right) \right\}$
$\left\{ -\frac{-a11+a22+\sqrt{a11^2-2a22a11+a22^2+4a12a21}}{2a21}, 1 \right\},$
$\left.\left. \left\{ -\frac{-a11+a22-\sqrt{a11^2-2a22a11+a22^2+4a12a21}}{2a21}, 1 \right\} \right\} \right\}$

2.8 Functions

In this section we will discuss how to define functions in the Wolfram Language. First we consider how we make assignments to variables, we will then review rules and transformation, before seeing how we can make function assignments.

2.8.1 SetDelayed

We have seen that we can create different variables by using the = assignment (or equivalently the **Set** function). Once such an assignment is made the evaluation on the right hand side (rhs) is assigned immediately to the left hand side (lhs). For a simple example, let us define a variable that is the sum of $1 + 1$:

In[146]:= sumExample=1+1
Out[146]= 2

We see that the evaluation was immediate, and our variable has a fixed value 2:

In[147]:= sumExample
Out[147]= 2

We sometimes do not want to fix a variable to have a finalized value. For example, if we want to make a list of 5 random integers between 1 and 100, we can think of first defining a number randomly chosen using **RandomInteger**:

In[148]:= randomInteger=**RandomInteger**[100]
Out[148]= 33

We might assume by creating a table we can get the 5 random integers we want:

In[149]:= **Table**[randomInteger,{i,1,5}]
Out[149]= {33,33,33,33,33}

Instead you will notice that the evaluated value is repeated and does not change. By the way, also note that if you try this yourself you will probably get a different random integer assigned. But you will see the same behavior described here.

There is another kind of evaluation, called a delayed assignment, and specified by the := operator. Here the variable to the left hand side(lhs) evaluates to the right hand side(rhs) only when the variable is called (lhs:=rhs). Let us see how it can be used:

In[150]:= randomIntegerDelayed:=**RandomInteger**[100]

We notice that no output is actually generated, since no assignment has been made. The **Head** of this expression is still an integer when the variable is called:

In[151]:= **Head**[randomIntegerDelayed]
Out[151]= **Integer**

In this case though, the evaluation is held and the assignment is only made when the variable is called:

In[152]:= randomIntegerList=**Table**[randomIntegerDelayed,{i,1,5}]
Out[152]= {82,49,29,93,34}

We now see that a new call to the variable is made as the **Table** iterator is incremented and we get the desired effect. A nice way to think about it is that you are calling the command to create a random integer either immediately with =, or when the variable is actually called with the := operator. The full form of the := operator is **SetDelayed**[lhs, rhs].

2.8.2 Transformation Rules

We have seen how to make assignments of variables, and briefly mentioned the use of rules. Let us create a variable z which is a list of three numbers $\{1, x, x^2\}$.

In[153]:= z={1,x,x²}
Out[153]= {1,x,x²}

In[154]:= ?z
       Global`z
       z={1,x,x²}

We see that the variable is defined in terms of x. We can evaluate z with different substitutions for x using transformation rules. This works by using the **ReplaceAll** command shorthand /. , and an assignment of variable to be replaced with a rule indicated by the arrow – > and formatted to → in Mathematica. For example let us replace x with 2 in our defined z expression:

In[155]:= z/.x -> 2
Out[155]= {1,2,4}

We could replace x with anything we want, say a symbol p:

In[156]:= z/.x -> p
Out[156]= {1,p,p²}

Or a more complicated expression, for example a list {p,q} :

In[157]:= z/.x->{p,q}
Out[157]= {1,{p,q},{p²,q²}}

We can have multiple rules in the case where we are using multiple variables. Let us create a new expression which takes x1,x2,x3 as inputs.

In[158]:= {x1∗x2,x1/x2,x3, **Exp**[−x1]}
Out[158]= {x1 x2, $\frac{x1}{x2}$, x3, e^{-x1}}}

We can have a multi assignment rule:

In[159]:= {x1∗x2,x1/x2,x3, **Exp**[−x1]}/.{x1->1,x2-> y,x3-> x}
Out[159]= {y, $\frac{1}{y}$, x, $\frac{1}{e}$}

We can define a variable for the original expression to make it more convenient to use:

In[160]:= expressionExample={x1∗x2,x1/x2,x3, **Exp**[−x1]}
Out[160]= {x1 x2, $\frac{x1}{x2}$, x3, e^{-x1}}

In[161]:= expressionExample/.{x1 -> 1,x2 -> y,x3 -> x}
Out[161]= {y, $\frac{1}{y}$, x, $\frac{1}{e}$}

We can also define the rule as a variable:

In[162]:= expressionSubRule={x1 -> 1,x2 -> y,x3 ->x}
Out[162]= {x→1,x2→y,x3→x}

2.8 Functions

Then we get a simplified version with the same result:

In[163]:= expressionExample /. expressionSubRule
Out[163]= $\{y, \frac{1}{y}, x, \frac{1}{e}\}$

We can also use **Thread** to create the rule more easily. **Thread** takes to lists to create a threaded assignment of variables to be replaced to their replacements:

In[164]:= **Thread**[{x1, x2, x3} -> {1, y, x}]
Out[164]= {x1→1, x2→y, x3→x}

Several rules can be applied one after the other. In our example, let us replace y and x after the first rule with values:

In[165]:= expressionExample /. expressionSubRule /. {x -> 1, y -> 2}
Out[165]= $\{2, \frac{1}{2}, 1, \frac{1}{e}\}$

There are other ways to user rules, for example by repeated replacement using **ReplaceRepeated** (which has the form expression//. rules). This will replace values repeatedly until there are no more changes to the expression. This is useful when replacing one symbol with another. Let us see how it works with our example substitution by creating a new rule:

In[166]:= expressionSubRuleRepeated={x1 -> 1, x2 ->x1, x3 -> x2}
Out[166]= {x1→1, x2→x1, x3→x2}

A simple replacement would give

In[167]:= expressionExampleReplaced1=expressionExample /. expressionSubRuleRepeated
Out[167]= $\{x1, \frac{1}{x1}, x2, \frac{1}{e}\}$

As you can see x1 and x2 still are parts of the expression as each replacement was only carried out once. We can repeat the substitution:

In[168]:= expressionExampleReplaced2=
           expressionExampleReplaced1 /. expressionSubRuleRepeated
Out[168]= $\{1, 1, x1, \frac{1}{e}\}$

And repeat again until no variables remain:

In[169]:= expressionExampleReplaced3=
           expressionExampleReplaced2 /. expressionSubRuleRepeated
Out[169]= $\{1, 1, 1, \frac{1}{e}\}$

Alternatively, using **ReplaceRepeated** (//.) we can get repeated replacement until the expression no longer changes in a single step:

In[170]:= expressionExample //. expressionSubRuleRepeated
Out[170]= $\{1, 1, 1, \frac{1}{e}\}$

2.8.3 Defining Functions

In the Wolfram Language we can define our own functions in a very general way. A function is typically defined as a left hand side with a dummy argument and a right hand side. Mathematically you can think of it as an assignment of a variable, say x to a more complicated expression. Let us first look at a simple example, and write a function called addOne that adds one to an input x.

In[171]:= addOne[x_]:=x+1

This is a delayed assignment, and again there is no **Output** cell. Also, the input dummy argument on the left has an underscore _ next to it, which is shorthand for **Blank**. This is actually called a pattern, which is matched to the right hand side (these are color coded green in a Notebook). The underscore is essential to the definition. The input argument x is then taken and 1 is added to it. The input can change every time as required:

In[172]:= ?addOne
           Global`addOne
           addOne[x_]:=x+1

We can now use the defined function with numbers, variables or expressions:

In[173]:= addOne[1]
Out[173]= 2

In[174]:= addOne[y]
Out[174]= 1+y

In[175]:= addOne[Sin[2x]]
Out[175]= 1+Sin[2 x]

We can define functions of several variables. For example let us define a function that adds two values:

In[176]:= addTwo[x_,y_]:= x+y
In[177]:= ?addTwo
           Global`addTwo
           addTwo[x_,y_]:=x+y

In[178]:= addTwo[1,2]
Out[178]= 3

In[179]:= addTwo[p,q]
Out[179]= p+q

Let us create a more complicated example. Here we integrate ax^2, and the user is asked to input the limits of integration and the constant a:

In[180]:= integratorFunction[a_,min_,max_]:=
              Integrate[-a x^2,{x,min,max}]
In[181]:= integratorFunction[3,-2,2]
Out[181]= -16

2.8 Functions

Our defined functions can be used as any Wolfram Language functions. For example, let us make a table where we evaluate our integrator function for different minima and maxima:

In[182]:= **Table[integratorFunction[2,minimum,maximum],**
{minimum,−3,0},{maximum,0,2}]
Out[182]= {{−18,−(56/3),−(70/3)},{−(16/3),−6,−(32/3)},
{−(2/3),−(4/3),−6},{0,−(2/3),−(16/3)}}

We can view this as a Matrix, which is essentially a list of lists:

In[183]:= **MatrixForm[%]**

Out[183]//**MatrixForm**= $\begin{pmatrix} -18 & -\frac{56}{3} & -\frac{70}{3} \\ -\frac{16}{3} & -6 & -\frac{32}{3} \\ -\frac{2}{3} & -\frac{4}{3} & -6 \\ 0 & -\frac{2}{3} & -\frac{16}{3} \end{pmatrix}$

Note how we used % as a shorthand to evaluate the previous output. Additionally, note how the **Out** cells has //**MatrixForm** printed with it - it will show up whenever you format **Out** cells in Mathematica using a postfix, similarly with **TableForm**.

2.8.4 Options

We have seen how functions have the form **FunctionHead** [argument$_1$, ..., argument$_n$]. In addition, function can take function-specific options that follow the standard arguments as a set of rules, **FunctionHead** [argument$_1$, argument$_2$..., argument$_n$, option$_1$ → optionValue$_1$, ..., option$_k$ → optionValue$_k$]. While all arguments need to be supplied in specific order to evaluate a function, options are optional, and can be in any number and in any order, and can also be omitted. For a given function we can find if it has options:

In[184]:= **Options[Table]**
Out[184]= { }

In[185]:= **Options[Integrate]**
Out[185]= {**Assumptions**:>$Assumptions,
  GenerateConditions→**Automatic**, **PrincipalValue**→**False**}

We notice that options have different pre-populated values. With the option **Assumptions** not set, we get for example:

In[186]:= **Integrate[$\sqrt{ax^2}$,{x,0,1}]**
Out[186]= $\frac{\sqrt{a}}{2}$

Now if we assume that a is less than zero we get a different form:

In[187]:= **Integrate[Sqrt[a x^2],{x,0,1},Assumptions→a<0]**
Out[187]= $\frac{i\sqrt{-a}}{2}$

2.9 Programming Essentials

The Wolfram Language allows for multiple programming paradigms, including procedural and functional approaches, or your favorite mixture therein.

2.9.1 Comments

Comments in the Wolfram Language are fairly easy to introduce, by simply enclosing them in brackets with *s: (* comments *).

```
In[188]:= 1+2 (*comments will not evaluate*)
Out[188]= 3

In[189]:= 1+(*comments can be placed anywhere – though here not
            recommended*)2
Out[189]= 3
```

2.9.2 Boolean Operations and Flow

The Wolfram Language has multiple Boolean operators that evaluate to True or False. This can be used to control code and involve decision making in code flow.

2.9.2.1 Boolean operators

Standard operators can evaluate to True of False. We can compare values using:

- smaller than <, smaller than or equal <=, greater than >, greater than or equal >=, not equal !=:

  ```
  In[190]:= exampleA=1;
            exampleB=2;
  In[192]:= 1<2
  Out[192]= True

  In[193]:= exampleA<exampleB
  Out[193]= True

  In[194]:= 1>2
  Out[194]= False

  In[195]:= 1>1
  Out[195]= False
  ```

2.9 Programming Essentials

```
In[196]:= 1>= 1
Out[196]= True

In[197]:= 1!= 2
Out[197]= True
```

- And (&&), Or (||)

```
In[198]:= 1 <2 && 1<3
Out[198]= True

In[199]:= 1<2 && 1>3
Out[199]= False

In[200]:= 1<2 || 1> 3
Out[200]= True
```

We note that numerical evaluations take precedent over the logical evaluations:

```
In[201]:= 1+1<2−1
Out[201]= False
```

When in doubt, use brackets:

```
In[202]:= (1+1)<(2−2)
Out[202]= False

In[203]:= 1+(1<2)−2
Out[203]= −1+True
```

Notice in our simplistic evaluation above we end up with a statement that has an unexpected form, as we have forced the inner Boolean evaluation to take precedent by enclosing it in brackets.

2.9.2.2 If Statements

The standard conditional statements of If and If...Else If are available:

```
In[204]:= ?If
          If[condition,t,f] gives t if condition evaluates
              to True, and f if it evaluates to False.
          If[condition,t,f,u] gives u if condition evaluates
              to neither True nor False. >>
```

For example, let us look at the following code defining a function checkerA that takes as input a single variable:

```
In[205]:= checkerA[a_]:=If[a, Print["Input is True"],
          Print["Input is False"],
          Print["Input is neither True nor False"]]
```

If we use a true value as input we get for obvious inputs that evaluate to True or False:

In[206]:= checkerA[1<2]

(*the line below is printed during the evaluation*)
Input is True

In[207]:= checkerA[1>2]

(*the line below is printed during the evaluation*)
Input is False

As you can see below, we get the third option when a Boolean value cannot be assessed (in this example this is because neither a nor b have been defined yet):

In[208]:= checkerA[a<b]

(*the line below is printed during the evaluation*)
Input is neither True nor False

2.9.2.3 Switch

A **Switch** statement allows us to carry out a multiple choice evaluation. In an expression matches one of the switch options an action is taken. Let us see an example:

In[209]:= ?Switch
    Switch [expr, form$_1$, value$_1$, form$_2$, value$_2$, ...]
        evaluates expr, then compares it with each of
        the form$_i$ in turn, evaluating and
        returning the value$_i$ corresponding to the first
        match found. >>

In[210]:= switcherEg[a_]:=**Switch**[a,"Match1",**Print**["Matched 1st"],
    "Match2",**Print**["Matched Second"],"Match3",
    Print["Matched Third"]]

In[211]:= switcherEg["Match3"]

(*the line below is printed during the evaluation*)
Matched Third

2.9.2.4 For

A for loop needs a starting value, a test that is evaluated, an incremental statement that changes the start value, and the body that keeps getting evaluated if the test is true.

In[212]:= ?For
    For[start, test, incr, body] executes start, then
        repeatedly evaluates body and incr until
        test fails to give **True**. >>

2.9 Programming Essentials

Let us write a simple table:

In[213]:= For[j=0,j <= 4, j=j+1,Print[{j,j^2}]]

    (*the lines below are printed during the evaluation*)
    {0,0}
    {1,4}
    {2,9}
    {3,16}

2.9.2.5 While

A while statements keeps evaluating its body code until the test fails.

In[214]:= ?While
    While[test,body] evaluates test, then body,
        repetitively, until test first fails to
        give **True**. >>

In[215]:= j=0;
    While[j<= 4, Print[{j,j^2}];j++]

    (*the lines below are printed during the evaluation*)
    {0,0}
    {1,4}
    {2,9}
    {3,16}

Note here how we have used ++ to post increment j by 1 (instead of $j = j + 1$ statements).

A trick often used is to set an operator to True explicitly in the test and to set it as False if a statement within the loop is achieved and we wish to terminate the loop. Let us use a random number generator for integers 1 to 1000, and see how many times it will take to evaluate to 1:

In[217]:= test=**True**;
    counter=0;
    While[test,x=**RandomInteger**[1000];
     counter++;
     If[x==1,**Print**[{"Number of Evaluations",counter}];
      test=**False**]]

    (*the line below is printed during the evaluation*)
    {Number of Evaluations, 172}

You will most likely get a different number printed. To generate the "random number", We have used **RandomInteger** before, and this is a useful function, so we take a brief look at its definition:

In[220]:= ?**RandomInteger**
    RandomInteger{i_{min}, i_{max}} gives a pseudorandom integer
        in the range {i_{min}, i_{max}}.
    RandomInteger[i_{max}] gives a pseudorandom integer
        in the range {$0, \ldots, i_{max}$}.

RandomInteger[] pseudorandomly gives 0 or 1.
RandomInteger[range,n] gives a list of n pseudorandom integers.
RandomInteger [range, {$n_1, n_2, \ldots$}] gives an $n_1 \times n_2 \times \ldots$ array of pseudorandom integers. >>

Keep in mind that random number generators are pseudorandom generators, as indicated in the Information above. These use an internal "seed" for pseudorandom number generation that changes continuously to simulate randomness (for example based on the date and time). You will get a different evaluation than above unless we set the seed explicitly, which we can do using **SeedRandom**:

```
In[221]:= test=True;
         counter=0;
         SeedRandom[1000];
         While[test, x=RandomInteger[1000]; counter++;
          If[x==1, Print[{"Number of Evaluations",
            counter}]; test=False]]
```

(∗the line below is printed during the evaluation∗)
{Number of Evaluations,1784}

If you evaluate the code above you will get the same result every time, as the seed will always be set to the same variable. While this is good for demonstration purposes, it takes away the "random" aspect of the numbers generated.

2.9.3 Modules, Blocks and With

Often we may want to code a sequence of commands to create a function, but not to have the intermediate variables defined globally in the section. The Module function can keep symbol names treated as local, and we typically use it in function definitions:

myfunction[x_,y_,...,z_]:=**Module**[{$local_1, local_2, \ldots, local_j = value_j, \ldots, local_n$}, expressions]

Here the function takes multiple arguments x through z. The module has two arguments, first, a list where local variable names are listed, and second an expression, or a series of expressions separated by semicolons that will be carried out using the variables. Additionally, in the local name declarations we can set the values of the local variables, e.g. to the input arguments or any values.

Let us look at a simple example that takes as inputs two numbers and adds them, takes the ratios and the differences between them.

```
In[225]:= myFunction[x_,y_]:=Module[{input1=x, input2=y,
         sum, difference, ratio, returnList},
         sum=input1+input2;
         difference=input1−input2;
         ratio=input1/input2;
```

2.9 Programming Essentials

```
            returnList={sum,difference,ratio};
            returnList];
In[226]:= myFunction[1,3]
```
Out[226]= $\{4,-2,\frac{1}{3}\}$

We can see that the internal assignments are not available globally:

```
In[227]:= returnList
Out[227]= returnList
```

Modules create new symbols for any local variables (with the naming convention being local name$number):

```
In[228]:= Module[{local},local]
Out[228]= local$6204
```

Block structures work with assigning local values to symbols, but the symbol names are available globally. **Block** is used for temporarily changing variables. The structure is:

```
In[229]:= ?Block
            Block[{x,y, ...,expr] specifies that expr is to be
               evaluated with local values for the
               symbols x, y, ....
            Block[{x=x0,...,expr] defines initial local values
               for x, ...]. >>
```

For a simple example:

```
In[230]:= a=30;
            f[x_]:=Block[{a,input1=x},a=40+x];
```

we evaluate f:

```
In[232]:= f[10]
Out[232]= 50
```

We will see that a is still as before:

```
In[233]:= a
Out[233]= 30
```

So a new assignment was temporarily used, but no new name is created by **Block** and values are not substituted:

```
In[234]:= Block[{a},a]
Out[234]= 30
```

Finally, **With** can be used to localize constants:

```
In[235]:= ?With
            With[{x=x0,y=y0,...,expr] specifies that all
               occurrences of the symbols x, y, ... in expr
               should be replaced by x0, y0, ...]. >>
In[236]:= With[{x=7,y=a+1},{x+y,x/y,x−y}]
Out[236]= {38,7/31,−24}
```

In[237]:= x
Out[237]= 1

If you are not manipulating variables, and just constants are local then you would use **With** instead of **Module**.

2.9.4 Patterns

The Wolfram Language has an extensive pattern matching ability. We have seen a simple example in the definition of functions, f [x_]:= **Expression**[x], where essentially a "dummy variable" x_ is matched to x on the right hand side and a delayed evaluation of **Expression**[x] is carried out on calling f[x]. The **Blank** following x on the left, _ , is the simplest pattern, and matches any expression. x_ means match any expression and name it x. **Blank** notation is used for pattern matching:

- _ : One blank, matches any expression.
- __ : Two blanks, matches a sequence of one or more expressions,
- ___: Three blanks, matches a sequence of zero or more expressions.

A useful function is **MatchQ**, which allows us to see if an expression matches a pattern - and good for practice too. A single _ will match anything:

In[238]:= **MatchQ**[q,_]
Out[238]= **True**

We can be more specific and follow a **Blank** with a **Head**. Since q has not been defined it matches a **Symbol**:

In[239]:= **MatchQ**[q,_Symbol]
Out[239]= **True**

In[240]:= **Head**[q]
Out[240]= **Symbol**

In[241]:= **MatchQ**[q,_Integer]
Out[241]= **False**

Let us now set q = 2 and try again

In[242]:= q=2;
         MatchQ[q,_Integer]
Out[243]= **True**

Let us try an algebraic expression:

In[244]:= **MatchQ**[r s^3 t, _^(_Integer)_]
Out[244]= **True**

We are matching two expression (r s to _ _) , raised to some expression that is an integer (3 to _Integer), followed by another expression (t to _). What if we want to check multiple expressions. We can use | to have an "OR" expression. Using q as before:

2.9 Programming Essentials

In[245]:= **MatchQ**[q,_Integer|_Real]
Out[245]= **True**

Whereas:

In[246]:= **MatchQ**[q,_Real]
Out[246]= **False**

We can use patterns in function definitions to define type. For example, let us define a function that only works if the first input is an integer and the second a real (decimal):

In[247]:= restrictedFunction[x_Integer,y_Real]:=x+y

We can see than when the inputs match the right pattern the function works as expected:

In[248]:= restrictedFunction[2,3.0]
Out[248]= 5.

However if either of the inputs does not match the pattern in the definition, we see that the function is left unevaluated, as it has not been defined for other patterns;

In[249]:= restrictedFunction[2,3]
Out[249]= restrictedFunction[2,3]

In[250]:= restrictedFunction[2.0,3.0]
Out[250]= restrictedFunction[2.,3.]

We can also see that the pattern persists in the information for our defined function - so we can always check our defined function:

In[251]:= ?restrictedFunction
         Global`restrictedFunction
         restrictedFunction[x_Integer,y_Real]:=x+y

We can often use patterns with conditions. Conditions are set as /; (you can think of this as saying "such that"). For example, let us match a pattern, so another pattern with a condition

In[252]:= **MatchQ**[{1,2,3},p_/;p[[3]]==3]
Out[252]= **True**

The above matches p such that the third element of p is identically equal to 3. Since {1,2,3} matches the pattern we get a True value returned. Pattern matching with conditions is very useful in substitutions. Let us say we have a list of expression values:

In[253]:= expressionValues={1,2,1.1,0.2,3,4,5,10,11.3,40,22.3};

Now let us substitute all values less than 5 with 5. We use a general pattern with a condition:

In[254]:= expressionValues/.p_/;p<5–> 5
Out[254]= {5,5,5,5,5,5,5,10,11.3,40,22.3}

In this expression, we match any number such that it is less than five, and then the rule assigns it the value 5.

We will use patterns throughout this book, and we will explain the matching when used. In addition you can get more familiar with patterns by checking out the following resources:

In[255]:= **SystemOpen**["paclet:tutorial/PatternsOverview"]
(*Execute the cell to open the help page, or type tutorial/PatternsOverview in the Wolfram Documentation Center*)

In[256]:= **SystemOpen**["paclet:guide/Patterns"]
(*Execute the cell to open the help page, or type guide/Patterns in the Wolfram Documentation Center*)

Additionally, patterns can be used extensively with the function **Cases**. This will return a list of elements in an expression matching a pattern:

In[257]:= Cases[expressionValues, 5]
Out[257]= {5}

In[258]:= Cases[expressionValues, p_/;p<5]
Out[258]= {1,2,1.1,0.2,3,4}

2.9.5 Pure Functions and Slots

We have seen so far how to define named functions, using f[x_]:= expression[x]. This gives the function the **Head** f and we can use it as such. The Wolfram Language additionally allows us to create pure functions that can be used "on the go" without having to define a name. You may find these are often referred to also as anonymous functions, or you may have heard the expression *lambda expression* in computation in the context of functional programming or the study of mathematical logic. We can use the function **Function** for this. For example, let us look at a function that multiplies a number by 2. The named function can be written as:

In[259]:= timesTwo[x_]:=2x

An equivalent pure function would look like:

In[260]:= **Function**[x,2x]
Out[260]= **Function**[x,2 x]

We can evaluate this on 3:

In[261]:= timesTwo[3]
Out[261]= 6

In[262]:= **Function**[x,2x][3]
Out[262]= 6

2.9 Programming Essentials

Notice how in the last version we have **Function**[x,body][argument supplied slot]. This may look cumbersome for now, but there are much more efficient ways to do this, through a shorthand use of slots. The supplied slot can be represented by # and the **Function** construct by &:

In[263]:= 2#&[3]
Out[263]= 6

2 # & will become 2 * 3 = 6.

Let us look at another example, say we want to raise a number to the 4th power. Let us look at equivalent constructions, first with a named function:

In[264]:= power4[x_]:= x^4
In[265]:= power4[3]
Out[265]= 81

Now with a pure function:

In[266]:= **Function**[x,x^4][3]
Out[266]= 81

And with the shorthand version:

In[267]:= #^4&[3]
Out[267]= 81

An even shorter way is to use a **Prefix** @ to represent the square brackets:

In[268]:= #^4&@3
Out[268]= 81

Slots can actually be numbered #1, #2, ..., #n if multiple arguments are supplied. For example:

In[269]:= (#1+#2)&[1,2]
Out[269]= 3

Here #1 refers to slot 1 and #2 refers to slot 2. Other pure function shorthands include ## to represent the sequence of all variables, and ##n for all variables starting from the nth one:

In[270]:= ##&[1,2,3,4]
Out[270]= **Sequence**[1,2,3,4]

In[271]:= **Plus**[##]&[1,2,3,4]
Out[271]= 10

In[272]:= ##2&[1,2,3,4]
Out[272]= **Sequence**[2,3,4]

In[273]:= **Plus**[##2]&[1,2,3,4]
Out[273]= 9

2.9.6 Function Manipulation

2.9.6.1 Map

The map function allows us to apply a function across elements in an expression at its first level, or any level desired. The full form is **Map**. The shorthand /@ is often used as well, and almost exclusively if applying a function at the first level.

```
In[274]:= ?Map
         Map[f,expr] or f/@expr applies f to each element on the
              first level in expr.
         Map[f,expr,levelspec] applies f to parts of expr
              specified by levelspec.
         Map[f] represents an operator form of Map that can be
              applied to an expression.>>
```

The easiest way to see how this works is to use a list {l1,l2,l3} and apply a function, g, to all elements in the list:

```
In[275]:= Map[g,{v1,v2,v3}]
Out[275]= {g[v1], g[v2], g[v3]}

In[276]:= g/@{v1,v2,v3}
Out[276]= {g[v1], g[v2], g[v3]}
```

Let us check how this works with a list of lists, which has more levels. If we do not specify a level we still apply to the first level, which is to all elements of the list:

```
In[277]:= Map[g,{{v1,1,{v13}},{v2,2,{v23}},{v3,3,{v33}}}]
Out[277]= g[{v1,1,{v13}}],g[{v2,2,{v23}}],g[{v3,3,{v33}}]
```

To apply to a specific level, we add in the level at the end:

```
In[278]:= Map[g,{{v1,1,{v13}},{v2,2,{v23}},{v3,3,{v33}}},{2}]
Out[278]= {{g[v1],g[1],g[{v13}]},
          {g[v2],g[2],g[{v23}]},
          {g[v3],g[3],g[{v33}]}}

In[279]:= Map[g,{{v1,1,{v13}},{v2,2,{v23}},{v3,3,{v33}}},{3}]
Out[279]= {{v1,1,{g[v13]}},
          {v2,2,{g[v23]}},
          {v3,3,{g[v33]}}}
```

If we want the function applied to all levels up to a specified level we do not give the curly brackets. For example we may want to apply g to all levels up to 3:

```
In[280]:= Map[g,{{v1,1,{v13}},{v2,2,{v23}},{v3,3,{v33}}},3]
Out[280]= {g[{g[v1],g[1],g[{g[v13]}]}],
          g[{g[v2],g[2],g[{g[v23]}]}],
          g[{g[v3],g[3],g[{g[v33]}]}]}
```

2.9 Programming Essentials

An extension of **Map** is **MapThread**:

In[281]:= ?**MapThread**
    MapThread[f,{{$a_1, a_2, \ldots$},{$b_1, b_2, \ldots$},...}] gives
        {$f(a_1, b_1, \ldots), f(a_2, b_2, \ldots), \ldots$}.
    MapThread[f,{$expr_1, expr_2, \ldots$},n] applies f to the parts
        of the $expr_i$ at level n.
    MapThread[f] represents an operator form of **MapThread**
        that can be applied to an expression. >>

MapThread will take the elements in the list and create a list where the function g is applied to corresponding levels across input expressions

In[282]:= **MapThread**[g,{{v1,1,{v13}},{v2,2,{v23}},{v3,3,{v33}}}]
Out[282]= {g[v1,v2,v3],g[1,2,3],g[{v13},{v23},{v33}]}

2.9.6.2 Apply

Apply is also used to conveniently replace headers in an expression. The forms are:

In[283]:= ?**Apply**
    Apply[f,expr] or f@@expr replaces the head of expr by f
    Apply[f,expr,{1}] or f@@@expr replaces heads at level 1
        of expr by f.
    Apply[f,expr,levelspec] replaces heads in parts of expr
        specified by levelspec.
    Apply[f] represents an operator form of **Apply** that can
        be applied to an expression. >>

The short forms function @ @ or function @ @ @ are used most often. The easiest example to understand how **Apply** works is again a list:

In[284]:= **Apply**[g,{{v1,1,{v13}},{v2,2,{v23}},{v3,3,{v33}}}]
Out[284]= g[{v1,1,{v13}},{v2,2,{v23}},{v3,3,{v33}}]

We see that the **Head** of the list (which is **List**) was replaced by g. As with map this can be done at a specific level by specifying with a curly bracket:

In[285]:= **Apply**[g,{{v1,1,{v13}},{v2,2,{v23}},{v3,3,{v33}}},{2}]
Out[285]= {{v1,1,g[v13]},{v2,2,g[v23]},{v3,3,g[v33]}}

If we want g replacing the Head at multiple locations up to and including a level:

In[286]:= **Apply**[g,{{v1,1,{v13}},{v2,2,{v23}},{v3,3,{v33}}},2]
Out[286]= {g[v1,1,g[v13]],
        g[v2,2,g[v23]],
        g[v3,3,g[v33]]}

2.9.6.3 Nest and NestList

Nest can be used to apply a function repeatedly to an expression. For example we can apply g to expression h 3 times:

```
In[287]:= Nest[g,h,3]
Out[287]= g[g[g[h]]]
```

NestList will return a list of the function's repeat applications from 0 to n:

```
In[288]:= NestList[g,h,3]
Out[288]= {h,g[h],g[g[h]],g[g[g[h]]]}

In[289]:= NestList[Cos,Sin[y],4]
Out[289]= {Sin[y],Cos[Sin[y]],Cos[Cos[Sin[y]]],
           Cos[Cos[Cos[Sin[y]]]],Cos[Cos[Cos[Cos[Sin[y]]]]]}
```

2.9.6.4 FoldList and Fold

FoldList takes three arguments, a function g, an expression h and a list v, say with elements v_i. It returns a list, where the first element is h. The subsequent elements are the function with arguments the previous value of the expression in the output list, and the next element in the input list v_i. This is repeated until the input list is traversed completely. This is easier to evaluate for an explanation:

```
In[290]:= FoldList[g,h,{v1}]
Out[290]= {h,g[h,v1]}

In[291]:= FoldList[g,h,{v1,v2}]
Out[291]= {h,g[h,v1],g[g[h,v1],v2]}

In[292]:= FoldList[g,h,{v1,v2,v3}]
Out[292]= {h,g[h,v1],g[g[h,v1],v2],g[g[g[h,v1],v2],v3]}
```

Fold is just the last element that you would get if you applied **FoldList**.

```
In[293]:= Fold[g,h,{v1,v2,v3,v4}]
Out[293]= g[g[g[g[h,v1],v2],v3],v4]
```

2.10 Strings

We can carry out a lot of string operations directly in the Wolfram Language:

```
In[294]:= Head["ACGT"]
Out[294]= String

In[295]:= stringExampleA="ACCTTACGGT";
          stringExampleB="CGCGgggTTACT";
```

There are different ways to manipulate strings. To change a string to upper case:

```
In[297]:= ToUpperCase[stringExampleB]
Out[297]= CGCGGGGTTACT
```

2.10 Strings

We can join strings either by using the **StringJoin** command, or the shortcut <> :

In[298]:= stringExampleA<>stringExampleB
Out[298]= ACCTTACGGTCGCGgggTTACT

In[299]:= **StringJoin**[stringExampleA, stringExampleB]
Out[299]= ACCTTACGGTCGCGgggTTACT

We can calculate the length of a string:

In[300]:= **StringLength**[stringExampleB]
Out[300]= 12

And we can break a string into its characters:

In[301]:= **Characters**[stringExampleB]
Out[301]= {C,G,C,G,g,g,g,T,T,A,C,T}

Let us check if our string contains Cs:

In[302]:= **StringContainsQ**[stringExampleA,"C"]
Out[302]= True

We can use patterns for strings, in this case we use "*C*", where the * indicates there can be characters before or after C.

In[303]:= **StringMatchQ**[stringExampleA,"*C*"]
Out[303]= True

We recommend reading the Patterns tutorial by typing "tutorial/StringPatterns" in the Wolfram Documentation Center in Mathematica to get familiar with the matching of string patterns. Patterns are a powerful feature of the Wolfram Language and will be used throughout this book.

We can count the different patterns or substrings within a string:

In[304]:= ?**StringCount**
> **StringCount**["string","sub"] gives a count of the number of times "sub" appears as a substring of "string".
> **StringCount**["string",patt] gives the number of substrings in "string" that match the general string expression patt.
> **StringCount**["string",{patt$_1$, patt$_2$,...}] counts the number of occurrences of any of the patt$_i$.
> **StringCount**[{$s_1, s_2,...$},p]] gives the list of results for each of the s_i. >>

In[305]:= **StringCount**[stringExampleA,"A"]
Out[305]= 2

Also we can replace strings. For example, we can replace all Ts with Us, for example if we were interested in the mRNA corresponding sequence:

In[306]:= **StringReplace**[stringExampleA, "T"→ "U"]
Out[306]= ACCUUACGGU

We can also take different parts of a string, like the 3rd letter:

```
In[307]:= StringTake[stringExampleA, {3}]
Out[307]= C
```

Or the 2nd letter from the end

```
In[308]:= StringTake[stringExampleA, {-2}]
Out[308]= G
```

We can take the first 5 letters:

```
In[309]:= StringTake[stringExampleA, 5]
Out[309]= ACCTT
```

Or we can take the last 5 letters:

```
In[310]:= StringTake[stringExampleA, -5]
Out[310]= ACGGT
```

And we can take strings pieces between locations:

```
In[311]:= StringTake[stringExampleA, {2, 5}]
Out[311]= CCTT
```

2.10.1 Character Codes

In the Wolfram Language we can use convert between characters and encodings:

```
In[312]:= ?ToCharacterCode
          ToCharacterCode["string"] gives a list of the integer
               codes corresponding to the characters in a
               string.
          ToCharacterCode["string","encoding"] gives integer
               codes according to the specified encoding. >>

In[313]:= ?FromCharacterCode
          FromCharacterCode[n] gives a string consisting of the
               character with integer code n.
          FromCharacterCode[{n_1, n_2, ...}] gives a string consisting
               of the sequence of characters with codes $n_i$.
          FromCharacterCode[{{n_{11}, n_{12}, ...}, {n_{21}, ...}, ...}] gives a list of
               strings.
          FromCharacterCode[..., "encoding"] uses the specified
               character encoding. >>

In[314]:= ToCharacterCode["C"]
Out[314]= {67}

In[315]:= FromCharacterCode[67]
Out[315]= C
```

2.10.2 Sequences as Strings

We have seen that the Wolfram Language has extensive string capabilities. When we think of nucleotide sequences, we can think of them as strings and manipulate them as such. For example, given a short sequence, we can get the complement:

In[316]:= stringExampleC="ACCttACGGT"
Out[316]= ACCttACGGT

In[317]:= **StringReplace**[stringExampleC,
{"A"→ "T","T"→"A","G"→ "C","C"→ "G"},
IgnoreCase→**True**]
Out[317]= TGGAATGCCA

If we wanted, we could also get the reverse complement by reversing the string:

In[318]:= **StringReverse**[**StringReplace**[ACGCAGTT,
{"A"→ "T","T"→"A",
"G"→ "C","C"→ "G"},
IgnoreCase→**True**]]
Out[318]= ACCGTAAGGT

We can even write a short function for this:

In[319]:= nucleotideSequenceManipulation[x_,**OptionsPattern**[]]:=
Module[{inputString=x,
operation=OptionValue["Operation"]},
Print[inputString];
Return[**Switch**[operation,
"None",inputString,
"UpperCase",**ToUpperCase**[inputString],
"Complement",
StringReplace[inputString,{"A"→"T",
"T"→"A","G"→"C","C"→"G"},
IgnoreCase→ **True**],
"ReverseComplement",**StringReverse**[**StringReplace**[
inputString,{"A"→"T","T"→"A","G"→"C","C"→"G"},
IgnoreCase→**True**]]]]];
Options[nucleotideSequenceManipulation]=
{"Operation"→ "None"};

Notice that we have given our function optional arguments, an **OptionsPattern**. The **Options** for the function are specified in the second statement. Internally, the function takes the value of an option and assigns it to a local variable. If an option is not specified, the assigned value in the **Options** rule for the option is used as the default. Let us check our simple function applied to a string:

In[321]:= nucleotideSequenceManipulation[stringExampleA,
"Operation"→ "ReverseComplement"]

(∗the following line is printed during the evaluation∗)
ACCTTACGGT
Out[321]= ACCGTAAGGT

2.11 Importing and Exporting Data

2.11.1 Importing Files

We can use the Wolfram Language to import data. We can also look at directory structures and change the directory. We can first print our current working directory:

In[322]:= **Directory**[]
Out[322]= /Users/user

Additionally we can check what directory our current Notebook is in:

In[323]:= **NotebookDirectory**[]
Out[323]= /Users/user/myMathematicaBookFolder/

We can set the directory using the **SetDirectory**["directory path string"]. And we can use the following to set the directory to the **NotebookDirectory**[].

In[324]:= **SetDirectory**[**NotebookDirectory**[]]
Out[324]= /Users/user/myMathematicaBookFolder/

Assuming you have downloaded the data accompanying the manuscript and are working through the Chap. 2 notebook, we can look at the file names in the current **Directory**:

In[325]:= **FileNames**[]

There is a small text file we can import:

In[326]:= **Import**["textfile"]
Out[326]= Hello World!

You will notice that the file content is displayed in the output. This can be a problem for big files, so you might want to suppress the output using ;. Let us import another file, with tab separated values content. This file is actually part of a bigger file from RNA-sequencing results, using TopHat [5] and Cufflinks.

In[327]:= inputTSV=**Import**["textData.tsv"]
Out[327]= {{tracking_id, class_code, nearest_ref_id, gene_id,
       gene_short_name, tss_id, locus, length, coverage,
       FPKM, FPKM_conf_lo, FPKM_conf_hi, FPKM_status},
      {MIR1302−2,−,−,MIR1302−2,MIR1302−2,TSS10595,
        chr1:30365−30503,−,−,0,0,0,OK},{FAM138A,−,−,FAM138A,
        FAM138A,TSS10184,chr1:34610−36081,
        −,−,0,0,0,OK},{OR4F5,−,−,OR4F5,OR4F5,
        TSS17749,chr1:69090−70008,−,−,0,0,0,OK},
       {DDX11L1,−,−,DDX11L1,DDX11L1,TSS18303,
        chr1:11873−14409,−,−,0.485822,0.362285,0.609358,OK},
       {WASH7P,−,−,WASH7P,WASH7P,TSS9156,chr1:14361−29370,
        −,−,2.14079,1.87911,2.40247,OK},
       {MIR6859−1,−,−,MIR6859−1,MIR6859−1,TSS27452,
        chr1:17368−17436,−,−,0,0,0,OK},
       {MIR6859−1,−,−,MIR6859−1,MIR6859−1,TSS34555,
        chr1:187890−187958,−,−,0,0,0,OK},

2.11 Importing and Exporting Data

```
{FAM138D,−,−,FAM138D,FAM138D,TSS27427,
 chr1:205128−206597,−,−,0,0,0,OK},
{OR4F29,−,−,OR4F29,OR4F29,TSS18830,
 chr1:450739−451678,−,−,0,0,0,OK},
{LOC729737,−,−,LOC729737,LOC729737,TSS22631,
 chr1:134772−140566,−,−,39.3387,38.7348,39.9426,OK},
{LOC101928626,−,−,LOC101928626,LOC101928626,
 TSS18498,chr1:627379−629009,−,−,0,0,0,OK},
{LOC100132287,−,−,LOC100132287,LOC100132287,TSS27248,
 chr1:490755−495445,−,−,13.5586,13.1669,13.9502,OK},
{MIR6723,−,−,MIR6723,MIR6723,TSS9033,
 chr1:632324−632413,−,−,0,0,0,OK},
{OR4F29,−,−,OR4F29,OR4F29,TSS33796,chr1:685715−686654,
 −,−,0,0,0,OK},{FAM87B,−,−,FAM87B,FAM87B,
 TSS11238,chr1:817370−819834,−,−,0,0,0,OK},
{LINC00115,−,−,LINC00115,LINC00115,TSS33282,
 chr1:826205−827522,−,−,0.0559651,0.00962327,0.102307,
 OK},{FAM41C,−,−,FAM41C,FAM41C,TSS15687,
 chr1:868070−876802,−,−,0.0280826,0,0.0577769,OK},
{LOC100130417,−,−,LOC100130417,LOC100130417,
 TSS15106,chr1:916817−919692,−,−,0,0,0,OK}}
```

We can also export the data. Notice that after our input, we have comma separated values. So, an easy and natural export format for these data would be a comma separated values (csv) file. We use **Export**, and add a .csv extension to the output name.

In[328]:= **Export**["outputCSV.csv",inputTSV]
Out[328]= outputCSV.csv

The Wolfram Language will recognize multiple file formats automatically. Alternatively we can specify the format if we know what it is explicitly. We can get a list of the file import formats that the Wolfram Language has inbuilt parsing for:

In[329]:= **$ImportFormats**
Out[329]= {3DS,ACO,Affymetrix,AgilentMicroarray,AIFF,ApacheLog,
 ArcGRID,AU,AVI,Base64,BDF,Binary,Bit,BMP,Byte,BYU,
 BZIP2,CDED,CDF,Character16,Character8,CIF,Complex128,
 Complex256,Complex64,CSV,CUR,DAE,DBF,DICOM,DIF,DIMACS,
 Directory,DOT,DXF,EDF,EML,EPS,ExpressionJSON,
 ExpressionML,FASTA,FASTQ,FCS,FITS,FLAC,GenBank,
 GeoJSON,GeoTIFF,GIF,GPX,Graph6,Graphlet,GraphML,GRIB,
 GTOPO30,GXL,GZIP,HarwellBoeing,HDF,HDF5,HIN,HTML,ICC,
 ICNS,ICO,ICS,Ini,Integer128,Integer16,Integer24,
 Integer32,Integer64,Integer8,JavaProperties,
 JavaScriptExpression,JCAMP−DX,JPEG,JPEG2000,JSON,JVX,
 KML,LaTeX,LEDA,List,LWO,M4A,MAT,MathML,MBOX,MDB,MESH,
 MGF,MIDI,MMCIF,MOL,MOL2,MP3,MPS,MTP,MTX,MX,MXNet,
 NASACDF,NB,NDK,NetCDF,NEXUS,NOFF,OBJ,ODS,OFF,OGG,
 OpenEXR,Package,Pajek,PBM,PCX,PDB,PDF,PGM,PHPIni,PLY,
 PNG,PNM,PPM,PXR,PythonExpression,QuickTime,Raw,
 RawBitmap,RawJSON,Real128,Real32,Real64,RIB,RSS,RTF,
 SCT,SDF,SDTS,SDTSDEM,SFF,SHP,SMILES,SND,SP3,Sparse6,
 STL,String,SurferGrid,SXC,Table,TAR,TerminatedString,

TeX, Text, TGA, TGF, TIFF, TIGER, TLE, TSV, UBJSON,
UnsignedInteger128, UnsignedInteger16,
UnsignedInteger24, UnsignedInteger32, UnsignedInteger64,
UnsignedInteger8, USGSDEM, UUE, VCF, VCS, VTK, WAV, Wave64,
WDX, WebP, WLNet, WMLF, XBM, XHTML, XHTMLMathML, XLS, XLSX,
XML, XPORT, XYZ, ZIP}

There is a similar list for export formats.

In[330]:= **$ExportFormats**
Out[330]= {3DS, ACO, AIFF, AU, AVI, Base64, Binary, Bit, BMP, Byte, BYU,
BZIP2, C, CDF, Character16, Character8, Complex128,
Complex256, Complex64, CSV, CUR, DAE, DICOM, DIF, DIMACS,
DOT, DXF, EMF, EPS, ExpressionJSON, ExpressionML, FASTA,
FASTQ, FCS, FITS, FLAC, FLV, GeoJSON, GIF, Graph6, Graphlet,
GraphML, GXL, GZIP, HarwellBoeing, HDF, HDF5, HTML,
HTMLFragment, ICNS, ICO, Ini, Integer128, Integer16,
Integer24, Integer32, Integer64, Integer8,
JavaProperties, JavaScriptExpression, JPEG, JPEG2000,
JSON, JVX, KML, LEDA, List, LWO, M4A, MAT, MathML, Maya, MGF,
MIDI, MOL, MOL2, MP3, MTX, MX, MXNet, NASACDF, NB, NetCDF,
NEXUS, NOFF, OBJ, OFF, OGG, Package, Pajek, PBM, PCX, PDB,
PDF, PGM, PHPIni, PICT, PLY, PNG, PNM, POV, PPM, PXR,
PythonExpression, QuickTime, RawBitmap, RawJSON,
Real128, Real32, Real64, RIB, RTF, SCT, SDF, SND, Sparse6,
STL, String, SurferGrid, SVG, SWF, Table, TAR,
TerminatedString, TeX, TeXFragment, Text, TGA, TGF, TIFF,
TSV, UBJSON, UnsignedInteger128, UnsignedInteger16,
UnsignedInteger24, UnsignedInteger32,
UnsignedInteger64, UnsignedInteger8, UUE, VideoFrames,
VRML, VTK, WAV, Wave64, WDX, WebP, WLNet, WMLF, X3D, XBM,
XHTML, XHTMLMathML, XLS, XLSX, XML, XYZ, ZIP, ZPR}

Going back to our examples we could have specified the import or export format explicitly. We can look at the first three elements:

In[331]:= **Import**["textData.tsv","TSV"][[1;;3]]
Out[331]= {{tracking_id, class_code, nearest_ref_id, gene_id,
gene_short_name, tss_id, locus, length, coverage, FPKM,
FPKM_conf_lo, FPKM_conf_hi, FPKM_status},
{MIR1302−2,−,−,MIR1302−2,MIR1302−2,TSS10595,
chr1:30365−30503,−,−,0,0,0,OK},
{FAM138A, −, −, FAM138A,FAM138A, TSS10184,
chr1:34610−36081,−,−,0,0,0,OK}}

If we do the import, and specify the wrong format we get an error:

In[332]:= **Import**["textData.tsv","XLS"]

(∗the line below is output during the evaluation∗)
...**Import**::fmterr: Cannot import data as XLS format.
Out[332]= **$Failed**

We can also import the same information in different formats. We may want a continuous text:

2.11 Importing and Exporting Data 51

Fig. 2.6 Out [336]= Open Stream for Writing, and, **Out** [337]= Open Streams

In[336]:= `temporaryFile = OpenWrite["temp"]`

Out[336]= OutputStream[⋯▸ Name: temp Unique ID: 3]

In[337]:= `Streams[]`

Out[337]= {OutputStream[⋯▸ Name: stdout Unique ID: 1],

OutputStream[⋯▸ Name: stderr Unique ID: 2],

OutputStream[⋯▸ Name: temp Unique ID: 3]}

In[333]:= `importText=`**Import**`["textData.tsv","Text"];`

This is a string now

In[334]:= **Head**[importText]
Out[334]= **String**

We can extract parts of the string:

In[335]:= **StringTake**[importText,{1,200}]
Out[335]= tracking_id class_code
      nearest_ref_id   gene_id   gene_short_name
      tss_id   locus   length   coverage   FPKM
      FPKM_conf_lo   FPKM_conf_hi   FPKM_status
      MIR1302–2 – – MIR1302–2 MIR1302–2
      TSS10595   chr1:30365–3050 – – 0

In this case the importing does not give us the structured list we expected. Sometimes you may have to try multiple importer formats, or build an importer yourself using low level file operations described in the next section.

2.11.2 Streams

In the Wolfram Language a lot of the importing and writing of files is done on whole files. While this works well for small files and known file formats, for larger files, or unknown formats it may be necessary to use low level streams to process file data.

To be able to write to a file we use the **OpenWrite** command, to open a named stream, which is represented graphically: **Streams**[] tells us what open streams exist, Fig. 2.6.

Fig. 2.7 Out [341]= Open Streams

```
In[341]:= Streams[]

Out[341]= {OutputStream[  Name: stdout, Unique ID: 1 ],

           OutputStream[  Name: stderr, Unique ID: 2 ]}
```

There are 3 **OutputStream** elements in the list, Fig. 2.6, the stdout and stderr ones are default. The last one, with the "Name: temp" and "Unique ID: 3" is the one we have opened. Once a stream is open we can write to it directly:

In[338]:= ?**Write**
    Write[channel, expr$_1$, expr$_2$, ...] writes the expressions expr$_1$ in sequence, followed by a newline, to the specified output channel. >>

Let us write a few expressions to our stream:

In[339]:= **Write**[temporaryFile,
    Integrate[y^2,{y,xmin,xmax}], "Test", z]

We can close the stream:

In[340]:= **Close**[temporaryFile]
Out[340]= temp

We can verify what streams are still open using **Streams**, Fig. 2.7.

Note that the variable defined previously still is the stream, even though it is now a closed stream.

In[342]:= temporaryFile
Out[342]= **OutputStream**[temp,3]

Let us check what we wrote in our file by importing the entire file:

In[343]:= **Import**["temp"]
Out[343]= xmax^3/3 − xmin^3/3"Test"{1, 1, 1}

Note that the information was simply concatenated together. Additionally, the first function evaluated prior to being written to the stream. When passing expressions to a stream we must be careful in terms of how we plan to distribute the file.

If we want to append to an existing file we can use **OpenAppend** to append to the file. Also we can write strings directly to a file:

In[344]:= temporaryFileWrite=**OpenAppend**["temp"];
    WriteString[temporaryFileWrite,"OutputString"];
    Close[temporaryFileWrite];

We can see that our string was written in the last two lines:

In[347]:= **Import**["temp"]
Out[347]= xmax^3/3 − xmin^3/3"Test"{1, 1, 1}
    OutputString

2.11 Importing and Exporting Data

In[348]:= **temporaryReadFile = OpenRead["temp"]**

Out[348]= **InputStream**[Name: temp Unique ID: 3]

Fig. 2.8 Out [348]= OpenRead Stream

We can also open a file for reading, Fig. 2.8.
We can read an expression in one piece at a time from a line in a specified format:

In[349]:= **Read**[temporaryReadFile , **Word**]
Out[349]= xmax^3/3

Or we can specify how to read in a **String**:

In[350]:= **Read**[temporaryReadFile , **String**]
Out[350]= − xmin^3/3 "Test" {1, 1, 1}

We can read a **Character** at a time:

In[351]:= **Read**[temporaryReadFile , **Character**]
Out[351]= O

In[352]:= **Read**[temporaryReadFile , **Character**]
Out[352]= u

Or a line from where we are in the stream:

In[353]:= ReadLine[temporaryReadFile]
Out[353]= tputString

In[354]:= ReadLine[temporaryReadFile]
Out[354]= **EndOfFile**

Note that we have reached the **EndOfFile** expression, and no more lines can be read. The stream position is at the end of the file.

In[355]:= **StreamPosition**[temporaryReadFile]
Out[355]= 47

We can set the position in the stream to a particular location:

In[356]:= **SetStreamPosition**[temporaryReadFile ,10]
Out[356]= 10

Now reading will continue from our current position in the stream:

In[357]:= **Read**[temporaryReadFile ,**Word**]
Out[357]= xmin^3/3 "Test" {1,

Always remember to close your streams once done:

In[358]:= **Close**[temporaryReadFile]
Out[358]= temp

2.12 Associations and Datasets

The Wolfram Language focuses on the use of lists, and more recently associations. Associations are similar to dictionaries in Python [4], and offer an effective structure for storing information with keys and values. An association has the form $< |key_1 \to value_1, key_2 \to value_2, ..., key_n \to value_n| >$. Here is an example, say we have:

```
In[359]:= patients=<|"subject1"→ "cancer",
              "subject2"→ "healthy",
              "subject3"→ "healthy",
              "subject4"→ "healthy",
              "subject5"→ "cancer"|>;
```

We can access the subjects by using single brackets and the keys to get the values (health status)

```
In[360]:= patients["subject5"]
Out[360]= cancer
```

We can also just use indexing like with lists, as Associations also preserve the order of the assignments:

```
In[361]:= patients[[2]]
Out[361]= healthy
```

We can extract **Keys** and **Values** of an association:

```
In[362]:= Keys[patients]
Out[362]= {subject1, subject2, subject3, subject4, subject5}
```

```
In[363]:= Values[patients]
Out[363]= {cancer, healthy, healthy, healthy, cancer}
```

For dealing with large datasets it is very useful to have annotations or query capabilities to search data. For example, consider the input TSV file from earlier when we imported sample data from a transcriptome experiment.

```
In[364]:= inputTSV[[1;;3]]
Out[364]= {{tracking_id, class_code, nearest_ref_id, gene_id,
           gene_short_name, tss_id, locus, length, coverage,
           FPKM, FPKM_conf_lo, FPKM_conf_hi, FPKM_status},
          {MIR1302-2,-,-,MIR1302-2,MIR1302-2,TSS10595,
           chr1:30365-30503,-,-,0,0,0,OK},
          {FAM138A,-,-,FAM138A,FAM138A,TSS10184,
           chr1:34610-36081,-,-,0,0,0,OK}}
```

We may be interested in the tracking_id column, the gene_short_name and the FPKM values. These are columns 1, 5 and 10 respectively:

```
In[365]:= inputTSVGeneFPKM=inputTSV[[All,{1,5,10}]]
Out[365]= {{tracking_id, gene_short_name, FPKM},
          {MIR1302-2,MIR1302-2,0},{FAM138A,FAM138A,0},
          {OR4F5,OR4F5,0},{DDX11L1,DDX11L1,0.485822},
          {WASH7P,WASH7P,2.14079},{MIR6859-1,MIR6859-1,0},
          {MIR6859-1,MIR6859-1,0},{FAM138D,FAM138D,0},
```

2.12 Associations and Datasets

{OR4F29,OR4F29,0},{LOC729737,LOC729737,39.3387},
{LOC101928626,LOC101928626,0},
{LOC100132287,LOC100132287,13.5586},
{MIR6723,MIR6723,0},{OR4F29,OR4F29,0},
{FAM87B,FAM87B,0},{LINC00115,LINC00115,0.0559651},
{FAM41C,FAM41C,0.0280826},
{LOC100130417,LOC100130417,0}}

Now imagine we had a full list of 20,000+ components and we want to look up data for one gene. That would be a bit cumbersome. We can instead make an association computationally by using **AssociationThread**:

In[366]:= ?**AssociationThread**

    AssociationThread[{key_1, key_2, ...} → {val_1, val_2, ...}] gives the association <|key_1 → val_1, key_2 → val_2, ...|>.
    AssociationThread[{key_1, key_2, ...}, {val_1, val_2, ...}] also gives the association <|key_1 → val_1, key_2 → val_2, ...|>. >>

In[367]:= inputTSVGeneAssociation=**AssociationThread**[inputTSV[[**All**,1]]→inputTSV[[**All**,{5,10}]]]
Out[367]= <|tracking_id→{gene_short_name,FPKM},
MIR1302−2→{MIR1302−2,0},FAM138A→{FAM138A,0},
OR4F5→{OR4F5,0},DDX11L1→{DDX11L1,0.485822},
WASH7P→{WASH7P,2.14079},
MIR6859−1→{MIR6859−1,0},FAM138D→{FAM138D,0},
OR4F29→{OR4F29,0},
LOC729737→{LOC729737,39.3387},
LOC101928626→{LOC101928626,0},
LOC100132287→{LOC100132287,13.5586},
MIR6723→{MIR6723,0},
FAM87B→{FAM87B,0},LINC00115→{LINC00115,0.0559651},
FAM41C→{FAM41C,0.0280826},
LOC100130417→{LOC100130417,0}|>

Now we can easily get information by the tracking id if we know it:

In[368]:= inputTSVGeneAssociation["WASH7P"]
Out[368]= {WASH7P,2.14079}

Another way to extract information form associations or structured lists is **Query**:

In[369]:= ?**Query**

    Query[$operator_1$, $operator_2$, ...], represents a query that can be applied to a **Dataset** object, in which the successive $operator_i$ are applied at successively deeper levels. >>

Query is typically used using a prefix form. Let us get the second element of the association:

In[370]:= **Query**[2]@inputTSVGeneAssociation
Out[370]= {MIR1302−2,0}

We can get the second, third and fifth element:

```
In[371]:= Query[{2,5}]@inputTSVGeneAssociation
Out[371]= <|MIR1302-2→{MIR1302-2,0},
           DDX11L1→{DDX11L1,0.485822}|>
```

Query is a very powerful function that can apply operators at deep levels of our association, either descending into it our ascending out of it. First descending into the association, the operators are applied without changing the structure of the data in terms of levels. Parts of the association are descending operators. Ascending operators apply to the results and can modify the data structure. They are applied after descending operators are done. A list of descending and ascending operators are given in the discussion of the details in "ref/Query" in the Wolfram Documentation Center.

Let us get the same elements as before and now extract the second value from each:

```
In[372]:= Query[{2,3,5},2]@inputTSVGeneAssociation
Out[372]= <|MIR1302-2→0,FAM138A→0,DDX11L1→0.485822|>
```

Now let us get the maximum:

```
In[373]:= Query[{2,3,5}/*Max,2]@inputTSVGeneAssociation
Out[373]= 0.485822
```

Here **Max** is an ascending operator, applied at the first level after the other operations.

The syntax for operators is $operator_1$/*$operator_2$/*... where successive ascending operators are applied one after the other.

Notice that this is a long list, and if we would like to find information in the list we need to search it. We can make a **Dataset**, from structured data:

```
In[374]:= datasetInputTSV=Dataset[inputTSV]
Out[374]= (*See Figure [2.9]*)
```

The output is well structured and can be navigated in the Notebook - see Fig. 2.9.

Additionally, we can make a **Dataset** from any nested list or association. Datasets from an association have clickable keys, and are easy to navigate. Similarly to associations, datasets can facilitate visualizing and quickly accessing information and we can use **Query** in similar fashion with an **Association**, with **Values** extracted using **Keys**, and we can perform complex operations can be performed. For more detail and powerful advanced features please consult the documentation:

```
In[375]:= SystemOpen["paclet:ref/Dataset"]
```

2.13 Exceptions

In the Wolfram Language we can use **Message** to raise exceptions to the usage of a function.

2.13 Exceptions

In[374]:= `datasetInputTSV = Dataset[inputTSV]`

tracking_id	class_code	nearest_ref_id	gene_id
MIR1302-2	-	-	MIR1302-2
FAM138A	-	-	FAM138A
OR4F5	-	-	OR4F5
DDX11L1	-	-	DDX11L1
WASH7P	-	-	WASH7P
MIR6859-1	-	-	MIR6859-1
MIR6859-1	-	-	MIR6859-1
FAM138D	-	-	FAM138D
OR4F29	-	-	OR4F29
LOC729737	-	-	LOC729737
LOC101928626	-	-	LOC101928626
LOC100132287	-	-	LOC100132287
MIR6723	-	-	MIR6723
OR4F29	-	-	OR4F29
FAM87B	-	-	FAM87B
LINC00115	-	-	LINC00115
FAM41C	-	-	FAM41C
LOC100130417	-	-	LOC100130417

Fig. 2.9 Out [374]= An Imported **Dataset**. We can import structured data and convert them to a Dataset - see **In**[374].

In[376]:= ?**Message**
    Message[symbol::tag] prints the message symbol::tag unless it has been switched off.
    Message[symbol::tag,$e_1, e_2, \ldots$] prints a message, inserting the values of the e_i as needed. >>

2.14 Graphical Capabilities

We have not discussed yet one of the best aspects of the Wolfram Language and Mathematica which is the ability to produce high-quality graphical output. Throughout this book we will see many examples. Here we only offer a very brief introduction. The simplest plot usually shown in a plot of a sine or cosine function. We can use **Plot** to generate a plot of a function f[x] evaluating it with the argument x taking values between a and b, entered in the form {x,a,b}.

In[377]:= ?**Plot**
    Plot[$f, \{x, x_{\min}, x_{\max}\}$] generates a plot of f as a function of x from x_{min} to x_{max}.
    Plot[$\{f_1, f_2, \ldots\}, \{x, x_{\min}, x_{\max}\}$] plots several functions f_i.
    Plot[$\{\ldots, w(f_i), \ldots\}, \ldots$] plots f_i with features defined by the symbolic wrapper w.

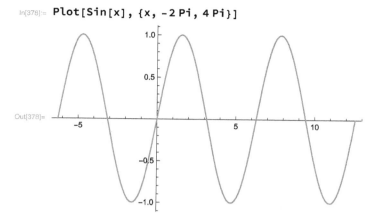

Fig. 2.10 Out [378]= Plot of Sine

> Plot[..., {x} ∈ reg] takes the variable x to be in the geometric region *reg*. >>

We can look at the plot from −2Pi to 4Pi in Fig. 2.10.

Plot takes extensive optional arguments which we can use to customize the look of graphics:

In[379]:= **Options[Plot]**
Out[379]= {AlignmentPoint→Center, AspectRatio→1/GoldenRatio,
Axes→True, AxesLabel→None, AxesOrigin→Automatic,
AxesStyle→{}, Background→None,
BaselinePosition→Automatic, BaseStyle→{},
ClippingStyle→None, ColorFunction→Automatic,
ColorFunctionScaling→True, ColorOutput→Automatic,
ContentSelectable→Automatic,
CoordinatesToolOptions→Automatic,
DisplayFunction:>$DisplayFunction, Epilog→{},
Evaluated→Automatic, EvaluationMonitor→None,
Exclusions→Automatic, ExclusionsStyle→None,
Filling→None, FillingStyle→Automatic,
FormatType:>TraditionalForm, Frame→False,
FrameLabel→None, FrameStyle→{}, FrameTicks→Automatic,
FrameTicksStyle→{}, GridLines→None, GridLinesStyle→{},
ImageMargins→0., ImagePadding→All,
ImageSize→Automatic, ImageSizeRaw→Automatic,
LabelStyle→{}, MaxRecursion→Automatic, Mesh→None,
MeshFunctions→{#1&}, MeshShading→None,
MeshStyle→Automatic, Method→Automatic,
PerformanceGoal:>$PerformanceGoal, PlotLabel→None,
PlotLabels→None, PlotLegends→None,
PlotPoints→Automatic, PlotRange→{Full, Automatic},
PlotRangeClipping→True, PlotRangePadding→Automatic,
PlotRegion→Automatic, PlotStyle→Automatic,
PlotTheme:>$PlotTheme, PreserveImageOptions→Automatic,

2.14 Graphical Capabilities 59

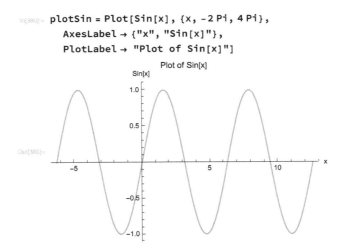

Fig. 2.11 Out [380]= Plot of Sine with labels and plot title

Prolog→ { } , RegionFunction→(**True**&) , **RotateLabel**→**True** ,
ScalingFunctions→**None** , TargetUnits→**Automatic** ,
Ticks→**Automatic** , TicksStyle→{ } ,
WorkingPrecision→**MachinePrecision** }

The options are provided to the function through a set of rules. For example we can add axes labels and a plot title, which is the minimum usually required for scientific plots, Fig. 2.11.

We can also define a **PlotTheme**, here chosen as "Scientific", as shown in Fig. 2.12.

Notice how we actually had to change from giving the option of **AxesLabel** to providing **FrameLabel** values. Plot can actually take many inputs as a list, and we can add legends as shown in Fig. 2.13.

We often need to plot numerical values, for example a list of paired values

In[383]:= exampleDataX={−5.74,−4.77,−3.47,−2.78,−1.59,−0.50,
  0.07,1.08,1.82,3.36,4.14,5.40,6.29,
  7.15,8.29,9.57,10.43,11.32,11.97};
  exampleDataY={0.36,1.10,1.18,0.74, 0.14, 0.84, 0, 34,
  1.08,1.20,−0.50, −0.41, −0.25, 1.23,
  1.32, 0.98, 0.50, −0.79, −0.55};

We can use **Transpose** to put the pairs together:

In[385]:= pairedXY=**Transpose**[{exampleDataX,exampleDataY}]
Out[385]= {{−5.74,0.36},{−4.77,1.1},{−3.47,1.18},
  {−2.78,0.74},{−1.59,0.14},{−0.5,0.84},{0.07,0},
  {1.08,34},{1.82,1.08},{3.36,1.2},{4.14,−0.5},
  {5.4,−0.41},{6.29,−0.25},{7.15,1.23},{8.29,1.32},
  {9.57,0.98},{10.43,0.5},{11.32,−0.79},{11.97,−0.55}}

In[381]:= `plotSin2 = Plot[Sin[x], {x, -2 Pi, 4 Pi},`
　　　　　`FrameLabel → {"x", "Sin[x]"},`
　　　　　`PlotLabel → "Plot of Sin[x]",`
　　　　　`PlotTheme → "Scientific"]`

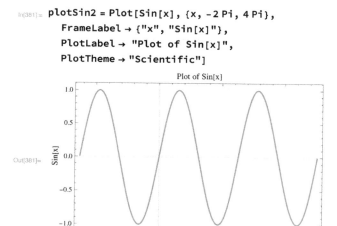

Fig. 2.12 Out [381]= Plot of Sine with PlotTheme

In[382]:= `plotSinCos = Plot[{Sin[x], Cos[x]}, {x, -2 Pi, 4 Pi},`
　　　　　`FrameLabel → {"x", "Sin[x]"},`
　　　　　`PlotLabel → "Plot of Sin[x]",`
　　　　　`PlotTheme → "Scientific",`
　　　　　`PlotLegends → {"Sin[x] Label", "Cos[x] Label"}]`

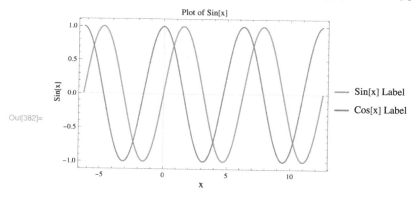

Fig. 2.13 Out [382]= Plot of list of functions

Now we can plot the points using **ListPlot**, as in Fig. 2.14.

You may have noticed that we have been assigning the plots to variables. We can use the variables and the **Show** function to combine graphics, as in Fig. 2.15.

2.15 Additional Capabilities Through Packages

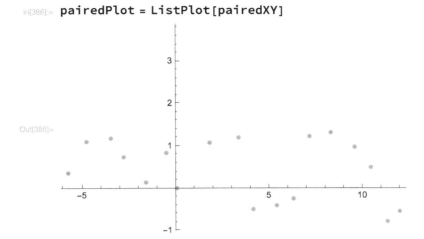

Fig. 2.14 Out [386]= ListPlot of paired data

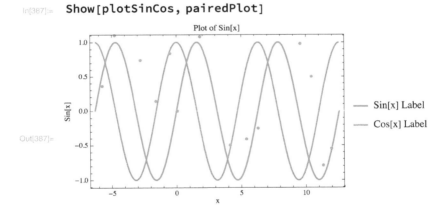

Fig. 2.15 Out [387]= Showing multiple plots together

2.15 Additional Capabilities Through Packages

2.15.1 Importing Packages

The Wolfram Language has many packages that have functions already defined to assist in various operations. Packages can be imported in 2 ways, with a **Needs** statement, or a **Get** statement (shorthand: << packageName`). As an example we will use the MathIOmica package in this manuscript as introduced below.

2.15.2 MathIOmica

In our discussions in various chapters we will make use of the package MathIOmica [3] created by the Mias Lab. This is an omics analysis package written in the Wolfram Language, and particularly designed to be used for dynamics (time series/longitudinal data). MathIOmica has functions for importing mapped omics data from text files to created collated analysis objects, and also standard normalization and filtering routines. Additionally MathIOmica offers visualization with combined dendrogram/heatmap plots based on cluster results, representation on biological pathways, and a mass spectrometry spectral viewer. Finally, MathIOmica offers basic enrichment analysis and annotation for Gene Ontology (GO) [1] and biological pathways [2].

2.15.2.1 Installation

The MathIOmica package can be downloaded from

In[388]:= **SystemOpen**[
    "https://github.com/gmiaslab/mathiomica/releases"]

The home page is:

In[389]:= **SystemOpen**["https://github.com/gmiaslab/mathiomica"]

And also:

In[390]:= **SystemOpen**["http://mathiomica.org."]

Once you have downloaded a release you can install it though the following steps:

1. Open Mathematica (version 11.2 and higher)
2. In the File menu select Install...
3. In the popup menu choose the following for each item:
    a. "Type of Item to Install:" Application
    b. "Source": From Directory... - Navigate to and select the "MathIOmica" directory containing the installation instructions document – Choose "Install for this User Only (username)"
4. Press OK.
5. Quit and restart Mathematica.
6. The package should now available with <<**MathIOmica`**
7. Documentation should be available in Mathematica's help browser (search for MathIOmica)

You can simply evaluate the following after installation to open the MathIOmica guide:

2.15 Additional Capabilities Through Packages   63

In[391]:= **SystemOpen**["paclet:MathIOmica/guide/MathIOmicaGuide"]

In[392]= << **MathIOmica`**

>   (*the following line is printed during evaluation*)
>   MathIOmica (http://mathiomica.org), by G. Mias Lab.

We anticipate to continue to add functionality to the MathIOmica package, and help Wolfram Language coders perform their own analysis. If you have requests for functionality you can e-mail us at mathiomica@gmail.com - we do not respond to requests directly, but will make our best effort to update the package if it is a popular request.

2.15.2.2 OmicsObject

MathIOmica uses internally associations extensively. To facilitate this usage the data from different omics is summarized in an **OmicsObject**. This is a structured nested association of the form shown in Fig. 2.16.

The **OmicsObject** is used typically with an external (outer) association to denote samples and an internal (inner) association for annotation. In the Fig. 2.16 structure example, the outer association has M outer labels as keys, corresponding to M samples. Across the samples there are N inner labels (e.g. identifiers for genes/proteins /small molecules), and inner labels are the same across samples. For a given jth outer label, OuterLabel$_j$, the kth inner label, InnerLabel$_k$ has a value of:
InnerLabel$_k$ → $\{\{$Measurements$_{jk}\}, \{$Metadata$_{jk}\}\}$

For any jth outer label OuterLabel$_j$, if the mth inner label, InnerLabel$_m$ is missing then it is assigned a value of **Missing**[]: InnerLabel$_m$ → **Missing**[].

Missing data are a problem in a wide range of omics data experiments (particularly in proteomics and mass spectrometry), and the MathIOmica **OmicsObject** approach was designed to handle such missing values.

Here is an example from the MathIOmica documentation, consisting of a list of 3 samples using protein data (UniProt [6] accessions). The measurements list has only a single value (corresponding to a relative intensity compared to a control), and the metadata is an integer that corresponds in this case to the number of unique peptides identified by the search engine.

In[394]:= omicsObjectExample=
    <|"FirstSample"→<|{"A0AVT1"}→{{0.937},{17}},
        {"A0MZ66"}→{{1.059},{9}},
        {"A1A4S6"}→{{1.03},{11}},
        {"A1L0T0"}→{{1.268},{4}},{"A0FGR8"}→**Missing**[]|>,
        "SecondSample"→<|{"A0AVT1"}→{{1.003},{17}},
        {"A0MZ66"}→ **Missing**[],{"A1A4S6"}→{{0.779},{11}},
        {"A1L0T0"}→{{0.917},{4}},
        {"A0FGR8"}→{{0.921},{24}}|>,
        "ThirdSample"→<|{"A0AVT1"}→{{1.064},{19}},
        {"A0MZ66"}→**Missing**[],{"A1A4S6"}→{{0.545},{5}},
        {"A1L0T0"}→**Missing**[],
        {"A0FGR8"}→ {{0.87}, {23}}|>|>;

Fig. 2.16 OmicsObject is one of the main constructs used in MathIOmica to organize information in associations

```
<|
 OuterLabel_1 -> <|InnerLabel_1 -> {{Measurements_{11}}, {Metadata_{11}}},
                  InnerLabel_2 -> {{Measurements_{12}}, {Metadata_{12}}},
                  InnerLabel_3 -> {{Measurements_{13}}, {Metadata_{13}}},
                  ...,
                  InnerLabel_k -> {{Measurements_{1k}}, {Metadata_{1k}}},
                  ...,
                  InnerLabel_N -> {{Measurements_{1N}}, {Metadata_{1N}}}|>,
 OuterLabel_2 -> <|
                  InnerLabel_1 -> {{Measurements_{21}}, {Metadata_{21}}},
                  InnerLabel_2 -> {{Measurements_{22}}, {Metadata_{22}}},
                  InnerLabel_3 -> {{Measurements_{23}}, {Metadata_{23}}},
                  ...,
                  InnerLabel_k -> {{Measurements_{2k}}, {Metadata_{2k}}},
                  ...,
                  InnerLabel_N -> {{Measurements_{2N}}, {Metadata_{2N}}}|>,
 ...,
 OuterLabel_j -> <|InnerLabel_1 -> {{Measurements_{j1}}, {Metadata_{j1}}},
                  InnerLabel_2 -> {{Measurements_{j2}}, {Metadata_{j2}}},
                  InnerLabel_3 -> {{Measurements_{j3}}, {Metadata_{j3}}},
                  ...,
                  InnerLabel_k -> {{Measurements_{jk}}, {Metadata_{jk}}},
                  ...,
                  InnerLabel_N -> {{Measurements_{jN}}, {Metadata_{jN}}}|>,
 ...,
 OuterLabel_M -> <|InnerLabel_1 -> {{Measurements_{M1}}, {Metadata_{M1}}},
                  InnerLabel_2 -> {{Measurements_{M2}}, {Metadata_{M2}}},
                  InnerLabel_3 -> {{Measurements_{M3}}, {Metadata_{M3}}},
                  ...,
                  InnerLabel_k -> {{Measurements_{Mk}}, {Metadata_{Mk}}},
                  ...,
                  InnerLabel_N -> {{Measurements_{MN}}, {Metadata_{MN}}}|>
|>
```

These are essentially multilevel associations, and can be used with **Query**. For example we can extract the measurements for the protein "AOMZ66" in the samples. This has the key **Key**[{"AOMZ66"}] in the **OmicsObject**. Note that the **Key** construct allows us to include an entire list as a key, which allows for generalized indexing and recall as necessary by our application.

In[394]:= **Query**[**All** ,**Key**[{"AOMZ66"}],1]@omicsObjectExample
Out[394]= <|FirstSample –>{1.059},SecondSample–>**Missing**[] ,
         ThirdSample–>**Missing**[]|>

Additional methods for handling **OmicsObject** data are included in MathIOmica and will be also used in some of the later chapters in the book.

References

1. Ashburner, M., Ball, C.A., Blake, J.A., Botstein, D., Butler, H., Cherry, J.M., Davis, A.P., Dolinski, K., Dwight, S.S., Eppig, J.T.: Gene ontology: tool for the unification of biology. Nat. Genet. **25**(1), 25–29 (2000)
2. Kanehisa, M., Goto, S.: Kegg: kyoto encyclopedia of genes and genomes. Nucl. Acids Res. **28**(1), 27–30 (2000)
3. Mias, G.I., Yusufaly, T., Roushangar, R., Brooks, L.R., Singh, V.V., Christou, C.: Mathiomica: an integrative platform for dynamic omics. Sci. Rep. **6**, 37,237 (2016)
4. Rossum, G.: Python reference manual. Tech. rep, Amsterdam, The Netherlands, The Netherlands (1995)
5. Trapnell, C., Pachter, L., Salzberg, S.L.: Tophat: discovering splice junctions with rna-seq. Bioinformatics **25**(9), 1105–11 (2009)
6. UniProt, C.: Uniprot: a hub for protein information. Nucl. Acids Res. **43**(Database issue), D204–12 (2015)
7. Wolfram, S.: An Elementary Introduction to the Wolfram Language, Wolfram Media (2015)
8. Wolfram Alpha LLC: Wolfram|Alpha (2017). Accessed Nov 2017
9. Wolfram Research, Inc.: Mathematica, Version 11.2. Champaign, IL (2017)

Chapter 3
Statistics

3.1 A Primer for Statistics with the Wolfram Language

In this chapter we will review some statistical approaches using the Wolfram Language. The Wolfram Language has extensive capabilities for statistics, including visualization tools. The abilities to generate distributions and intermix code with mathematical statistics greatly facilitate prototyping various analysis approaches. Additionally, the new features of data sets and entities of information (e.g. country information) allow us to actually combine multimodal information in statistical analysis. In this chapter we briefly touch on various examples of using the Wolfram Language [28, 29] for standard statistical analysis, and describe some examples with the inbuilt example datasets. Our aim here is not to cover the whole of statistics, for which there are excellent expositions, as for example in the non-exhaustive references [1, 3, 7, 8, 10, 11, 17, 27].

3.2 Descriptive Statistics

The Wolfram Language has many standard functions for descriptive statistics. Let us consider a set of numbers, for example say we have a class of ten students and we have their heights in meters:

In[1]:= dataHeights = {1.50, 1.73, 1.53, 1.85,
1.80, 1.60, 1.68, 1.55, 1.72, 1.84}
Out[1]= {1.5, 1.73, 1.53, 1.85, 1.8, 1.6, 1.68, 1.55, 1.72, 1.84}

We can see how many measurements we have:

In[2]:= Length[dataHeights]
Out[2]= 10

Electronic supplementary material The online version of this chapter (https://doi.org/10.1007/978-3-319-72377-8_3) contains supplementary material, which is available to authorized users.

© Springer International Publishing AG 2018
G. Mias, *Mathematica for Bioinformatics*,
https://doi.org/10.1007/978-3-319-72377-8_3

We can calculate the means, medians, and variances, standard deviations, maxima and minima of these data:

```
In[3]:= Mean [dataHeights]
Out[3]= 1.68

In[4]:= Median [dataHeights]
Out[4]= 1.7

In[5]:= Variance [dataHeights]
Out[5]= 0.0168

In[6]:= StandardDeviation [dataHeights]
Out[6]= 0.129615

In[7]:= Max [dataHeights]
Out[7]= 1.85

In[8]:= Min [dataHeights]
Out[8]= 1.5
```

We can plot the data to take a quick look, Fig. 3.1.

```
In[9]: dataPlot = ListPlot[dataHeights,
         PlotLabel -> "Example Height Data",
         AxesLabel -> {"Student", "Height/m"}]
```

We can also plot the mean and median at the same time, Fig. 3.2.

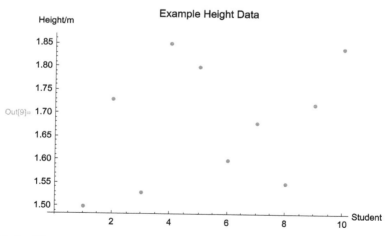

Fig. 3.1 Out[9]= ListPlot of data

3.3 A Short Probabilistic Sequence of Events

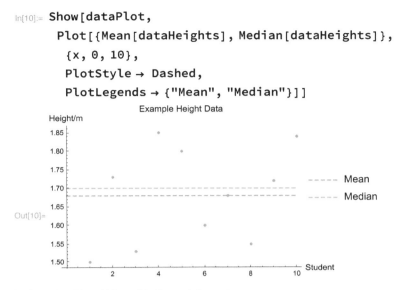

Fig. 3.2 Out[10]= Plot of Mean, Median and data points

3.3 A Short Probabilistic Sequence of Events

When we talk about statistics, the first thing that comes to mind is probability. Our everyday intuition gives us a notion of what a probable or improbable event is - and often misleads us with our human biases. The simplest probabilistic model everyone has probably thought about is the coin flip: what will you get when you flip a coin - heads or tails? For a coin that is not biased we expect a fifty-fifty chance to get either heads or tails. So we talk of the theoretical probability of getting Heads to be 0.5 or the probability of getting Tails to be 0.5. What if we wanted to calculate the empirical probability of this happening? In practice, if we were to flip 1000 coins we would expect about 500 of them to be heads, give or take. If we wanted to test this experimentally we could probably flip a coin manually 1000 times. That would be a bit tiresome, and would only yield one possible set of outcomes for 1000 flips, i.e. just one realization. We would want to repeat the experiment multiple times, and then maybe average the results across experiments. To facilitate this model probabilistic process we can use a simple algorithm to generate simulations of the experiment. The function **RandomChoice** will allow us to select repeatedly from the set { "H","T" } as many times as we want:

In[11]:= ?**RandomChoice**
    RandomChoice[{$e_1, e_2, \ldots$}] gives
        a pseudorandom choice of one of the e_i.
    RandomChoice[list,n] gives a list of n pseudorandom
        choices.

RandomChoice[list, $\{n_1, n_2, \ldots\}$] gives an $n_1 \times n_2 \times \ldots$ array of pseudorandom choices.
RandomChoice[$\{w_1, w_2, \ldots\} \to \{e_1, e_2, \ldots\}$] gives a pseudorandom choice weighted by the w_i.
RandomChoice[wlist $\to$ elist, n] gives a list of n weighted choices.
RandomChoice[wlist $\to$ elist, $\{n_1, n_2, \ldots\}$] gives an $n_1 \times n_2 \times \ldots$ array of weighted choices. >>

In[12]:= randomFlips=**RandomChoice**[{"H","T"},1000];

For brevity, let us look at the first 10 elements of this set:

In[13]:= **Take**[randomFlips,10]
Out[13]= {T,H,T,T,T,H,T,T,H,H}

We can see how many counts we get:

In[14]:= **Tally**[randomFlips]
Out[14]= {{T,484},{H,516}}

Now let us repeat this experiment 2000 times and tally each time. The code will first create a set of 2000 lists of 1000 flips, and the **Tally** function will calculate the tally for each list. The **SortBy** will sort the results of **Tally** based on the first element matching (**First**):

In[15]:= talliesCoinFlip=**SortBy**[**Tally**[#],**First**]&/@
         RandomChoice[{"H","T"},{2000,1000}];

We can check the dimensions:

In[16]:= **Dimensions**[talliesCoinFlip]
Out[16]= {2000,2,2}

And we can see the first few results:

In[17]:= talliesCoinFlip[[1;;10]]
Out[17]= {{{H,501},{T,499}},{{H,522},{T,478}},{{H,482},
         {T,518}},{{H,488},{T,512}},{{H,498},{T,502}},
         {{H,499},{T,501}},{{H,478},{T,522}},{{H,495},
         {T,505}},{{H,523},{T,477}},{{H,499},{T,501}}}

Basically, we have the result of 2000 experiments. Because we sorted Heads is always first in the list. Let us now calculate the mean value across the experiments:

In[18]:= N[**Mean**[talliesCoinFlip]]
Out[18]= {{H,499.654},{T,500.347}}

We see that the mean approaches what our intuition tells us for an unbiased coin, that over a large number of coin flips we will get approximately equal numbers of Heads or Tails.

In statistics we talk of having random variables. After performing an experiment, such as our coin flip or randomly selecting say cards from a card deck, or rolling a dice we get possible outcomes. Random variables assign these outcomes to numerical values. For our coin flip we can define a random variable F:

3.3 A Short Probabilistic Sequence of Events

$$F = \begin{cases} 0 \to H \\ 1 \to T \end{cases} \tag{3.1}$$

The set of values a random variable can take is called its range. We can then assign a probability to each of these possible values that the random variable can take, and in our example we would have:

$$P = \begin{cases} 0.5 \text{ if } F = 0 \\ 0.5 \text{ if } F = 1 \end{cases} \tag{3.2}$$

This function is the probability mass function of the discrete random variable P, which maps real numbers to a probability value between 0 and 1.

Let us see another example, in terms of DNA and creating strings corresponding to nucleotides. We can think, of these as sequences of words made up of the random letters {A,C,G,T}. We can then create sequences of words by joining random choices over a specified length:

In[19]:= randomWord=
   StringJoin[RandomChoice[{"A","C","G","T"},
   100]]
Out[19]= TGCCACGTCGGTAGGAGGCTATGCGGTGCGGCCAGTTACATAGAAGTTTTGCAAC
   GATTGACCAACGTTGGATAGCTTTTCGAGTGGAGTGGGCGGTGCC

We can then define a *discrete random variable N* that takes values, i.e. has range, {0,1,2,3}:

$$N = \begin{cases} 0 \to A \\ 1 \to C \\ 2 \to G \\ 3 \to T \end{cases} \tag{3.3}$$

RandomChoice samples equally between the input selection set. We can assign a weighted probability for each element in the set by using for example the frequencies of nucleotides in the human genome, say for chromosome 1 [30]:

In[20]:= randomWeightedSequence =
   StringJoin[
   RandomChoice[{0.29, 0.21, 0.21, 0.29} ->
   {"A", "C", "G", "T"}, 100]]
Out[20]= AAGACACACTTCAAGAATGCGACCAGGGGCTGTGATCAATCTTTTTAGACGTACG
   ACTCACTTTTCCGCTATATTCGAATCTCATCAGCCTTCTATTATA

In terms of a probability mass function we can write this as:

$$P(N) = \begin{cases} 0.29 \text{ if } N = 0 \\ 0.21 \text{ if } N = 1 \\ 0.21 \text{ if } N = 2 \\ 0.29 \text{ if } N = 3 \end{cases} \tag{3.4}$$

In a similar fashion, we can also simulate numerical experiments. How about rolling two dice? We can do this in different ways. We can ask to get 2 random integers from 1 to 6:

In[21]:= diceRoll=**RandomInteger**[{1,6},2]
Out[21]= {6,6}

Or we can again use **RandomChoice** from the set {1,2,3,4,5,6}:

In[22]:= diceRoll2=**RandomChoice**[**Range**[6],2]
Out[22]= {5,6}

Additionally, we can also create continuous random variables. For example we can generate a random real number, from 0 to 1.

In[23]:= exampleContinuous=**RandomReal**[1,1]
Out[23]= {0.490825}

We should note that the random generator is actually using an internal seed for the computation so this is a pseudorandom number as we discussed in Chap. 2. We can effect the choice of random seed using **SeedRandom**[] to reset the pseudorandom generator, or use **SeedRandom**[n] with an integer n.

Continuous random variables can vary continuously in terms of their range. In terms of continuous random variables, these are defined as functions from a real number to another real number. The probability is not defined for a specific value, but rather for a range. For example, the probability of some value for a continuous random variable X taking values between x_1 and x_2 is:

$$P(x_1 \leq X \leq x_2) = \int_{x_1}^{x_2} \text{pdf}(x)\, dx \tag{3.5}$$

In the integral we have defined the probability density function (pdf). The area under this function between x_1 and x_2 gives us the probability that the continuous random variable X takes values in this range.

3.4 Examples of Discrete Distributions

3.4.1 Bernoulli Distribution

The Bernoulli distribution is the model for a binary experiment, such as our single coin flip, or any two outcome experiment, where the outcomes can be assigned to be 0 or 1. Typically, one of the events is assigned probability parameter p, which mean the other event has probability 1–p. We can write the probability mass function $p_{\text{Bernoulli}}$,

$$p_{\text{Bernoulli}(X=x)} = \begin{cases} p & \text{if } x = 1 \\ 1 - p & \text{if } x = 0 \end{cases} \tag{3.6}$$

3.4 Examples of Discrete Distributions

```
In[25]:= DiscretePlot[PDF[BernoulliDistribution[0.5],x],
    {x, 0, 1}, Frame → True, PlotLabel →
    "Probability Mass Function for BernoulliDistribution[0.5]",
    FrameLabel → {"X", "Probability"}]
```

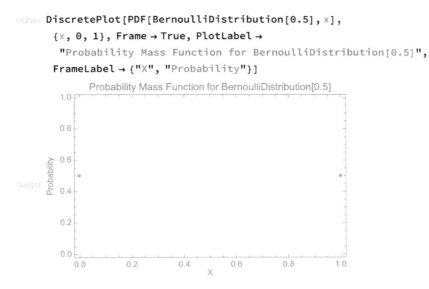

Fig. 3.3 Out[25]= PDF for Bernoulli distribution

The function **BernoulliDistribution** can encode this information:

```
In[24]:= ?BernoulliDistribution
    BernoulliDistribution[p] represents a Bernoulli
    distribution with probability parameter p.   >>
```

We can plot the probability mass function for our coin toss, where we assign p = 0.5 using the PDF function, Fig. 3.3. Note that in the Wolfram Language the PDF built-in symbol is used both for the probability mass function (pmf) for discrete random variables, as well as for the probability density function (pdf) for continuous random variables.

Notice that we have used **DiscretePlot** and that we have only two points defined in the plot, corresponding to the binary nature of the outcome. The sum of the two probabilities will be 1 as expected.

We can also obtain the distribution function, or cumulative distribution function (CDF). The CDF is defined as the probability that the observed value is less than a cutoff c:

$$\text{CDF}[c] = P(X \leq c) \tag{3.7}$$

for c a real number. In the Wolfram Language we use:

```
In[26]:= ?CDF
    CDF[dist,x  gives  the  cumulative  distribution  function  for
        the  symbolic  distribution  dist  evaluated  at  x.
    CDF[dist, {x₁, x₂,...}]  gives  the  multivariate  cumulative
        distribution  function  for  the  symbolic
        distribution  dist  evaluated  at  {x₁, x₂,...}.
    CDF[dist]  gives  the  CDF  as  a  pure  function.     >>
```

In[27]:= `DiscretePlot[CDF[BernoulliDistribution[0.5], x],
 {x, 0, 1}, ExtentSize → Right,
 ExtentMarkers → {"Filled", "Empty"},
 Frame → True,
 PlotLabel →
  "Cumulative Distribution Function for
 BernoulliDistribution[0.5]",
 FrameLabel → {"X", "Probability"}]`

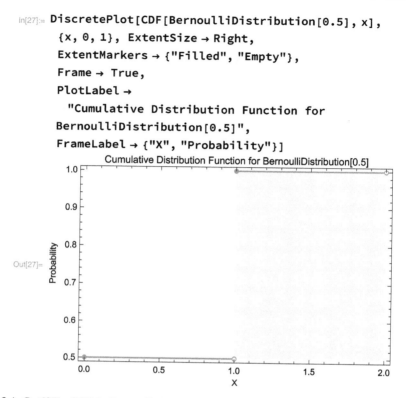

Fig. 3.4 Out[27]= CDF for Bernoulli distribution

We can also plot the CDF over the same range using **DiscretePlot**, Fig. 3.4.

Notice that in Fig. 3.4 we use points and open points to distinguish that when X=1, we have that point excluded from the lower line plotted, i.e. the interval [0,1), but it is included in the upper line, i.e. in the interval [1,2).

We can also obtain a formula for both the PDF and CDF:

In[28]:= **PDF[BernoulliDistribution [0.5], x]**

Out[28]= $\begin{cases} 0.5 & x == 0 \;||\; x == 1 \\ 0 & \text{True} \end{cases}$

In[29]:= **CDF[BernoulliDistribution [0.5], x]**

Out[29]= $\begin{cases} 0 & x < 0 \\ 0.5 & 0 \leq x < 1 \\ 0 & \text{True} \end{cases}$

More generally, we can get symbolic expressions for each of these as a function of the parameter p:

3.4 Examples of Discrete Distributions

In[30]:= **PDF[BernoulliDistribution[p],x]**

Out[30]= $\begin{cases} 1-p & x==0 \\ p & x==1 \\ 0 & \text{True} \end{cases}$

In[31]:= **CDF[BernoulliDistribution[p],x]**

Out[31]= $\begin{cases} 0 & x<0 \\ 1-p & 0 \le x < 1 \\ 0 & \text{True} \end{cases}$

We can use the function **RandomVariate** to give a list of pseudorandom variates from any given distribution:

In[32]:= **RandomVariate[BernoulliDistribution[0.5],10]**
Out[32]= {1,0,1,1,1,1,0,1,0,1}

3.4.2 Binomial Distribution

In the case where we have multiple, n, Bernoulli trials, i.e. experiments with binary outcomes, then we talk about having a binomial distribution. For example tossing a coin twenty times in a row.

We now have two variables: (1) the success probability given by p, and (2) the number of trials n. The probability of getting a value x in n trials is given by:

$$p_{\text{Binomial}}(x) = \binom{n}{x} p^x (1-p)^{n-x} \tag{3.8}$$

We can think of this as multiplying the probability of x successes and $n-x$ failures in a total of n trials, times the number of ways these can occur, $\binom{n}{x}$. Here $\binom{n}{x}$ represents the binomial coefficient (function **Binomial[n,x]** in the Wolfram Language), which is the the number of different ways one can choose x out of n possible outcomes,

$$\binom{n}{x} = \frac{n!}{x!(n-x)!} \tag{3.9}$$

When n=1 this reduces to a single Bernoulli trial:

$$p_{\text{Binomial}}(x)\mid_{n=1} = \binom{1}{x} p^x (1-p)^{1-x} \tag{3.10}$$

since we have:

$$p_{\text{Binomial}}(0)\mid_{n=1} = 1-p \tag{3.11}$$

and also,

$$p_{\text{Binomial}}(1)\mid_{n=1} = p \tag{3.12}$$

```
In[34]:= DiscretePlot[PDF[BinomialDistribution[20, 0.5], x],
         {x, 0, 20}, Frame → True,
         PlotLabel →
          "Probability Mass Function
           for BinomialDistribution[20,0.5]",
         FrameLabel → {"X", "Probability"}]
```

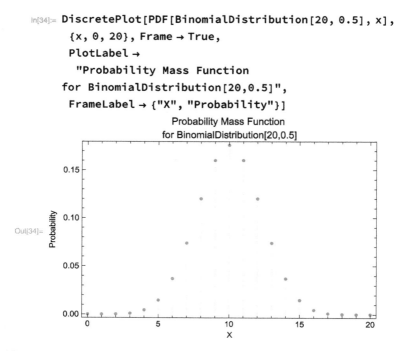

Fig. 3.5 Out[34]= PDF for Binomial distribution

The **BinomialDistribution** can be used directly in the Wolfram Language by:

```
In[33]:= ?BinomialDistribution
         BinomialDistribution[n,p] represents a binomial
           distribution with n trials and success
           probability p.  >>
```

We can plot the probability mass function, Fig. 3.5

If we want a specific value we can evaluate it. For example, what is the probability to get 5 successes, i.e. 5 heads in twenty flips:

```
In[35]:= PDF[BinomialDistribution[20,0.5],5]
Out[35]= 0.0147858

In[36]:= Table[PDF[BinomialDistribution[20,0.5],x],{x,0,20}]
Out[36]= {9.53674×10⁻⁷,
          0.0000190735,0.000181198,
          0.00108719,0.00462055,0.0147858,0.0369644,
          0.0739288,0.120134,0.160179,0.176197,0.160179,
          0.120134,0.0739288,0.0369644,0.0147858,0.00462055,
          0.00108719,0.000181198,0.0000190735,9.53674×10⁻⁷}
```

As we can see the probability at the tails approaches zero as expected, and we have higher number in the middle of the list, where the peak is at 0.176.

3.4 Examples of Discrete Distributions

Fig. 3.6 Out[37]= PDF, Out[38]= CDF for a Binomial distribution

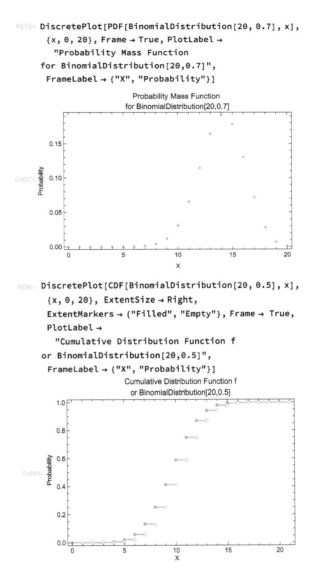

So far we have considered unbiased coins. The assumption has been that there is an equal probability for the two outcomes for heads and tails. What if our coin is biased? Let us say the probability of success is 0.7. Then we would expect that we will get more hits, and that the peak of the probability mass function will shift to the right. We can plot both the new PDF and CDFs, Fig. 3.6.

Note again that we can get symbolic forms for the mass function:

In[39]:= **PDF[BinomialDistribution[n,p],x]**
Out[39]= $\begin{cases} (1-p)^{n-x} p^x \text{Binomial}[n,x] & 0 \leq x \leq n \\ 0 & \text{True} \end{cases}$

In[39]:= **CDF[BinomialDistribution[n,p],x]**
Out[39]= $\begin{cases} \text{BetaRegularized}[1\text{-}p, n\text{- Floor}[x], 1\text{+Floor}[x]] & 0 \leq x < n \\ 1 & x \geq n \\ 0 & \text{True} \end{cases}$

We can also calculate the mean and the variance:

In[41]:= **Mean[BinomialDistribution[n,p]]**
Out[41]= n p

This is simply the number of trials times the probability of success as we would intuitively expect.

In[42]:= **Variance[BinomialDistribution[n,p]]**
Out[42]= n (1−p) p

The variance is the number of trials times the probability of success times the probability of failure.

We can again use **RandomVariate** to get a list of pseudorandom variates from a **BinomialDistribution**

In[43]:= **RandomVariate[BinomialDistribution[20,0.25],10]**
Out[43]= {7,9,10,1,2,3,6,8,3,6}

3.4.3 Geometric Distribution

The Geometric Distribution describes the probability of having a fixed number of failures prior to a success. It is parametrized with a probability of success p. The probability mass function for x number of failed trials up to success (x+1 trials total) is:

In[44]:= **PDF[GeometricDistribution[p],x]**
Out[44]= $\begin{cases} (1-p)^x p & x \geq 0 \\ 0 & \text{True} \end{cases}$

We can think of this as a series of x Bernoulli trials that fail until we get a successful outcome.

In[45]:= **?GeometricDistribution**
       GeometricDistribution[p] represents a geometric distribution with probability parameter p.   ≫

We can plot the probability mass function, let's say for a probability of 0.25, in Fig. 3.7. As we can see from the plot, this looks like a discrete analog of an exponential distribution.

3.4 Examples of Discrete Distributions

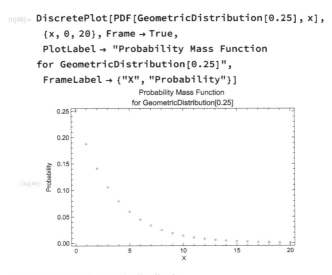

Fig. 3.7 Out[46]= PDF for a Geometric distribution

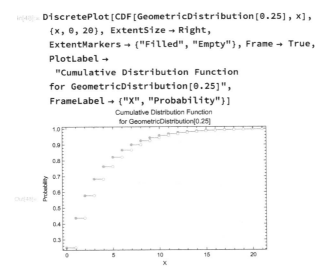

Fig. 3.8 Out[48]= CDF for a Geometric distribution

If we want a specific value we can evaluate the PDF function. For example, what is the probability to get success on the fourth trial, i.e. 3 failures first:

In[47]:= **PDF[GeometricDistribution [0.25],3]**
Out[47]= 0.105469

We can also plot the cumulative distribution function, Fig. 3.8.

We can sum the PDF directly for the various points:

In[49]:=Sum[PDF[GeometricDistribution[0.25], x], {x, 0, 21}]
Out[49]= {0.4375,0.578125,0.683594,0.762695,0.822021,
0.866516,0.899887,0.924915,0.943686,0.957765,0.968324,
0.976243,0.982182,0.986637,0.989977,0.992483,0.994362,
0.995772,0.996829,0.997622,0.998216}

Note again that we can get symbolic forms for CDF as well:

In[50]:= CDF[GeometricDistribution[n],x]
Out[50]= $\begin{cases} 1-(1-p)^{\text{Floor}[x]+1} & x \geq 0 \\ 0 & \text{True} \end{cases}$

3.4.4 Poisson Distribution

The Poisson Distribution is a model for random processes, where points occur randomly in temporal or spatial locations (see for example Modern Introduction to Probability and Statistics by Dekking et al. [7]). It is used extensively to model radioactive particle processes (as measured by counters), or hits on two dimensional targets, or digital droplet PCR (polymerase chain reaction) [13, 26].

For example, in one dimension we can think of events occurring at sequential times. The parameter for the Poisson distribution (frequently represented by λ), is the characterization of the rate of event occurrence, which remains a constant average rate over time - i.e. per time interval. Additionally, each event can take place independently of how long it has been since a previous event. The probability mass function is given by:

In[51]:= PDF[PoissonDistribution[λ],x]
Out[51]= $\begin{cases} 1\frac{e^{-\lambda}\lambda^x}{x!} & x \geq 0 \\ 0 & \text{True} \end{cases}$

One characteristic of the Poisson distribution is its mean is equal to its variance:

In[52]:= Mean[PoissonDistribution[λ]]
Out[52]= λ

In[53]:= Variance[PoissonDistribution[λ]]
Out[53]= λ

As an example for a mean of 1 we plot the PDF in Fig. 3.9 and the CDF in Fig. 3.10. We can also look for comparison at the the PDF, Fig. 3.11, and CDF, Fig. 3.12 for a mean of 10.

3.4.5 Hypergeometric Distribution

We can think of a Binomial distribution as the series of n independent Bernoulli trials that have the same probability of success. What if the trials are not independent? This

3.4 Examples of Discrete Distributions

```
In[54]:= DiscretePlot[PDF[PoissonDistribution[1], x],
   {x, 0, 20}, Frame → True,
   PlotLabel →
     "Probability Mass Function
   for PoissonDistribution[1]",
   FrameLabel → {"X", "Probability"},
   PlotRange -> Full]
```

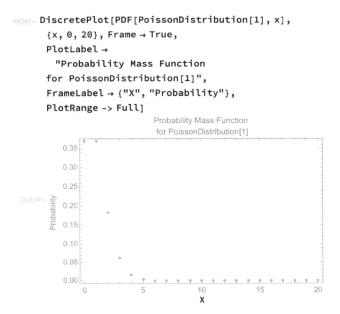

Fig. 3.9 Out[54]= PDF for a Poisson distribution

Fig. 3.10 Out[55]= CDF for a Poisson distribution

```
In[55]:= DiscretePlot[CDF[PoissonDistribution[1], x],
   {x, 0, 20}, Frame → True,
   PlotLabel →
     "Cumulative Distribution Function
   for PoissonDistribution[1]",
   FrameLabel → {"X", "Probability"},
   PlotRange → Full]
```

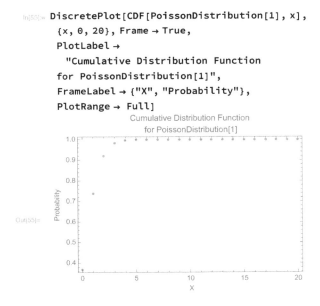

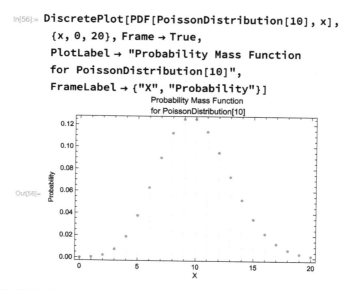

Fig. 3.11 Out[56]= PDF for a Poisson distribution with mean 10

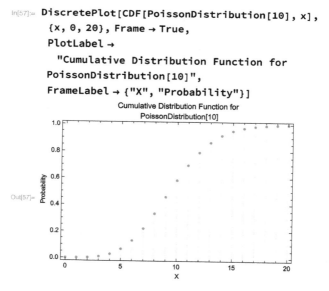

Fig. 3.12 Out[57]= CDF for a Poisson distribution with mean 10

3.4 Examples of Discrete Distributions

is for example when we sample without replacement from a population. We can use a hypergeometric distribution for the number of successes from n draws (without replacement), given a population of size n_{tot}, having n_{succ} possible successes. The prototypical model is that of an urn model, where n balls are drawn from an urn that contains n_{succ} black balls out of a total of n_{tot} balls, and n_{tot}-n_{succ} white balls.

In[58]:= **?HypergeometricDistribution**
HypergeometricDistribution[n,n_{succ},n_{tot}] represents a hypergeometric distribution. >>

The mean of the distribution is:

In[59]:= **Mean[HypergeometricDistribution[n,n_{succ},n_{tot}]]**
Out[59]= $\frac{n n_{\text{succ}}}{n_{\text{tot}}}$

The hypergeometric distribution is used for example in pathway and gene ontology(GO) over-representation (enrichment) calculations [2]. For an enrichment model we have a pathway that contains n_P total of entities (usually proteins) out of a population of n_{Total} proteins, and we are testing against this a list of say n proteins. We are interested in the possibility of getting n or more hits in the pathway by chance alone.

We assume here we are considering a population (possible measurements) of N total genes, and a given pathway or GO term T has M_T genes associated to it. For any set of genes/proteins we are interested in (e.g. as the result of a differential expression analysis), that contains a subset of $n_{\text{group}} < N$ genes, we can identify which x_T genes are associated to term T. Then we would like to calculate the probability, p_T, that at least x_T genes could have an association to term T randomly (i.e. more than x_T number of successes):

$$p_T \text{ of at least } x_T \text{ successes} = \sum_{i=x_T}^{n_{\text{group}}} \frac{\binom{M_T}{i}\binom{N-M_T}{n_{\text{group}}}}{\binom{N}{n_{\text{group}}}} \quad (3.13)$$

This is essentially a sum of HypergeometricDistribution terms:

In[60]:= **PDF[HypergeometricDistribution[n_{group},M_T,N],i]**

Out[60]= $\begin{cases} \frac{\text{Binomial}[N-M_T,-i+n_{\text{group}}]\text{Binomial}[M_T,i]}{\text{Binomial}[N,n_{\text{group}}]} & \begin{array}{l} 0 \leq i \leq M_T \&\& \\ -N + M_T + n_{\text{group}} \leq i \leq M_T \&\& \\ 0 \leq i \leq n_{\text{group}} \&\& \\ -N + M_T + n_{\text{group}} \leq i \leq n_{\text{group}} \end{array} \\ 0 & \text{True} \end{cases}$

For example, for 10 selections, in a population of 100 that has 5 in the "success" group, we can calculate the PDF at successive hits:

In[61]:= **Table[
   N[PDF[HypergeometricDistribution[10, 5, 100],
      i]], {i, 1, 5}]**

Out[20]= {0.339391, 0.0702188,
        0.00638353, 0.000251038, 3.34717×10^{-6}}

84   3 Statistics

We will see additional examples in the following chapters, but let us assume we have carried out an experiment and found a set of genes showing up-regulation in their expression.

In[62]:= genesUpregulated =
  {"TAB1","TNFSF13B","MALT1","TIRAP",
   "CHUK","TNFRSF13C","PARP1","CSNK2A1",
   "IKBA","CSNK2B","LTBR","LYN","MYD88",
   "GADD45B","ATM","NFKB1","NFKB2","NFKBIA",
   "AKT3","PIAS4","FOS","JUN"};

Here we are using official gene symbols. We can carry out an over-representation analysis in MathIOmica [21]:

In[63]:= **Needs**["MathIOmica`"]

  (* the line below is printed during the evaluation *)
  MathIOmica (http://mathiomica.org), by G. Mias Lab.

In[64]:= exampleORA=**GOAnalysis**[genesUpregulated];

Let us get the first few results:

In[65]:= exampleORA[[1;;3]]
Out[65]= <|GO:0038095→
  {{6.75904*10^−14,3.56877*10^−11,**True**},
   {20,251,47250,8},
   {{Fc-epsilon receptor signaling pathway,
     biological_process},
    {{TAB1},{MALT1},{CHUK},{LYN},{NFKB1},
     {NFKBIA},{FOS},{JUN}}}},GO:0005515→
  {{2.19898×10^{-13},4.69303×10^{-11},**True**},
   {20,8801,47250,19},
   {{protein binding,molecular_function},
    {{TAB1},{TNFSF13B},{MALT1},{TIRAP},{CHUK},
     {PARP1},{CSNK2A1},{CSNK2B},{LTBR},
     {LYN},{GADD45B},{ATM},{NFKB1},{NFKB2},
     {NFKBIA},{AKT3},{PIAS4},{FOS},{JUN}}}},
  GO:0051092→{{2.6665×10^{-13},4.69303×10^{-11},**True**},
   {20,155,47250,7},
   {{positive regulation of NF-kappaB
     transcription factor activity,
     biological_process},{{TAB1},{MALT1},
     {TIRAP},{CHUK},{NFKB1},
     {NFKB2},{NFKBIA}}}}|>

The output is a dictionary which has the following form for keys → values

<|
GO:Term$_1$]→{{p-value$_1$,
     multiple hypothesis adjusted p-value$_1$,
     True/False for statistical significance},
    {{number of members in group being tested,
     number of successes for term$_1$ in population,
     total number of members in population,

3.4 Examples of Discrete Distributions 85

```
              number of members (or more) in current group being
                tested associated to term₁},
   {{GO term₁ description,
      ontology category for term₁},
      {input IDs associated to Term₁}}}|>
```

We see that the results are statistically significant (see further below for a discussion of p-values), and multiple hypothesis correction has been performed using Benjamini-Hochberg [5] approach. A list of genes is included for each GO term. For example in the GO:0038095, the term is a Biological Process called "Fc-epsilon receptor signaling pathway". We had 8 hits out of the 20 input genes with annotation associated to this term. The term has 251 genes associated to it.

3.5 Examples of Continuous Distributions

3.5.1 Uniform Distribution

The simplest continuous distribution is a uniform distribution. The idea is that in a given interval, [a,b], a random variable can take any value in the interval with equal probability.

In[66]:= `PDF[UniformDistribution[{a,b}],x]`

Out[66]= $\begin{cases} \frac{1}{b-a} & a \leq x \leq b \\ 0 & \text{True} \end{cases}$

We look at an example for the PDF for our variable being able to take values between 1 and 2, Fig. 3.13. Also, we can get the corresponding CDF, Fig. 3.14.

The density function in Fig. 3.13 is a straight line, and has zero slope (i.e. does not change as we move across the interval by increasing x). The area under the PDF is equal to 1, as we can see from the plot, Fig. 3.13, and also verify directly by integration:

In[69]:= `Integrate[PDF[UniformDistribution[{1,3}],x],`
`         {x,-Infinity,Infinity}]`
Out[69]= 1

Equivalently, we notice how the CDF, Fig. 3.14, is ramping up at a constant rate between 1 and 3. We can take the derivative with respect to x:

In[70]:= `D[CDF[UniformDistribution[{1,3}],x],x]`

Out[70]= $\begin{cases} 0 & x < 1 \\ \frac{1}{2} & 1 < x < 3 \\ 0 & x > 3 \\ \text{Indeterminate} & \text{True} \end{cases}$

The mean and variance of the distribution are:

In[71]:= `Mean[UniformDistribution[{a,b}]]`
Out[71]= $\frac{a+b}{2}$

```
In[67]:= Plot[PDF[UniformDistribution[{1, 3}], x],
         {x, 0, 4}, PlotRange → All, Frame → True,
         FrameLabel → {"x", "P(X=x)"},
         PlotLabel → "Probability Density Function
         for UniformDistribution[{1,2}]"]
```

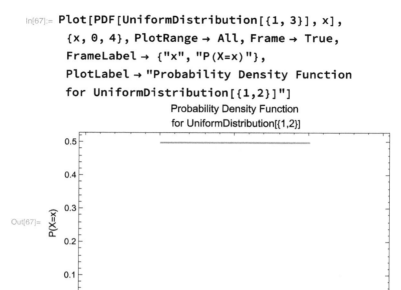

Fig. 3.13 Out[67]= PDF for a Uniform distribution

In[72]:= **Variance[UniformDistribution[{a,b}]]**
Out[72]= $\frac{1}{12}(-a+b)^2$

The Wolfram Language also has functions for multivariate uniform distributions. Notice below how the PDF takes now as arguments two or more variables listed in a list, {x,y,z,...}. For example, say we have 3 variables:

In[73]:= **PDF[UniformDistribution[{{x_{min},x_{max}},{y_{min},y_{max}},
         {z_{min},z_{max}}}],{x,y,z}]**

Out[73]= $\begin{cases} \frac{1}{(x_{max}-x_{min})(y_{max}-y_{min})(z_{max}-z_{min})} & \begin{array}{l} x - x_{min} \geq 0 \&\& y - y_{min} \geq 0 \&\& \\ z - z_{min} \geq 0 \&\& -x + x_{max} \geq 0 \&\& \\ -y + y_{max} \geq 0 \&\& -z + z_{max} \geq 0 \end{array} \\ 0 & \text{True} \end{cases}$

Let us pick some values, say x is uniform across [1,3], y across [1,4] and z across [0,2].

In[74]:= **PDF[UniformDistribution[{{1,3},{1,4},{0,2}}],{x,y,z}]**
Out[74]=
$\begin{cases} \frac{1}{12} & -1+x \geq 0 \&\& -1+y \geq 0 \&\& z \geq 0 \&\& 3-x \geq 0 \&\& 4-y \geq 0 \&\& 2-z \geq 0 \\ 0 & \text{True} \end{cases}$

If we integrate out 2 of the variables, x and y, we still get the correct symbolic expression for z:

3.5 Examples of Continuous Distributions 87

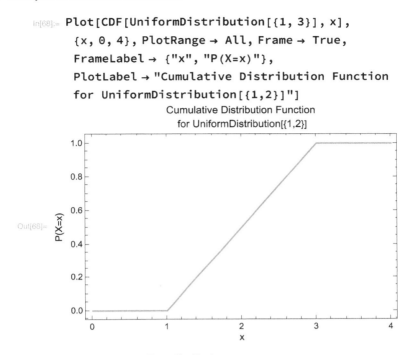

```
In[68]:= Plot[CDF[UniformDistribution[{1, 3}], x],
         {x, 0, 4}, PlotRange → All, Frame → True,
         FrameLabel → {"x", "P(X=x)"},
         PlotLabel → "Cumulative Distribution Function
         for UniformDistribution[{1,2}]"]
```

Fig. 3.14 Out[68]= CDF for a Uniform distribution

```
In[75]:= Integrate[
         PDF[UniformDistribution[
         {{1,3},{1,4},{0,2}}],{x,y,z}],
         {x,-Infinity, Infinity},
         {y,-Infinity, Infinity}]
```
$$\text{Out}[75]= \begin{cases} \frac{1}{2} & 0 \le z \le 2 \\ 0 & \text{True} \end{cases}$$

3.5.2 Exponential Distribution

We have seen above how the Poisson distribution can be used to describe events over particular time intervals. One can think of the exponential distribution as a continuous version of the Poisson process.

```
In[76]:= ?ExponentialDistribution
         ExponentialDistribution [λ] represents an
         exponential distribution with scale inversely
         proportional to parameter λ. >>
```

The PDF is given by:

In[77]:= **PDF[ExponentialDistribution[λ,x]**

Out[77]= $\begin{cases} e^{-x\lambda}\lambda & x \geq 0 \\ 0 & \text{True} \end{cases}$

λ here is again used as the parameter to measure the density of events over a unit time interval.

We can plot the PDF and CDF, Fig. 3.15. We can also see the effect of changing the rate to say $\lambda = 10$ in Fig. 3.16.

As you can see, the probability drop is much sharper than for smaller λ. In contrast to the Poisson Distribution, the Mean and Variance are not equal, but still have a simple dependence on λ.

In[82]:= **Mean[ExponentialDistribution[λ]**
Out[82]= $\frac{1}{\lambda}$

In[83]:= **Variance[ExponentialDistribution[λ]**
Out[83]= $\frac{1}{\lambda^2}$

3.5.3 Normal Distribution

The Normal Distribution is the most used distribution in the majority of gene expression analysis, and possibly computational biology. Processed expression data, is typically transformed or analyzed towards assuming a Normal distribution. This is also called a Gaussian distribution, or equivalently described by its shape, the bell distribution. The distribution is described by two parameters, the mean of the data μ and the standard deviation σ. In terms of a representation in the Wolfram Language we have:

In[84]:= **PDF[NormalDistribution[μ,σ],x]**

Out[84]= $\frac{e^{-\frac{(x-\mu)^2}{2\sigma^2}}}{\sqrt{2\pi}\sigma}$

This continuous distribution is defined across all real numbers. Notice the denominator $\sqrt{2\pi}\sigma$ which assures us of the proper normalization so that the integral over the entire set of real numbers for the distribution is unity:

In[85]:= **Integrate[PDF[NormalDistribution[μ,σ],x],**
            {x,-Infinity,Infinity},Assumptions$\to \sigma \geq 0$]
Out[85]= 1

As an aside, we note that in the Wolfram Language when we deal with symbolic evaluations certain assumptions must be noted. If we had not included the **Assumptions** option above to specify that the standard deviation should be a number we would instead get a **ConditionalExpression**:

In[86]:= **Integrate[PDF[NormalDistribution[μ,σ],x],**
            {x,-Infinity,Infinity}]

3.5 Examples of Continuous Distributions

```
In[78]:= Plot[PDF[ExponentialDistribution[1], x],
        {x, 0, 4}, PlotRange → All, Frame → True,
        FrameLabel → {"x", "P(X=x)"},
        PlotLabel → "Probability Density Function
        for ExponentialDistribution[1]"]
```

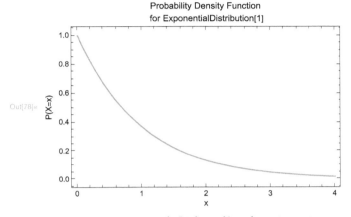

```
In[79]:= Plot[CDF[ExponentialDistribution[1], x],
        {x, 0, 4}, PlotRange → All, Frame → True,
        FrameLabel → {"x", "P(X=x)"},
        PlotLabel → "Cumulative Distribution Function
        for ExponentialDistribution[1]"]
```

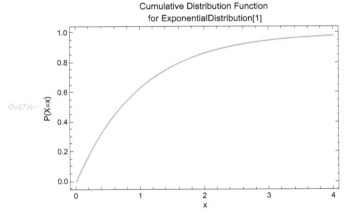

Fig. 3.15 **Out**[78]= PDF, and, **Out**[79]= CDF, for an Exponential [1] distribution.

In[80]:= `Plot[PDF[ExponentialDistribution[10], x],`
`  {x, 0, 4}, PlotRange → All, Frame → True,`
`  FrameLabel → {"x", "P(X=x)"},`
`  PlotLabel → "Probability Density Function`
`    for ExponentialDistribution[10]"]`

Out[80]=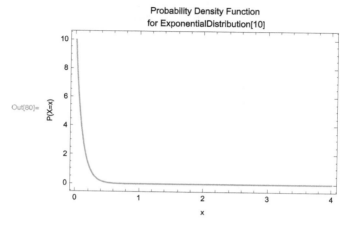

In[81]:= `Plot[CDF[ExponentialDistribution[10], x],`
`  {x, 0, 4}, PlotRange → All, Frame → True,`
`  FrameLabel → {"x", "P(X=x)"},`
`  PlotLabel → "Cumulative Distribution Function`
`    for ExponentialDistribution[10] "]`

Out[81]=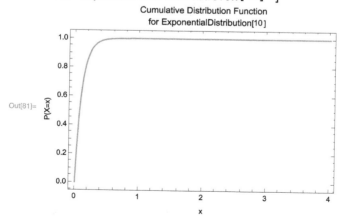

Fig. 3.16 **Out**[80]= PDF, and, **Out**[81]= CDF, for an Exponential [10] distribution.

3.5 Examples of Continuous Distributions

```
In[89]:= Plot[PDF[NormalDistribution[1, 1], x],
    {x, -3, 4}, PlotRange → All, Frame → True,
    FrameLabel → {"x", "P(X=x)"},
    PlotLabel → "Probability Density Function
    for NormalDistribution[1,1]"]
```

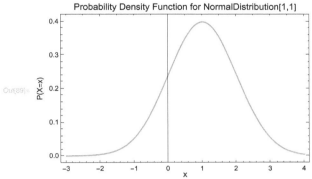

Fig. 3.17 Out[89]= PDF for a Normal distribution

Out[86]= ConditionalExpression$[\frac{1}{\sqrt{\frac{1}{\sigma^2}}\sigma}, \text{Re}[\sigma^2] \geq 0]$

The CDF has the form containing the complementary error function, **Erfc**. We can evaluate this for any set of parameters (e.g. $\mu = 1, \sigma = 1$):

In[87]:= CDF[NormalDistribution[μ, σ], x]
Out[87]= $\frac{1}{2}$**Erfc**$\left[\frac{-x+\mu}{\sqrt{2}\sigma}\right]$

And we can also get a table for any numeric values we want, as a list:

In[88]:= Table[N[CDF[NormalDistribution[1,1],x]],{x,0,5,0.5}]
Out[88]= {0.158655,0.308538,0.5,0.691462,0.841345,0.933193,
 0.97725,0.99379,0.99865,0.999767,0.999968}

Both the **NormalDistribution** PDF and CDF can be plotted in Figs. 3.17 and 3.18 respectively.

Now let us generate some set of random variates. For example, what if we generate 1000 data points from the same distribution:

In[93]:= sampleNormal=**RandomVariate[NormalDistribution**
 [0,1],1000];

We can generate a histogram of our random variates, and see that it follows a normal distribution, Fig. 3.19.

RandomVariate is very useful in programming and statistical computations. We can use random variates in simulations for any kind of distribution. The **RandomVariate** function takes as input a distribution and will generate a pseudorandom variate

In[90]:= `Plot[CDF[NormalDistribution[1, 1], x],`
`{x, -3, 4}, PlotRange → All, Frame → True,`
`FrameLabel → {"x", "P(X=x)"},`
`PlotLabel → "Cumulative Distribution Function`
`for NormalDistribution[1,1]"]`

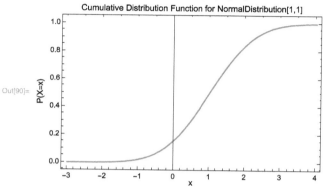

Out[90]=

Fig. 3.18 Out[90]= CDF for a Normal distribution

In[94]:= `Histogram[sampleNormal, Frame → True,`
`FrameLabel → {"x", "count"},`
`PlotLabel → "Histogram of 1000 Random Variates`
`from NormalDistribution[0,1]"]`

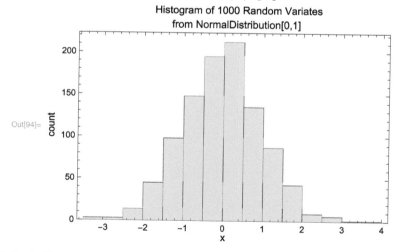

Out[94]=

Fig. 3.19 Out[94]= **Histogram** of random variates

3.5 Examples of Continuous Distributions

```
In[97]:= QuantilePlot[sampleNormal,
    ReferenceLineStyle → Red,
    FrameLabel → {"Normal Distribution Quantiles",
      "Random Variate Quantiles"},
    PlotLabel → "Quantile Plot for 1000 Random Variates
    from NormalDistribution[0,1]"]
```

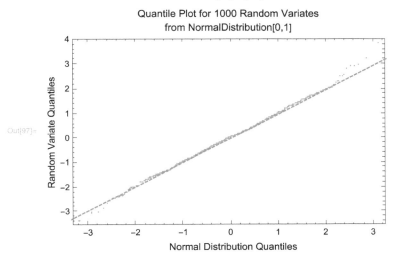

Fig. 3.20 Out[97]= **QuantilePlot** for random variates

from the symbolic distribution. This works for continuous and discrete distributions. **SeedRandom** can be used to initialize the pseudorandom sequence if you would like to use a particular seed.

In addition, as for any distributions we can calculate Quantiles. For example the 0.95 quantile for a Normal distribution with mean 0 and unit standard deviation is:

In[95]:= **Quantile**[**NormalDistribution**[0,1],0.95]
Out[95]= 1.64485

We can also calculate lower and upper quartiles:

In[96]:= **Quantile**[**NormalDistribution**[0,1],{0.25,0.75}]
Out[96]= {−0.67449,0.67449}

A quantile quantile plot can also be created (by default against the quantiles of a normal distribution). We can use our random variates generated above to obtain a **QuantilePlot**, Fig. 3.20.

We can also look at a box plot of our data, in Fig. 3.21. This is a simple example, let us generate a second set with a different mean, say 1.5.

In[100]:= sampleNormal2=**RandomVariate**[**NormalDistribution**
    [1.5,1],1000];

Now let us plot both in Fig. 3.22.

The mean is represented by the center of the square, 25 and 75% quantiles are the edges of the rectangle, and outliers are represented by dots, with far outliers shown in lighter colors.

3.5.4 Chi-Squared Distribution

A chi-square (χ^2) distribution is commonly used in various calculations. One typical source is that the sum of the squares of identically normally distributed random variables $\{X_1, X_2, \ldots, X_n\}, S_n = \{X_1^2 + X_2^2 + \cdots + X_n^2\}$ follows a χ^2 distribution with n degrees of freedom. The distribution has only one parameter which is the degrees for freedom, ν. We can get symbolic forms of both the PDF and CDF:

In[101]:= **PDF[ChiSquareDistribution[ν, x]**

Out[101]= $\begin{cases} \frac{2^{-\frac{\nu}{2}} e^{-\frac{x}{2}} x^{-1+\frac{\nu}{2}}}{\text{Gamma}\left[\frac{\nu}{2}\right]} & x > 0 \\ 0 & \text{True} \end{cases}$

In[102]:= **CDF[ChiSquareDistribution[ν], x]**

Out[102]= $\begin{cases} \text{GammaRegularized}\left[\frac{\nu}{2}, 0, \frac{x}{2}\right] & x > 0 \\ 0 & \text{True} \end{cases}$

Sums of squares typically arise when comparing statistical errors.

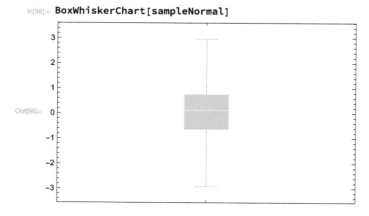

Fig. 3.21 Out[98]= Box plot from random variates

3.5 Examples of Continuous Distributions 95

```
In[100]:= BoxWhiskerChart[{sampleNormal, sampleNormal2},
    "Outliers", ChartLegends → {"1", "2"},
    ChartStyle → {Blue, Orange}]
```

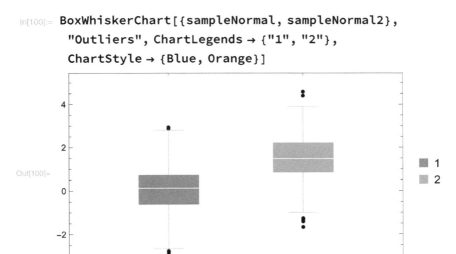

Fig. 3.22 Out[100]= Box plots from two samples

3.5.5 Student-t Distribution

The standard Student-t distribution is parametrized by the number of degrees of freedom v. For a normal distribution if we have a sample size v, we can look at the deviation of the mean of this sample compared to the true mean. These deviations follow a standard Student t-distribution.

In[105]:= **PDF[StudentTDistribution[v], x]**

Out[105]= $\dfrac{\left(\frac{v}{x^2+v}\right)^{\frac{1+v}{2}}}{\sqrt{v}\,\text{Beta}\left[\frac{v}{2}, \frac{1}{2}\right]}$

In[106]:= **CDF[StudentTDistribution[v], x]**

Out[106]= $\begin{cases} \frac{1}{2}\text{BetaRegularized}\left[\frac{v}{x^2+v}, \frac{v}{2}, \frac{1}{2}\right] & x \leq 0 \\ \frac{1}{2}\left(1 + \text{BetaRegularized}\left[\frac{x^2}{x^2+v}, \frac{1}{2}, \frac{v}{2}\right]\right) & \text{True} \end{cases}$

The degrees of freedom determine the variance, while for the standard Student t distribution the mean is zero:

In[107]:= **Variance[StudentTDistribution[v]]**

Out[107]= $\begin{cases} \frac{v}{-2+v} & v > 2 \\ \text{Indeterminate} & \text{True} \end{cases}$

In[108]:= **Mean[StudentTDistribution[v]]**

Out[108]= $\begin{cases} 0 & v > 1 \\ \text{Indeterminate} & \text{True} \end{cases}$

3.6 Other Distributions

The Wolfram Language documentation has multiple Discrete and Continuous Distributions included. You can refer to the documentation tutorial type either:

- "tutorial/DiscreteDistributions"
- "tutorial/ContinuousDistributions"

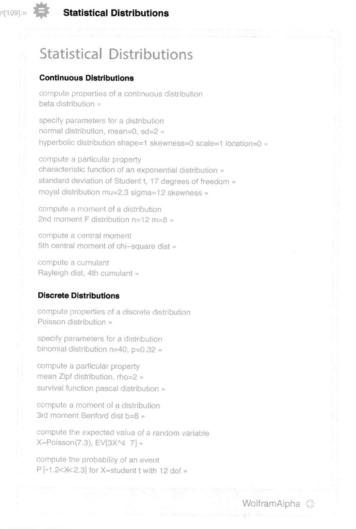

Fig. 3.23 Out[109]= Wolfram|Alpha can be used to obtain further statistical distribution information and computation

3.6 Other Distributions

in the Wolfram Documentation help bar) to see additional examples. The documentation contains additional distributions and multi-variate and mixed distribution possibilities which we have not covered here.

Furthermore, Wolfram|Alpha can provide additional access as indicated in Fig. 3.23.

Additionally, the Wolfram Language has multivariable distributions, such as

In[110]:= ?MultinormalDistribution
MultinormalDistribution[μ, Σ] represents a
    multivariate normal (Gaussian) distribution
    with mean vector μ and covariance matrix Σ.
MultinormalDistribution[Σ] represents a multivariate
    normal distribution with zero mean and covariance
    matrix Σ. >>

In[111]:= ?MultinomialDistribution
MultinomialDistribution[$n, \{p_1, p_2, \ldots, p_m\}$]
    represents a multinomial distribution
    with n trials and probabilities p_i. >>

3.7 Data Sets

Data analysis is driven by, well, data. We need data to test algorithms, to make hypotheses to validate results. The better the quality of the data, the easier the analysis. A brilliant algorithm still needs good data to give good results. The Wolfram Language has a lot of example data that can be used, and in addition **ResourceData** which may be useful for testing code and algorithms and **Entity** types that contain information and will be briefly described below. Additionally, a plethora of data is available for omics experiments hosted on websites such as Gene Expression Omnibus (GEO) [4, 9], ArrayExpress [15], UCSC Browser [14, 16], and many more. We also include a subset of data together with the notebooks accompanying the book, as well as data incorporated into the MathIOmica package which are used for illustrative examples of workflows in our examples. Here we simply introduce the various data and we will use these in later parts of the book for our coding examples.

3.7.1 ExampleData

We will use some of the **ExampleData** in the Wolfram Language for this chapter. Mathematica has multiple example data distributed with the language software, and we can get a full list of the different topics:

In[112]:= **ExampleData**[]
Out[112]= {AerialImage, Audio, ColorTexture,
      Dataset, Geometry3D, LinearProgramming,
      MachineLearning, Matrix, NetworkGraph,
      Sound, Statistics, TestAnimation, TestImage,
      TestImage3D, TestImageSet, Text, Texture}

Within each topic we can search for additional descriptions. Let us retrieve all the data of the "Statistics" topic:

In[113]:= exampleDataStatistics=**ExampleData**["Statistics"];

There are 111 sets as of this writing:

In[114]:= **Length**[exampleDataStatistics]
Out[114]= 111

In[115]:= exampleDataStatistics[[1;;10]]
Out[115]= {{Statistics, AirlinePassengerMiles},
      {Statistics, AirplaneGlass},
      {Statistics, AnimalWeights},
      {Statistics, AnorexiaTreatment},
      {Statistics, AnscombeRegressionLines},
      {Statistics, AustraliaAIDS},
      {Statistics, AustraliaRainfall},
      {Statistics, Baboon},
      {Statistics, BatchChemicalProcessYields},
      {Statistics, BeaverBodyTemperatures}}

Each of the data sets carries additional properties. For example let us check the "DenmarkMelanoma"" dataset.

In[116]:= **ExampleData**[
      {"Statistics","DenmarkMelanoma"},
      "Properties"]
Out[116]= {ApplicationAreas, ColumnDescriptions,
      ColumnHeadings, ColumnTypes, DataElements,
      DataType, Description, Dimensions,
      EventData, EventSeries, LongDescription,
      Name, ObservationCount, Source, TimeSeries}

For each of the elements we can get the information. We can see what the data is about:

In[117]:= **ExampleData**[
      {"Statistics","DenmarkMelanoma"},
      "Description"]
Out[117]= Survival from malignant melanoma.

Also, we can get the source of the data:

In[118]:= **ExampleData**[{"Statistics","DenmarkMelanoma"},"Source"]
Out[118]= P. K. Andersen, O. Borgan, R. D. Gill
      and N. Keiding (1993) Statistical Models
      based on Counting Processes. Springer.

3.7 Data Sets

Let us use the `"BoneMarrowTransplants"` data as an example to look further into. The description is again fairly short:

In[119]:= **ExampleData**[
    {"Statistics","BoneMarrowTransplants"},
    "Description"]

Out[119]= Autologous and allogeneic bone marrow transplants.

The data has again other properties we can check, as the `"DenmarkMelanoma"` dataset did:

In[120]:= **ExampleData**[
    {"Statistics","BoneMarrowTransplants"},
    "ApplicationAreas"]

Out[120]= Medicine

If we plan to use the data, we should take a look at the longer description:

In[121]:= **ExampleData**[
    {"Statistics","BoneMarrowTransplants"},
    "LongDescription"]

Out[121]= A sample of 101 patients with advanced acute myelogenous leukemia reported to the International Bone Marrow Transplant Registry. Fifty-one of these patients had received an autologous (auto) bone marrow transplant in which, after high doses of chemotherapy, their own marrow was reinfused to replace their destroyed immune system. Fifty patients had an allogeneic (allo) bone marrow transplant where marrow from an HLA-matched sibling was used to replenish their immune systems. These data are right censored."

And also we need to cite our sources:

In[122]:= **ExampleData**[
    {"Statistics","BoneMarrowTransplants"},
    "Source"]

Out[122]= Klein and Moeschberger (1997) Survival Analysis Techniques for Censored and truncated data, Springer."

It is good practice to check the dimensions of the data:

In[123]:= **ExampleData**[
    {"Statistics","BoneMarrowTransplants"},
    "Dimensions"]

Out[123]= {101,3}

And of course, we can get the `"DataElements"` themselves. We store the data in a variable:

100   3   Statistics

In[124]:= dataBoneMarrow=
           ExampleData[
             {"Statistics","BoneMarrowTransplants"},
             "DataElements"];

In[125]:= dataBoneMarrow // **Short**
Out[125]//**Short**= {{0.03,1,0},<<99>>,{56.086,2,0}}

Note how we used **//Short** above to display a shorter dialog. This is done using the **Postfix** notation //:

In[126]:= ?**Postfix**
           Postfix[f[expr]] prints with f[expr]
              given in default postfix form: expr//f.
           Postfix[f[expr],h] prints as exprh. >>

Additionally, we are interested in the type of data for all the columns:

In[127]:= **ExampleData**[
             {"Statistics","BoneMarrowTransplants"},
             "ColumnTypes"]
Out[127]= {Numeric, Numeric, Numeric}

And the columns themselves would not be that useful without "ColumnDescriptions":

In[128]:= **ExampleData**[{"Statistics",
              "BoneMarrowTransplants"},#]&@ "ColumnDescriptions"
Out[128]= {Time to death or relapse in months,
            Type of transplant (1 =
              allogeneic, 2 = autologous),
            Leukemia–free survival indicator (1 = alive
              without relapse, 0 = dead or relapse)}

Let us look at the distributions of the "Time to death or relapse" separately for allogeneic or autologous transplants. We use the **Gather** function to separate the data using a test.

In[129]:= ?**Gather**
           Gather[list] gathers the elements of list into sublists
              of identical elements.
           Gather[list,test] applies test to pairs of elements to
              determine if they should be considered
              identical. >>

In[130]:= dataBoneMarrowSeparated=
           Gather[dataBoneMarrow, #1[[2]]==#2[[2]]&];

In[131]:= dataBoneMarrowSeparated // **Short**
Out[131]//**Short**= {{{0.03,1,0},<<48>>,{60.625,1,1}},{<<1>>}}

Notice the form of our test: #1[[2]]==#2[[2]]&. This is basically a function that takes 2 inputs to compare, argument #1 and #2 (i.e. pairwise comparisons are carried out within the data). The test returns True or False if the 2nd parts of the arguments are

3.7 Data Sets

```
In[132]:= BoxWhiskerChart[{dataBoneMarrowSeparated[[1, All, 1]],
        dataBoneMarrowSeparated[[2, All, 1]]},
    PlotTheme → "Scientific", ChartStyle → "Pastel",
    FrameLabel → { "Patient Group",
        "Time to Death or Relapse (Months)"}, PlotLabel →
    "Distributions of Time to Death or Relapse for
    Patients Receiving Allogeneic or Autologous Transplants",
    ChartLabels → {"Allogeneic", "Autologous"},
    ChartLegends → {"Allogeneic", "Autologous"}]
```

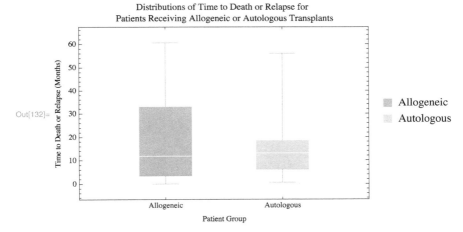

Fig. 3.24 Out[132]= Box plots

identical, i.e. in our case whether the transplants are 1: allogeneic or 2: autologous. Now we can make a box plot, Fig. 3.24.

We can see from Fig. 3.24 that the distributions have similar means, but that the spread is much bigger for the allogeneic data. We can compare the means and standard deviation:

```
In[133]:= Mean[N[#]]&/@
            {dataBoneMarrowSeparated[[1,All,1]],
             dataBoneMarrowSeparated[[2,All,1]]}
Out[133]= {18.5519,16.7317}

In[134]:= StandardDeviation[N[#]]&/@
            {dataBoneMarrowSeparated[[1,All,1]],
             dataBoneMarrowSeparated[[2,All,1]]}
Out[134]= {17.7627,13.9388}
```

Notice here how we used **Map** /@ to apply the function to each element and shorten our code. We need to further separate the patients based on the third column

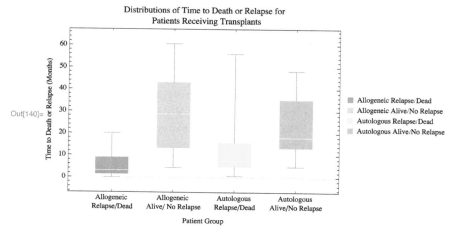

Fig. 3.25 Out[140]= Box plots for Allogenic and Autologous categorization

of information: 1 = alive without relapse or 0 = dead or relapse. We can further split our lists:

In[135]:= dataBoneMarrowSeparatedAliveDead=**Gather**[dataBoneMarrow,
 #1[[2]]==#2[[2]]&&#1[[3]]==#2[[3]]&];
In[136]:= dataBoneMarrowSeparatedAliveDead[[1]]//**Short**
Out[136]//**Short**= {{0.03,1,0},{0.493,1,0},<<19>>,{20.066,1,0}}

In[137]:= dataBoneMarrowSeparatedAliveDead[[2]]//**Short**
Out[137]//**Short**= {{4.441,1,1},<<26>>,{60.625,1,1}}

In[138]:= dataBoneMarrowSeparatedAliveDead[[3]]//**Short**
Out[138]//**Short**= {{0.658,2,0},<<26>>,{56.086,2,0}}

In[139]:= dataBoneMarrowSeparatedAliveDead[[4]]//**Short**
Out[139]//**Short**= {{4.836,2,1},<<21>>,{48.322,2,1}}

Now we can create a box plot, Fig. 3.25 for the results:

In[140]:= **BoxWhiskerChart**[
 dataBoneMarrowSeparatedAliveDead[[**All**,**All**,1]],
 PlotTheme-> "Scientific",ChartStyle ->"Pastel",
 FrameLabel->{ "Patient Group",
  "Time to Death or Relapse (Months)"},
 PlotLabel->
  "Distributions of Time to Death or Relapse for
  Patients Receiving Transplants",
 ChartLabels ->
 {"Allogeneic Relapse/Dead",
  "Allogeneic Alive/No Relapse",
  "Autologous Relapse/Dead",
  "Autologous Alive/No Relapse"},
 ChartLegends ->

3.7 Data Sets

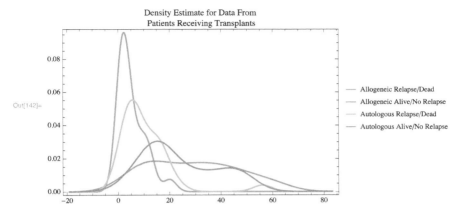

Fig. 3.26 Out[142]= SmoothHistogram plots

```
{"Allogeneic Alive","Allogenic Dead",
 "Autologous Alive","Autologous Dead"}]
```

We can plot a density estimate of these data using **SmoothHistogram**, Fig. 3.26. This has multiple options which we can get:

```
In[141]:= Options[SmoothHistogram]
Out[141]= {AlignmentPoint→Center, AspectRatio→1/GoldenRatio,
    Axes→True, AxesLabel→None, AxesOrigin→Automatic,
    AxesStyle→{}, Background→None,
    BaselinePosition→Automatic, BaseStyle→{},
    ClippingStyle→None, ColorFunction→Automatic,
    ColorFunctionScaling→True,
    ColorOutput→Automatic,
    ContentSelectable→Automatic,
    CoordinatesToolOptions→Automatic,
    DisplayFunction:>$DisplayFunction, Epilog→{},
    Filling→None, FillingStyle→Automatic,
    FormatType:>TraditionalForm, Frame→False,
    FrameLabel→None, FrameStyle→{}, FrameTicks→Automatic,
    FrameTicksStyle→{}, GridLines→None,
    GridLinesStyle→{}, ImageMargins→0., ImagePadding→All,
    ImageSize→Automatic, ImageSizeRaw→Automatic,
    LabelStyle→{}, MaxRecursion→Automatic, Mesh→None,
    MeshFunctions→{#1&}, MeshShading→None,
    MeshStyle→Automatic, Method→Automatic,
    PerformanceGoal:>$PerformanceGoal, PlotLabel→None,
    PlotLegends→None, PlotPoints→Automatic,
    PlotRange→Automatic, PlotRangeClipping→True,
    PlotRangePadding→Automatic, PlotRegion→Automatic,
    PlotStyle→Automatic, PlotTheme:>$PlotTheme,
    PreserveImageOptions→Automatic, Prolog→{},
    RegionFunction→(True&), RotateLabel→True,
    ScalingFunctions→None, TargetUnits→Automatic,
    Ticks→Automatic, TicksStyle→{},
```

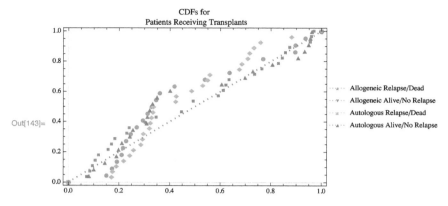

Fig. 3.27 Out[143]= ProbabilityPlot

WorkingPrecision→MachinePrecision }

We generate the plot in Fig. 3.26, to include a **PlotTheme**, labels and legends using the code below:

```
In[142]:= SmoothHistogram[
          dataBoneMarrowSeparatedAliveDead[[All,All,1]],
          PlotTheme->"Scientific",
          PlotLabel->"Density Estimate for Data From
          Patients Receiving Transplants",
          PlotLegends ->{"Allogeneic Relapse/Dead",
                        "Allogeneic Alive/No Relapse",
                        "Autologous Relapse/Dead",
                        "Autologous Alive/No Relapse"}]
```

We can also look at the CDF compared to the CDF of a normal distribution, Fig. 3.27, using a **ProbabilityPlot** :

```
In[143]:= ProbabilityPlot[
          dataBoneMarrowSeparatedAliveDead[[All,All,1]],
          PlotTheme->"Scientific",PlotLabel->
          "CDFs for \n Patients Receiving Transplants",
          PlotLegends ->{"Allogeneic Relapse/Dead",
                        "Allogeneic Alive/No Relapse",
                        "Autologous Relapse/Dead",
                        "Autologous Alive/No Relapse"},
          PlotMarkers ->Automatic]
```

3.7.2 ResourceData

ResourceData Following Mathematica 11, ResourceData can be accessed from the Wolfram Data Repository (Fig. 3.28).

3.7 Data Sets

In[145]:= **ResourceSearch["Cancer"]**

Out[145]=

Fig. 3.28 Out[145]= **ResourceSearch**["Cancer"]

In[146]:= **roEsophageal = ResourceData["Sample Data: Esophageal Cancer"]**

Out[146]=

AgeGroup	AlcoholConsumption	TobaccoConsumption	NumberOfCases	NumberOfControls
1	1	1	0	40
1	1	2	0	10
1	1	3	0	6
1	1	4	0	5
1	2	1	0	27
1	2	2	0	7
1	2	3	0	4
1	2	4	0	7
1	3	1	0	2
1	3	2	0	1
1	3	4	0	2
1	4	1	0	1
1	4	2	1	1
1	4	3	0	1
1	4	4	0	2
2	1	1	0	60
2	1	2	1	14
2	1	3	0	7
2	1	4	0	8
2	2	1	0	35

showing 1-20 of 88

Fig. 3.29 Out[146]= **ResourceData**["Sample Data: Esophageal Cancer"].

In[144]:= ?**ResourceData**
  ResourceData[resource] gives the primary content of the specified resource.
  ResourceData[resource, elem] gives element elem of the content of the resource. »

We can search for a data resource (Experimental feature in 11.2) using **ResourceSearch** as shown in Fig. 3.29.

The data can be loaded and used once we know the name, which is loaded as a **Dataset**, in Fig. 3.29.

3.7.3 Entity Types

The Wolfram Language has multiple entities that can be used in computations. This includes country, disease, chemical, file formats, genes, proteins, medical tests, and many more information, These are entered in normal form as **Entity**[`"type"`, `"name of entity"`]. The list of available entity types is:

In[147]:= entityValues=**EntityValue**[];

We only print a few as an example:

In[148]:= entityValues[[1;;10]]
Out[148]= {AdministrativeDivision, Aircraft, Airline, Airport, Alphabet, AmusementPark, AmusementParkRide, AnatomicalFunctionalConcept, AnatomicalStructure, Artwork}

Each type of entity has various members in its list, which can be obtained with **EntityList**. If you try to get the protein/gene information note that there is a lot of information/values so it might take a while to first download the data:

In[149]:= proteinList=**EntityList**[`"Protein"`];
In[150]:= **Length**[proteinList]
Out[150]= 27 479

We can look at the first few members of the list, shown as a graphical object in Fig. 3.30.

In[151]:= **proteinList[[1 ;; 10]]**

Out[151]= { alpha 1B-glycoprotein precursor , DEK oncogene , translocase of inner mitochondrial membrane 23 (yeast) homolog , allograft inflammatory factor 1 isoform 1 , allograft inflammatory factor 1 isoform 3 , absent in melanoma 1 , adenylate kinase 1 , adenylate kinase 2 isoform a , aminoacylase 1 , ATP-binding cassette, sub-family A, member 2 isoform b }

Fig. 3.30 Out[151]= Protein **EntityList**

3.7 Data Sets   107

In[154]:= `diseaseList[[1 ;; 10]]`

Out[154]= { Asthma , CerebrovascularDisease , IschemicHeartDisease ,
salmonella infections , unclassified salmonella infections ,
unspecified type of salmonella infection , shigellosis ,
unclassified specified shigella infections ,
bacterial food poisoning ,
unspecified type of bacterial food poisoning }

Fig. 3.31 Out[154]= Disease **EntityList**

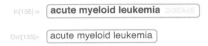

In[155]:= acute myeloid leukemia DISEASE

Out[155]= acute myeloid leukemia

Fig. 3.32 Out[155]= Use the Ctrl="Acute Myeloid Leukemia" method to obtain the entity

In[156]:= `amlEntity = Entity["Disease", "ICDNine205.0"]`

Out[156]= acute myeloid leukemia

Fig. 3.33 Out[156]= Use the ICD9 or ICD10 code to method to obtain the disease information

Other entities of interest include "Disease":

In[152]:= diseaseList=**EntityList**["Disease"];
In[153]:= **Length**[diseaseList]
Out[153]= 10886

Again, we can look at the frist few items as shown in Fig. 3.31.

Entities can be entered also using the Wolfram|Alpha control+= entry (ctrl+=) method. This will create a box you can type into and then will process the entry for matches. For example you can type ctrl = acute myeloid leukemia and enter to get the disease input, Fig. 3.32:

Alternatively, if you know the disease ICD9 code you can type it out, Fig. 3.33.

We can look at the **InputForm** for the ICD9:

In[157]:= **InputForm**[amlEntity]
Out[157]//**InputForm**= Entity["Disease", "ICDNine205.0"]

The entities have many useful properties, Fig. 3.34.

While these are formatted in Mathematica, we can get the Wolfram Language **InputForm**:

In[159]:= **InputForm**[#]&/@amlEntity["Properties"]
Out[159]= {EntityProperty["Disease", "AgeMean"],
          EntityProperty["Disease", "BodyMassIndexMean"],

In[158]:= `amlEntity["Properties"]`

Out[158]= { average patient age ,

average patient BMI , average temperature ,

average diastolic blood pressure , average patient height ,

ICD-9 code , ICD-10 code , name ,

average systolic blood pressure , average patient weight }

Fig. 3.34 Out[158]= Disease properties

Fig. 3.35 Out[161]= AML ICD-10 Code

In[161]:= `amlEntity["ICDTenCode"]`

Out[161]= ICD-10 C92.0

```
EntityProperty["Disease", "BodyTemperatureMean"],
EntityProperty["Disease", "DiastolicMean"],
EntityProperty["Disease", "HeightMean"],
EntityProperty["Disease", "ICDNineCode"],
EntityProperty["Disease", "ICDTenCode"],
EntityProperty["Disease", "Name"],
EntityProperty["Disease", "SystolicMean"],
EntityProperty["Disease", "WeightMean"]}
```

You can either use the ctrl= method to input these information, hover over the formatted entity boxes to get the information or cut and paste the formatted entity objects at the right locations. Here are a couple of examples of properties, such as the "AgeMean" and "ICDTenCode" for the disease, Fig. 3.35.

In[160]:= `amlEntity["AgeMean"]`
Out[160]= 53.1106 yr

3.7.4 Data Example: The Golub ALL AML Data Set

As part of our examples we are including information from a study performed by Golub et al. [12]. These data have become a standard for illustrating biological data analysis approaches. We have ported a version of the dataset together with the files that accompany this book. The study studied the possibility of using gene expression information to classify cancer types. It used data collected from 72 subjects, 25 with acute myeloid leukemia(AML) and 47 acute lymphoblastic leukemia (ALL). This set was initially used in two parts, with 38 subject data used for training purposes to come up with a classification between the cancer subtypes, and 34 used to test

3.7 Data Sets

the classification. The data consisted of gene expression results which were high-throughput at the time, for 7129 features/gene products on Affymetrix Hu8600 arrays.

The ALL/AML data were used extensively in followup work as an example dataset, and we have ported a version for use with the Wolfram Language, (golubAssociation). Additionally we have created an association containing a map of the Affymetrix accessions for the array to gene information (hu6800IDtoAnnotation). In Chap. 6 we will see how we obtain the normalized set used in the manuscript and the annotations. We will refer to this as the Golub dataset, which has been filtered to contain 3571 genes for 72 subjects by using all the data from the original study (obtained from: `"http://portals.broadinstitute.org/cgi-bin/cancer/publications/pub_paper.cgi?mode=view\&paper_id=43"`, Accessed on Nov 30, 2017).

If you are using the notebook and files accompanying this chapter, you can access the data by loading the association:

In[162]:= **SetDirectory[NotebookDirectory[]];**

In[163]:= golubAssociation=<<golubAssociation;

In[164]:= golubAssociation // **Short**
Out[164]//**Short**= <|AML–><|AFFX–BioDn–3_at –>{<<1>>},<<3570>>|>,
　　　　　　　ALL–><<1>>|>

Additionally, you can load the array annotation created for our purposes:

In[165]:= hu6800IDtoAnnotation=<<hu6800IDtoAnnotation;

In[166]:= hu6800IDtoAnnotation[[1;;4]]
Out[166]//**Short**= <|Probe Set ID"→"{Gene Symbol,
　　　　　　　Gene Title, Entrez Gene ID, Cytoband},
　　　　　　　AB000409_at"→"{MKNK1,MAP kinase
　　　　　　　　interacting serine/threonine kinase 1,
　　　　　　　　8569,1p33},
　　　　　　　AB000410_s_at"→"{OGG1,8–oxoguanine DNA
　　　　　　　　glycosylase,4968,3p26.2},
　　　　　　　AB000450_at"→"{VRK2,vaccinia related
　　　　　　　　kinase 2,7444,2p16.1}|>

In[167]:= hu6800IDtoAnnotation["M71243_f_at"]
Out[167]= {GYPA, glycophorin A (MNS blood group),2993,4q31.21}

Now, say we know the gene name we want, but do not really know the symbol. We can look for the string in the association and select the one that matches. There are 4 columns of information. We use a **Query** to select. For example, say we are interested in the information on Myosin light chain 6B, which is listed in the original paper as showing significant differences between ALL and AML:

In[168]:= **Query[**
　　　　　Select[
　　　　　　Or@@StringContainsQ[#,
　　　　　　　"myosin light chain 6B",
　　　　　　IgnoreCase->False]&]]@hu6800IDtoAnnotation

Out[168]= <|M31211_s_at"→"{MYL6B,
           myosin light chain 6B,140465,12q13.13}|>

If we known the **Key** we can get the information:

In[169]:= hu6800IDtoAnnotation["M31211_s_at"]
Out[169]= {MYL6B,myosin light chain 6B,140465,12q13.13}

Let us dissect the **Query** above: The **StringContainsQ** function returns true if the non-case dependent match of the string exists in any of the four lists in the values of the association. The Or@@ calculates a logical OR operation across the list of Boolean variables returned, and will yield True only if a True is contained in the list. Finally, **Select** will only return the values that are true. The **Query** acts (@) on the annotation so we get the entry that was selected. Here is some more information on **StringMatchQ**, which is very useful for matching strings in text files:

In[170]:= ?StringMatchQ
    StringMatchQ["string",patt] tests whether "string"
        matches the string pattern patt.
    StringMatchQ["string",**RegularExpression**["regex"]] tests
        whether "string" matches the specified regular
        expression.
    StringMatchQ[{$s_1,s_2,...$},p] gives the list
        of results for each of the s_i.
    StringMatchQ[patt] represents an operator form of
        StringMatchQ that can be applied to an expression.
    >>

Now we know the accession (the **Key** in the output), we can extract the data we are interested in:

In[171]:= myosinExample=**Query**[**All**,"M31211_s_at"]@golubAssociation
Out[171]= <|AML→{−0.929698,−1.21439,
           −1.36559,−1.03054,−1.29915,−0.401122,
           −0.263245,−1.27241,−0.651176,−0.323125,
           −1.43473,−0.203518,−1.04336,−1.22702,
           −1.28071,0.260207,−1.05801,−0.720331,
           −0.809905,−0.588208,−0.513632,
           −0.247296,−1.42036,−1.48163,−1.3111},
        ALL→{0.21673,−0.0512012,0.0512196,
           0.0658298, 0.355842,−0.0764156,−0.298932,
           −0.46567,0.490372,0.651373,0.715212,
           0.148149,0.354363,−0.466193,0.576175,
           0.417461,−0.407586,0.388138,0.266817,
           0.634913,0.707898,−0.775871,0.0765331,
           0.108658,0.224704,0.466898,−0.87136,
           0.193097,−0.37865,−0.0896119,0.284866,
           0.51995,−0.204948,0.560927,−0.0758743,
           0.726773,0.337038,0.374954,0.146782,
           −0.474574,0.170777,0.136991,0.107438,
           −0.55069,0.279121,−0.665758,0.545882}|>

We can plot the information using a **SmoothHistogram**, as in Fig. 3.36.

3.7 Data Sets   111

In[172]:= **SmoothHistogram[myosinExample]**

Out[172]=
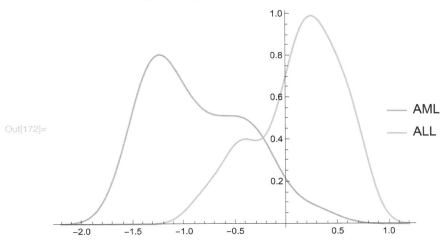

Fig. 3.36 Out[172]= Histogram of gene expression example

Additionally, we can look at a box plot for the data, Fig. 3.37. We notice that the expression in distinctly higher in the mean in ALL versus AML, as reported as well in the paper by Golub et al.

3.7.5 Example: Sandberg et al. Data by Pavlidis

In Chap. 6 we will analyze data from Sandberg et al. [25] and follow the analysis by Pavlidis [22, 23]. The data consists of 24 microarray samples, from two mouse strains (129, B6) and six different brain regions (amygdala, cerebellum, cortex, entorhinal cortex, hippocampus and midbrain) and include two duplicates for each tissue type. The data is available together with the notebooks for this book.

In[174]:= **SetDirectory[NotebookDirectory[]]**;

In[175]:= dataSandberg=**Import**["sandberg-sampledata.txt","TSV"];

We take just a brief look at the data prior to our analysis later in the book:

In[176]:= dataSandberg[[1;;3]]
Out[176]= {{clone,129Amygdala(1a),
         129Amygdala(1b),129Cerebellum(1),
         129Cerebellum(2),129Cortex(1),
         129Cortex(2),129EntorhinalCortex(1a)
         ,129EntorhinalCortex(1b),129Hippocampus(1),
         129Hippocampus(2),129Midbrain(1),
         129Midbrain(2),B6Amygdala(1a),B6Amygdala(1b),
         B6Cerebellum(1),B6Cerebellum(2),B6Cortex(1),

In[173]:= `BoxWhiskerChart[myosinExample,`
   `PlotTheme → "Scientific", ChartStyle → "Pastel",`
   `FrameLabel →`
    `{"Patient Group",`
     `"Normalized Gene Expression for MYL6B"},`
   `PlotLabel →`
    `"Normalized Gene Expression`
    `for MYL6Bin Golub et al. data",`
   `ChartLabels → {"AML", "ALL"},`
   `ChartLegends → {"AML", "ALL"}]`

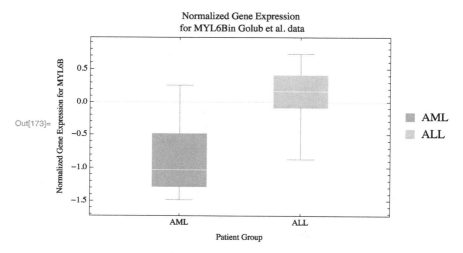

Fig. 3.37 Out[173]= Box plot of gene expression example for AML and ALL

```
B6Cortex(2),B6EntorhinalCortex(1a),
B6EntorhinalCortex(1b),B6Hippocampus(1),
B6Hippocampus(2),B6Midbrain(1),B6Midbrain(2)},
{aa000148.s.at,298,212,258,205,220,269,
167,294,218,239,220,157,257,236,256,
232,240,284,236,264,188,228,241,257},
{AA000151.at,243,154,228,162,192,173,
191,125,99,144,186,176,127,54,202,
166,135,251,185,198,102,145,202,218}}
```

3.7 Data Sets

3.7.6 Example: Marcobal et al.

In Chap. 8 we will analyze data from the study by Marcobal et al. [18]. The researchers carried out metabolomics profiling of urine in mice responding to inoculation by bacteria. Specifically, two groups of Swiss - Webster germ free (GF) mice (in gnotobiotic isolators) were used, with kept germ free, and the second group's gut colonized with two bacterial species, *Bacteroides thetaiotaomicron* and *Bifidobacterium longum*. Samples for up to 25 days at 5 day intervals are available. The data contain mass features (in Dalton) as identifiers from mass spectrometry, and the intensities for each sample per feature.

We can import the data from the notebooks directory - the file is called "GFMice.csv", and look at the first few entries:

In[177]:= **SetDirectory**[**NotebookDirectory**[]];

In[178]:= importGFMice = **Import**["GFMice.csv"];

In[179]:= importGFMice[[3;;6]]

Out[179]={{ ,0(GF) ,0(GF) ,0(GF) ,5 ,5 ,5 ,5 ,10 ,10 ,10 ,
10 ,15 ,15 ,15 ,15 ,20 ,20 ,20 ,25 ,25 ,25},
{269.125 ,4.77187×10^6 ,1.83739×10^6 ,
2.69081×10^6 ,1.78109×10^6 ,
1.95988×10^6 ,1.61111×10^6 ,1.84043×10^6 ,
3.46027×10^6 ,7.17564×10^6 ,4.24283×10^6 ,
5.18863×10^6 ,3.44216×10^6 ,5.37601×10^6 ,
5.75372×10^6 ,5.62317×10^6 ,5.60079×10^6 ,
4.82033×10^6 ,2.78184×10^6 ,2.68937×10^6 ,
3.73472×10^6 ,5.98949×10^6},
{225.083 ,741909. ,667996. ,555465. ,693806. ,
812922. ,690705. ,922265. ,955981. ,1.3267×10^6 ,
1.64057×10^6 ,1.43668×10^6 ,1.28518×10^6 ,
1.25653×10^6 ,1.46838×10^6 ,1.34728×10^6 ,
1.06292×10^6 ,1.09649×10^6 ,1.02028×10^6 ,
806857. ,1.21311×10^6 ,1.04819×10^6},
{323.134 ,533782. ,260343. ,218852. ,383337. ,
377515. ,341260. ,583339. ,249954. ,
356329. ,276584. ,263299. ,255125. ,
343460. ,472003. ,437638. ,199619. ,314219. ,
288501. ,109916. ,227727. ,169341.}}

The first line contains information on whether the mice are germ free and the day, and the next few lines are examples of intensities. The first entry in each list of intensities corresponds to a mass feature measurement in Daltons. We will use these data as an example analysis in metabolomics.

3.7.7 Example Data: MathIOmica

The MathIOmica package comes with multiple example data. The data can be found in the **ConstantMathIOmicaExamplesDirectory**. You can get list all the files of the available Example Data in MathIOmica by evaluating:

In[180]:= filesMathIOmicaExamples=
        FileNames[__, ConstantMathIOmicaExamplesDirectory];

The data consists of raw and processed results that have been analyzed in various examples in the MathIOmica Documentation (please note that you should not move or alter these files from the installed directory as this may change the integrity of the package and its documentation). The data contain information form the integrative Personal Omics Profile [6] pilot described below.

Data Description	File Name(s) located in the ConstantMathIOmicaExamplesDirectory.
iPOP Transcriptome. The transcriptomic data included was obtained from mapping of the originally RNA Sequencing raw data using the Tuxedo suite. The data corresponds to transcriptome from peripheral blood mononuclear cells (PBMCs).	iPOP_07_genes.fpkm_tracking
	iPOP_08_genes.fpkm_tracking
	iPOP_09_genes.fpkm_tracking
	iPOP_10_genes.fpkm_tracking
	iPOP_11_genes.fpkm_tracking
	iPOP_12_genes.fpkm_tracking
	iPOP_13_genes.fpkm_tracking
	iPOP_14_genes.fpkm_tracking
	iPOP_15_genes.fpkm_tracking
	iPOP_16_genes.fpkm_tracking
	iPOP_17_genes.fpkm_tracking
	iPOP_18_genes.fpkm_tracking
	iPOP_19_genes.fpkm_tracking
	iPOP_20_genes.fpkm_tracking
	iPOP_21_genes.fpkm_tracking
iPOP Proteome. The Proteomics data from analysis of mass spectrometry data using the Sequest algorithm implemented by ProteomeDiscoverer. The data corresponds to proteome from PBMCs. The names of the files provide a correspondence of samples to Tandem Mass Tag labels in order of increasing m/z values from 126 to 131 amu. 6 TMT labels were used in each experiment. The data has been adapted from the original to UniProt accessions.	8_7_9_10_11_14_MulticonsensusReports_3Replicates.csv
	8_12_13_15_16_14_MulticonsensusReports_3Replicates.csv
	8_17_19_20_21_14_MulticonsensusReports_3Replicates.csv
iPOP Metabolome. The Metabolomics data from analysis of mass spectrometry data. The data corresponds to small molecule metabolomics from plasma ran with technical triplicates. The names of the files provide a correspondence of samples ran in positive or negative mode.	metabolomics_positive_mode.csv
	metabolomics_negative_mode.csv

Fig. 3.38 Description of Example iPOP original datasets and corresponding files in the ConstantMathIOmicaExamplesDirectory

3.7 Data Sets

Data Description	File Name(s) located in the ConstantMathIOmicaExamplesDirectory.
iPOP transcriptome imported as an OmicsObject across all timepoints.	rnaExample
iPOP proteome data imported as an OmicsObject across all timepoints.	proteinExample
iPOP metabolome data imported as an OmicsObject across all timepoints and technical replicates for negative and positive mode aligned mass spectrometry features.	metabolomicsNegativeModeExample
	metabolomicsPositiveModeExample
Example time series from proteomics.	proteinTimeSeriesExample
Example classification results from proteomics.	proteinClassificationExample
Example classification results from proteomics.	proteinClusteringExample
Example combined clustering results from transcriptome, proteome and metabolome data.	combinedClustersExample
Example enrichment analysis results for Gene Ontology and KEGG pathway analysis for combined omics data in this tutorial.	combinedGOAnalysis
combinedKEGGAnalysis	combinedKEGGAnalysis
Spectra from proteomics mass spectrometry data examples.	small.pwiz.1.1.mzML
	exampleMS3.mzXML

Fig. 3.39 Description of Example MathIOmica analyzed datasets and corresponding files in the ConstantMathIOmicaExamplesDirectory

3.7.7.1 integrative Personal Omics

Data from the first integrative Omics Profiling (iPOP) in MathIOmica includes data from the dynamics of proteomics transcriptomics and metabolomics [6, 19, 20]. The data corresponds to a time series analysis of omics from blood components from a single individual. Different samples (from 7 to 21 included here) were obtained at different time points. The time points included here correspond to days ranging from 186th to the 400th day of the study:

```
In[181]:= sampleToDays= <|"7"->"186","8"->"255","9"->"289",
         "10"->"290","11"->"292","12"->"294","13"->"297",
         "14"->"301","15"->"307","16"->"311","17"->"322",
         "18"->"329","19"->"369","20"->"380","21"->"400"|>;
```

On day 289 the subject of the study had a Respiratory Syncytial Virus (RSV) infection. Additionally, after day 301, the subject displayed high glucose levels and was eventually diagnosed with type 2 diabetes (T2D). The data was mapped and analyzed and in used in simplified examples (additional data will be included in MathIOmica as they become available - see https://mathiomica.org).

Based on MathIOmica's documentation the raw data files available are listed in Fig. 3.38. Additionally, analyzed files are available, as listed in Fig. 3.39.

3.8 Hypothesis Testing

In many situations we want to test whether or not variables are similar, or if they are related in some way. We talk of the null hypothesis (H_0) and an alternative hypothesis (H_A). In the usual context you may want to compare a quantitative measure in a population, a random variable (e.g. height, or gene expression, levels of a metabolite) between different subsets of the population. For example, μ_0 can be the base healthy population, and perhaps a group associated with illness will have μ_1 as the mean measurement. Then the null hypothesis is that $\mu_0 = \mu_1$ and the alternative hypothesis that $\mu_0 \neq \mu_1$. Only one of the hypotheses can be true.

Often, $\mu_0 = 0$, or the data is adjusted that way so we are comparing against the possibility of matching this population baseline, or alternatively deviating from it. In hypothesis testing the object is whether or not to reject the null hypothesis.

In real testing we are likely to make errors. The two types of errors we can make in testing a null hypothesis are:

1. Type I errors, where we rejected the null hypothesis but we should not have (with probability α), and often referred to as false positive errors.
2. Type II errors: where we did not reject the null hypothesis, but should have (with probability β), and often referred to as false negative errors.

When carrying out a test to check the null hypothesis, β gives us a measure of how much error we make in rejecting the null hypothesis, and so $1 - \beta$ gives us a measure of the probability that the test correctly rejects the null hypothesis, called the power. Both the typical α selection for significance, $\alpha < 0.05$, and $\beta < 0.2$ (corresponding to power 0.8.) are traditional and arbitrary.

We usually use some statistic to decide whether or not to reject the null hypothesis. The statistic has its own null distribution and is a random variable. The probability of observing extreme values of the test statistic is what we call a p-value and it is an observed value given that the null hypothesis is true. A small p value means that there is a small probability to find this statistic value, and corresponds to extremes in observation. The significance level of the statistic is typically chosen as 0.05 corresponding to α. In the Wolfram Language many hypothesis tests are incorporated, that allow us to carry out traditional tests such as **TTest**, **ZTest**, **MannWhitneyTest**. Additionally, there are many functions that allow us to run immediately multiple tests at once and generate reports.

3.8.1 Location Tests

Location tests allow us to test the means or medians of a set of data, μ_1, as the location parameter, against a fixed value μ_0 (often set to zero, or obtained from another data set). The null hypothesis is that $\mu_1 = \mu_0$. Many location test statistics exist, and typically they return a probability (p-value) telling us how unlikely it is that the null hypothesis is true. We first look at a few examples, before applying the general **LocationTest** function:

3.8.1.1 Z Test

The Z-test location test assumes that the data is normally distributed and has a known variance. The statistic is defined as

$$Z = \frac{\tilde{X} - \mu_0}{\frac{\sigma}{\sqrt{n}}}, \qquad (3.14)$$

where μ_0 is the null distribution mean. Let us draw some data from a normal distribution with mean 1, variance 1:

In[182]:= **SeedRandom**[123];
randomNormalSet=
RandomVariate[**NormalDistribution**[1,1],100];

In[184]:= **ZTest**[randomNormalSet]
Out[184]= 2.3277×10^{-18}

In[185]:= ?**ZTest**
ZTest[*data*] tests whether the mean of the *data* is zero.
ZTest[{data$_1$, data$_2$}] tests whether the means of *data$_1$* and *data$_2$* are equal.
ZTest[*dspec*, σ^2] tests for zero or equal means assuming a population variance σ^2.
ZTest[dspec, σ^2, μ_0] tests the mean against μ_0.
ZTest[dspec, σ^2, μ_0, property] returns the value of "*property*". >>

The null hypothesis that this is from a normal distribution of mean 0 is rejected. We can see what the difference is if we have a zero mean set:

In[186]:= **SeedRandom**[123];
randomNormalSet0=
RandomVariate[**NormalDistribution**[0,1],100];
ZTest[randomNormalSet0]
Out[188]= 0.17115

In this case the value is greater than 0.05 so we do not reject the null hypothesis.

We can use **ZTest** to test whether the means of two data sets are equal. We can use as an example the expression of myosin between the AML and ALL populations as discussed above.

In[189]:= myosinExample=**Query**[**All**, "M31211_s_at"]@golubAssociation
Out[189]= <|AML→ { −0.929698, −1.21439,
−1.36559, −1.03054, −1.29915, −0.401122,
−0.263245, −1.27241, −0.651176, −0.323125,
−1.43473, −0.203518, −1.04336, −1.22702,
−1.28071, 0.260207, −1.05801, −0.720331,
−0.809905, −0.588208, −0.513632,
−0.247296, −1.42036, −1.48163, −1.3111},
ALL→ {0.21673, −0.0512012, 0.0512196,
0.0658298, 0.355842, −0.0764156, −0.298932,

$$\begin{aligned}&-0.46567, 0.490372, 0.651373, 0.715212,\\&0.148149, 0.354363, -0.466193, 0.576175,\\&0.417461, -0.407586, 0.388138, 0.266817,\\&0.634913, 0.707898, -0.775871, 0.0765331,\\&0.108658, 0.224704, 0.466898, -0.87136,\\&0.193097, -0.37865, -0.0896119, 0.284866,\\&0.51995, -0.204948, 0.560927, -0.0758743,\\&0.726773, 0.337038, 0.374954, 0.146782,\\&-0.474574, 0.170777, 0.136991, 0.107438,\\&-0.55069, 0.279121, -0.665758, 0.545882\}|>\end{aligned}$$

In[190]:= **ZTest**[Values@myosinExample]
Out[190]= 3.71152×10^{-18}

We see that the expression difference is significant. We can compare the means between the sets:

In[191]:= **Query**[**All**, **Mean**]@myosinExample
Out[191]= <|AML→ −0.873201, ALL→ 0.115926|>

We see that for myosin ALL has higher expression.

3.8.1.2 TTest

The Z-Test assumed that σ is known as a parameter. A t-test does not assume this and uses the sample variance s. So we have a statistic T:

$$T = \frac{\tilde{X} - \mu_0}{\frac{s}{\sqrt{n}}}, \qquad (3.15)$$

which also obeys a Student t-distribution. We can again apply the test to our myosin gene expression example:

In[192]:= **TTest**[Values@myosinExample]
Out[192]= 1.82388×10^{-13}

3.8.1.3 Mann-Whitney Non-parametric Test

Both the Z-test and T-test are examples of parametric testing. This parametric testing involves assumptions of the underlying distributions that the measurements came from, and very often assumes a normal distribution. Some parameter is involved in terms of the underlying distribution, that is used as a statistic for testing for significance. In non-parametric testing no assumption is made about the distribution of the data being tested. The data can often be ranked, and the rank is then used to formulate a testing statistic. The probability then is estimated whether or not values will have higher or lower ranks if compared across groups.

3.8 Hypothesis Testing

An example of non parametric testing is a Mann-Whitney U-Test for univariate data. It is based on ranking two sets of data, of length n_1 and n_2 respectively, after joining them. The test statistic is $U=Min[U_1,U_2]$ where [24]:
$U_1= n_1\ n_2+n_1(n_1+1)/2- R_1$,
$U_2= n_1\ n_2+n_2(n_2+1)/2- R_2$.

R_1 and R_2 correspond to sum of ranks from sample 1 and 2 respectively. In the Wolfram Language we can carry this out:

n_1 and n_2 are 25 and 47 respectively:

In[193]:= Query[All , Length]@myosinExample
Out[193]= <|AML→25, ALL→47|>

Ordering gives us the position of each set when we order them together:

In[194]:= ?Ordering
   Ordering[*list*] gives the positions in *list* at which each successive element of Sort[*list*] appears.
   Ordering[*list*,*n*] gives the positions in *list* at which the first n elements of Sort[*list*] appear.
   Ordering[*list*,−*n*] gives the positions of the last *n* elements of Sort[*list*].
   Ordering[*list*,*n*,*p*] uses Sort[*list*,*p*]. >>

In[195]:= order1=
   Ordering[Join[myosinExample[[1]],myosinExample[[2]]]]
Out[195]= {24,11,23,3,25,5,15,8,14,2,17,13,4,
   1,52,19,47,18,71,9,20,69,21,65,39,
   33,42,6,54,10,32,7,22,58,12,55,31,
   60,27,28,29,48,68,49,67,64,37,66,53,
   26,50,16,44,70,56,62,38,30,63,43,41,
   51,34,57,72,59,40,45,35,46,36,61}

If we order again we get the ranks in their original positions by locating the positions (i.e. 1 is at 14th position(i.e. rank), 2 is at position (i.e. rank) 10, 3 is at position 4, etc.):

In[196]:= order2=
   Ordering[
   Ordering[Join[myosinExample[[1]],
   myosinExample[[2]]]]]
Out[196]= {14,10,4,13,6,28,32,8,20,30,2,35,12,
   9,7,52,11,18,16,21,23,33,3,1,5,50,
   39,40,41,58,37,31,26,63,69,71,47,57,
   25,67,61,27,60,53,68,70,17,42,44,51,
   62,15,49,29,36,55,64,34,66,38,72,
   56,59,46,24,48,45,43,22,54,19,65}

We can now calculate R1 for the AML samples, the first 25, and R2 for the remaining ALL samples. We also calculate U1 and U2 to find the smallest value:

In[197]:= r1=Plus@@order2[[1;;25]];
   r2=Plus@@order2[[26;;]];
   U1=25*47+25*26/2−r1
   U2=25*47+47*48/2 −r2

Out[199]= 1087

Out[200]= 88

So the smallest value is U = 88. A faster way to do it is:

In[201]:= **MannWhitneyTest**[Values@myosinExample,
   Automatic,"TestStatistic"]
Out[201]= 88.

3.8.2 A High-Level Approach

The function **LocationTest** can perform many location tests on data, returning a p value.

In[202]:= **LocationTest**[Values@myosinExample, **Automatic**,
   {"TestDataTable", **All**}]

Out[202]=

	Statistic	P-Value
Mann-Whitney	88.	3.58778×10^{-9}
T	-9.08801	1.82388×10^{-13}
Z	-8.6873	3.71152×10^{-18}

The argument **Automatic** assumes that $\mu_1 - \mu_2 = 0$. The tests reported are what are deemed to be applicable by the software.

We can also get a Test Conclusion, which we can see is based on an α of 0.05.

In[203]:= **LocationTest**[Values@myosinExample, **Automatic**,
   "TestConclusion"]
Out[203]= The null hypothesis that
   the mean difference is 0 is rejected at the
   5 percent level based on the T test.

3.9 Additional Statistics

3.9.1 Expectations and Correlations

For a given distribution we can calculate the expectation of any variable. Essentially that is given by the probability times the variable of interest:

$$E[X] = \sum_i \text{pmf}[x_i] \times x_i, \qquad (3.16)$$

and in the continuous variable case,

$$E[X] = \int \text{PDF}[x] x \, dx. \qquad (3.17)$$

Fig. 3.40 Special characters for **Distributed** and **Conditioned** can be entered through multiple commands

The ≈ symbol stands for the **Distributed** function. Based on the Wolfram documentation: $x \approx dist$ can be entered as:

- x esc dist esc *dist*
- x\[Distributed] *dist*
- Distributed[x, *dist*].

Similarly, the **Conditioned** expression, *expr* ⫫ *cond*, can be entered as:

- *expr* esc cond esc *cond*
- *expr*\[Conditioned] *cond*.
- Conditioned[*expr,cond*].

For a function f(i) for some list of variable i we have:

$$E[f[X]] = \sum_i \text{pmf}[x_i] \times f[x_i], \tag{3.18}$$

and again in the case of a continuous variable we have:

$$E[X] = \int \text{PDF}[x] f[x] dx. \tag{3.19}$$

In the Wolfram Language we use the **Expectation** function to find the expectation of expression *expr*: **Expectation** [*expr*, **Distributed** [*x,dist*]], where x is distributed as *dist*, which is declared through the **Distributed** function, Fig. 3.40.

For example, the expectation of x for a normal distribution with parameters μ and σ would be:

In[205]:= **Expectation**[x, **Distributed**[x, **NormalDistribution**[μ,σ]]]
Out[205]= μ

We can also write down the covariance between two random variables:

$$\text{Covariance}[X,Y] = E[(X - \mu_X)(Y - \mu_Y)], \tag{3.20}$$

and the correlation,

$$\text{Correlation}[X,Y] = \text{Covariance}[X,Y]/\sigma_x \sigma_y \tag{3.21}$$

In[206]:= ?**Covariance**
           Covariance[v_1,v_2] gives the covariance between the
                   vectors v_1 and v_2.
           Covariance[m] gives the covariance matrix for the
                   matrix m.

Covariance[m_1, m_2] gives the covariance matrix
for the matrices m_1 and m_2.
Covariance[$dist$] gives the covariance matrix for the
multivariate symbolic distribution $dist$.
Covariance[$dist,i,j$] gives the $(i,j)^{th}$ covariance for the
multivariate symbolic distribution $dist$. »

In[207]:= ?**Correlation**
Correlation[v_1,v_2] gives the correlation between the
vectors v_1 and v_2.
Correlation[m] gives the correlation matrix for the
matrix m.
Correlation[m_1,m_2] gives the correlation matrix for the
matrices m_1 and m_2.
Correlation[$dist$] gives the correlation matrix for the
multivariate symbolic distribution $dist$.
Correlation[$dist,i,j$] gives the $(i,j)^{th}$ for the
multivariate symbolic distribution $dist$. »

We can create correlations plots between samples. For example we can look at correlations in the Golub dataset:

In[208]:= dataGolubFull=**Values** @
Merge[**Values**[golubAssociation] , **Flatten**];

Code notes: We first extract the values of the golubAssociation for AML and ALL using **Values**[golubAssociation]. The resulting two associations of genes to expressions for samples are merged with a **Flatten** function that will put together matching keys, i.e. gene accessions and create one list of all samples. Finally the **Values** externally retrieves a matrix we can use. We can visualize the correlations as an **ArrayPlot**, Fig. 3.41.

As an aside, we can see if any of the samples cluster together using MathIOmica's **MatrixClusters** to generate clustering based on the correlations:

In[210]:= <<**MathIOmica**'

In[211]:= clustersGolubCorrelation=
MatrixClusters[**Correlation**[dataGolubFull],
DistanceFunction->CorrelationDistance];

The result is from one set of data, and can be passed to the **MatrixDendrogramHeatmap** function that will generate a visualization using hierarchical clustering, Fig. 3.42.

We see that there is a clear division in two groups in Fig. 3.42. We can extract the information from the clusters results:

In[213]:= Query[All , "GroupAssociationsRows"]
@clustersGolubCorrelation
Out[213]= <|1→
<|H:G1→ { 1,9,21,3,8,4,11,13,14,15,6,10,
18,20,2,24,5,12,23,25,16,19,17},
H:G2→ { 35,34,36,31,48,28,39,42,70,
72,43,71,26,53,58,52,33,47,29,
32,54,62,63,61,64,68,57,38,45,

3.9 Additional Statistics

```
In[209]:= ArrayPlot[Correlation[dataGolubFull],
    Mesh → True,
    PlotTheme → "Scientific", PlotLabel →
    "Array Plot of Correlations Between Samples
    from AML/ALL Study"]
```

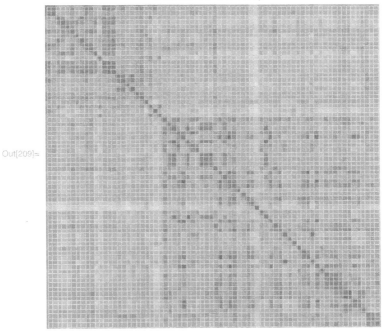

Fig. 3.41 Out[209]= ArrayPlot of correlations form AML/ALL study

$$67,59,30,40,49,56,41,51,44,66,$$
$$65,55,60,50,37,7,22,69,27,46\}|>|>$$

As we can see in the first group the samples are all from AML (the first 25). Only one non-AML sample is found in the clustering. We will present other methods for classification in the Machine Learning Chapter. For any pair of samples we can actually construct scatter plots and regression fits. Let us consider the first two samples.

We can carry out a linear regression using **LinearModelFit** :

In[214]:= ?**LinearModelFit**
  LinearModelFit[$\{y_1, y_2, \ldots\}, \{f_1, f_2, \ldots\}, x$] constructs a linear model of the form $\beta_0 + \beta_1 f_1 + \beta_2 f_2 + \ldots$ that fits the y_i for successive x values $1, 2, \ldots$.

In[212]:= `MatrixDendrogramHeatmap[clustersGolubCorrelation[[1]], ScaleShift → 1]`

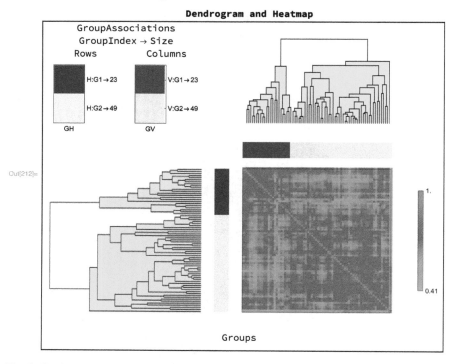

Fig. 3.42 Out[212]= MathIOmica MatrixDendrogramHeatmap of correlations in AML/ALL dataset

LinearModelFit[
 $\{\{x_{11}, x_{12}, \ldots, y_1\}, \{x_{21}, x_{22}, \ldots, y_2\}, \ldots\}, \{f_1, f_2, \ldots\}, \{x_1, x_2, \ldots\}\}$]
  constructs a linear model of the form
  $\beta_0 + \beta_1 f_1 + \beta_2 f_2 + \ldots$ where the f_i depend on the variables x_k.
LinearModelFit[m, v] constructs a linear model from the design matrix m and response vector v. »

In[215]:= modelGolub12=
    LinearModelFit[dataGolubFull[[All,1;;2]], x,x];

The model has many properties:

In[216]:= modelGolub12["Properties"]//**Short**
Out[216]//**Short**= {AdjustedRSquared,AIC,<<61>>, VarianceInflationFactors}

3.9 Additional Statistics 125

```
In[219]:= Show[ListPlot[dataGolubFull[[All, 1 ;; 2]],
    PlotTheme → "Scientific"],
  Plot[modelGolub12["BestFit"], {x, -10, 10},
    PlotLegends → Grid[{{"Best Fit:",
        Chop@modelGolub12["BestFit"]},
      {"R²:", ToString[ modelGolub12["RSquared"]]}}]],
  PlotLabel →
    "Linear Fit for 2 AML Datasets from Golub Study"]
```

Out[219]=

Best Fit: 0.831672 x
R^2: 0.691679

Fig. 3.43 Out[219]= Linear Fit for 2 AML Datasets from Golub Study. *Code notes*: We used **Show** above to show together the scatter plot and the model fit. For the Model we evaluate the line and provide legends that give information on the BestFit, and the RSquared statistic. For the "BestFit" formula we used **Chop** to remove the intercept value that is approaching zero

For our purposes we need the "BestFit" formula and the "RSquared" value:

In[217]:= modelGolub12["BestFit"]
Out[217]= $1.11811 \times 10^{-15} + 0.831672$ x

In[218]:= **ToString**[modelGolub12@"RSquared"]
Out[218]= 0.691679

Now we can put all the information together and create a linear model fit with a fit line and information regarding the model, shown in Fig. 3.43.

3.9.2 Moments of a Distribution

Expectations of the powers of a random variable are called moments of the distribution, and are defined as [see for example Weisstein, Eric W. "Moment." From MathWorld–A Wolfram Web Resource. http://mathworld.wolfram.com/Moment.html]:

$$\mu'_n = \mathrm{E}\left[X^n\right] \tag{3.22}$$

If we take the moment about the mean, then it is called a central moment:

$$\mu_n = \mathrm{E}\left[(X - \mu)^n\right] . \tag{3.23}$$

The moments give us different information: The first moment is the usual mean, the second moment is equal to the variance μ_2.

The third moment related the skewness(γ_1), normalized by $\mu_2^{3/2}$, which tells us how much a distribution is skewed from center:

$$\gamma_1 = \frac{\mu_3}{\mu_2^{3/2}} . \tag{3.24}$$

The fourth moment is related to the kurtosis (β_2), which is the fourth moment normalized by the variance.

$$\beta_2 = \frac{\mu_4}{\mu_2^2} . \tag{3.25}$$

The Wolfram language has functions to evaluate any **CentralMoment** of a distribution or data, as well as the **Skewness** and **Kurtosis** for any distribution or set of data:

In[220]:= data=N[**RandomVariate**[**BinomialDistribution**[100,0.1],50]]
Out[220]= {13.,9.,5.,8.,10.,12.,10.,2.,10.,6.,
12.,10.,10.,11.,9.,15.,8.,13.,5.,12.,
11.,8.,12.,6.,15.,9.,6.,7.,11.,10.,
9.,13.,9.,10.,14.,7.,10.,6.,9.,7.,
12.,15.,10.,7.,11.,8.,10.,11.,11.,10.}

In[221]:= **CentralMoment**[data,2]
Out[221]= 7.4976

In[222]:= **CentralMoment**[data,4]
Out[222]= 173.809

In[223]:= **Skewness**[data]
Out[223]= −0.230375

In[224]:= **Kurtosis**[data]
Out[224]= 3.09192

We can check the **Kurtosis** value by using the **CentralMoment** functions directly:

In[225]:= **CentralMoment**[data,4]/**CentralMoment**[data,2]^2
Out[225]= 3.09192

We can get the central moments of any distribution, for example the first 6 central moments of a normal distribution:

In[226]:= Table[**CentralMoment**[**NormalDistribution**[μ,σ],i],{i,6}]
Out[226]= {0,σ^2,0,3σ^4,0,15 σ^6}

3.9 Additional Statistics

```
In[229]:= TableForm[({ToString[#], Skewness[#], Kurtosis[#]} & /@ {NormalDistribution[μ, σ],
    BinomialDistribution[n, p], ExponentialDistribution[λ],
    ChiSquareDistribution[ν], StudentTDistribution[ν]}),
    TableHeadings → {None, {"Distribution", "Skewness", "Kurtosis"}}]
```

Distribution	Skewness	Kurtosis
NormalDistribution[μ, σ]	0	3
BinomialDistribution[n, p]	$\frac{1-2p}{\sqrt{n(1-p)p}}$	$3 + \frac{1-6(1-p)p}{n(1-p)p}$
ExponentialDistribution[λ]	2	9
ChiSquareDistribution[ν]	$2\sqrt{2}\sqrt{\frac{1}{\nu}}$	$3 + \frac{12}{\nu}$
StudentTDistribution[ν]	$\begin{bmatrix} 0 & \nu > 3 \\ \text{Indeterminate} & \text{True} \end{bmatrix}$	$\begin{bmatrix} 3 + \frac{6}{-4+\nu} & \nu > 4 \\ \text{Indeterminate} & \text{True} \end{bmatrix}$

Fig. 3.44 Out[229]= **TableForm** of **Skewness** and **Kurtosis** for various distributions

Or equivalently for a Poisson distribution:

In[227]:= **Table[CentralMoment[PoissonDistribution[**μ**], i], {i, 6}]**
Out[227]= $\{0, \mu, \mu, 3\mu^2 + \mu, 10\mu^2 + \mu, 15\mu^3 + 25\mu^2 + \mu\}$

We can also use the **Kurtosis** or **Skewness** directly.

In[228]:= **Kurtosis[NormalDistribution[**μ, σ**]]**
Out[228]= 3

Let us get the information for a few distributions as a rule list in a table format, Fig. 3.44.

3.9.3 Distribution Fit

Often with real data we are not sure that we have the right distribution. We can formulate a null hypothesis H₀ that our data is from a certain distribution and test it using the DistibutionFitTest . In this context a small p value means we reject the null hypothesis that the data came from this distribution. Let us create some data from an **ExponentialDistribution** and a **NormalDistribution**, and check against a **NormalDistribution**, **BinomialDistribution** and **ExponentialDistribution** . We use **RandomVariate** to create 2 sets, and visualize them in Fig. 3.45.

In[230]:= **SeedRandom[123]**
  exponentialSampleData=
    RandomVariate[ExponentialDistribution[1], 1000];
  normalSampleData=
    RandomVariate[NormalDistribution[2,3], 1000];

For example testing the exponential sample against a normal distribution we reject the null hypothesis.

In[234]:= **DistributionFitTest[exponentialSampleData,
    NormalDistribution[2,3]]**

In[233]:= `Histogram[#] & /@ {exponentialSampleData, normalSampleData}`

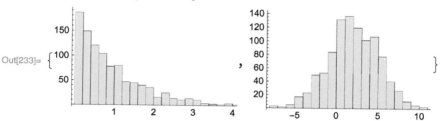

Fig. 3.45 Out[233]= Histogram of Random Variates from Exponential and Normal Distributions

Out[234]= 1.33227×10^{-15}

If we check against an **ExponentialDistribution** we get:

In[235]:= `DistributionFitTest[exponentialSampleData, ExponentialDistribution[1]]`
Out[235]= 0.489776

We can create a HypothesisTestData Object, which is the result of the **DistributionFitTest** can be created from the result to retrieve various properties from the testing:

In[236]:= `hFitExponentialSampleTests = DistributionFitTest[exponentialSampleData, NormalDistribution[2,3], "HypothesisTestData"];`

The object has many properties:

In[237]:= `hFitExponentialSampleTests["Properties"]`
Out[237]= {AllTests, AndersonDarling, AutomaticTest, BaringhausHenze, CramerVonMises, DegreesOfFreedom, DistanceToBoundary, FittedDistribution, FittedDistributionParameters, HypothesisTestData, JarqueBeraALM, KolmogorovSmirnov, Kuiper, MardiaCombined, MardiaKurtosis, MardiaSkewness, PearsonChiSquare, Properties, PValue, PValueTable, ShapiroWilk, ShortTestConclusion, SzekelyEnergy, TestConclusion, TestData, TestDataTable, TestEntries, TestStatistic, TestStatisticTable, WatsonUSquare}

We can obtain the statistic from all different tests:

In[238]:= `hFitExponentialSampleTests["TestDataTable", All]`

3.9 Additional Statistics   129

Out[238]=
	Statistic	P-Value
Anderson-Darling	273.125	0.
Baringhaus-Henze	327.828	0.
Cramér-von Mises	57.0741	1.33227×10^{-15}
Jarque-Bera ALM	1089.88	0.
Kolmogorov-Smirnov	0.38977	3.26584×10^{-133}
Kuiper	0.641303	$3.089077791341700 \times 10^{-359}$
Mardia Combined	1089.88	0.
Mardia Kurtosis	24.6349	5.33911×10^{-134}
Mardia Skewness	469.076	5.09234×10^{-104}
Pearson χ^2	3314.24	$3.030676237868480 \times 10^{-685}$
Shapiro-Wilk	0.847096	5.56647×10^{-30}
Watson U^2	38.8369	0.

We can also check the normal data for example against the same distribution:

In[239]:= hFitNormalSampleTests=
           DistributionFitTest[normalSampleData,
           NormalDistribution[2,3],"HypothesisTestData"];

In[240]:= hFitNormalSampleTests["TestDataTable",**All**]

Out[240]=
	Statistic	P-Value
Anderson-Darling	0.564789	0.681835
Baringhaus-Henze	1.00766	0.204813
Cramér-von Mises	0.0987282	0.591145
Jarque-Bera ALM	2.14485	0.332788
Kolmogorov-Smirnov	0.0224672	0.685025
Kuiper	0.033845	0.667026
Mardia Combined	2.14485	0.332788
Mardia Kurtosis	-0.497718	0.618683
Mardia Skewness	1.91942	0.165921
Pearson χ^2	27.904	0.626117
Shapiro-Wilk	0.998117	0.335309
Watson U^2	0.086213	0.362369

And we can get a test conclusion for each:

In[241]:= hFitExponentialSampleTests["TestConclusion"]
Out[241]= The null hypothesis that
             the data is distributed according to the
                NormalDistribution[2,3]
             is rejected at the 5 percent level
             based on the Cramér—von Mises test.

In[242]:= hFitNormalSampleTests["TestConclusion"]
Out[242]= The null hypothesis that
             the data is distributed according to the
                NormalDistribution[2,3]
             is not rejected at the 5 percent level
             based on the Cramér—von Mises test.

We can also go back and check the golubData set. Let us extract all the values in the first sample. We get first all the first **Association** (AML) and from All genes the first sample and then we flatten the list:

In[244]:= `Histogram[golubFirstSet,`
`  PlotLabel → "Gene Expression Distribution`
`  Golub et al. AML Example",`
`  FrameLabel → {"Normalized Expression", "Count"}]`

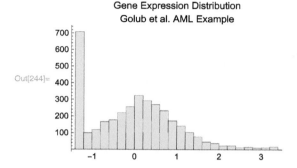

Fig. 3.46 Out[244]= Histogram of normalized gene expression in Golub et al. example

In[243]:= golubFirstSet=
 (**Query**["AML" , **All**/∗**Values** ,1]@
 golubAssociation);

We can plot the data, in Fig. 3.46. You will notice a bimodal distribution, because of how the data was processed.

We could think these data fit a normal distribution (for the most part, and this is widely assumed in expression papers). Let us see what a fit test would actually give:

In[245]:= hFitTestsGolubFirst=**DistributionFitTest**[golubFirstSet,
 NormalDistribution[], "HypothesisTestData"];

In[246]:= hFitTestsGolubFirst["TestDataTable",**All**]

Out[246]=

	Statistic	P-Value
Anderson-Darling	40.4247	0.
Baringhaus-Henze	68.2693	0.
Cramér-von Mises	3.83399	1.0965×10^{-9}
Jarque-Bera ALM	154.672	0.
Mardia Combined	154.672	0.
Mardia Kurtosis	0.395616	0.692389
Mardia Skewness	154.238	2.05421×10^{-35}
Pearson χ^2	6558.29	$3.903781409875469 \times 10^{-1362}$
Shapiro-Wilk	0.944707	1.20641×10^{-34}

In[247]:= hFitTestsGolubFirst["TestConclusion"]
Out[247]= The null hypothesis that the data is distributed according to the **NormalDistribution**[0,1] is rejected at the 5 percent level based on the Cramér–von Mises test.

References

1. Abell, M.L., Braselton, J.P., Rafter, J.A.: Statistics with Mathematica, vol. 1. Academic Press, New York (1999)
2. Ashburner, M., Ball, C.A., Blake, J.A., Botstein, D., Butler, H., Cherry, J.M., Davis, A.P., Dolinski, K., Dwight, S.S., Eppig, J.T., Harris, M.A., Hill, D.P., Issel-Tarver, L., Kasarskis, A., Lewis, S., Matese, J.C., Richardson, J.E., Ringwald, M., Rubin, G.M., Sherlock, G.: Gene ontology: tool for the unification of biology. The gene ontology consortium. Nat. Genet. **25**(1), 25–9 (2000)
3. Baglivo, J.A.: Mathematica Laboratories for Mathematical Statistics: Emphasizing Simulation and Computer Intensive Methods. SIAM, Philadelphia (2005)
4. Barrett, T., Wilhite, S.E., Ledoux, P., Evangelista, C., Kim, I.F., Tomashevsky, M., Marshall, K.A., Phillippy, K.H., Sherman, P.M., Holko, M., Yefanov, A., Lee, H., Zhang, N., Robertson, C.L., Serova, N., Davis, S., Soboleva, A.: NCBI GEO: archive for functional genomics data sets–update. Nucleic Acids Res. **41**(D1), D991–D995 (2013)
5. Benjamini, Y., Hochberg, Y.: Controlling the false discovery rate: a practical and powerful approach to multiple testing. J. R. Stat. Soc. Ser. B (Methodol.) **57**, 289–300 (1995)
6. Chen*, R., Mias*, G.I., Li-Pook-Than*, J., Jiang*, L., Lam, H.Y., Chen, R., Miriami, E., Karczewski, K.J., Hariharan, M., Dewey, F.E., Cheng, Y., Clark, M.J., Im, H., Habegger, L., Balasubramanian, S., O'Huallachain, M., Dudley, J.T., Hillenmeyer, S., Haraksingh, R., Sharon, D., Euskirchen, G., Lacroute, P., Bettinger, K., Boyle, A.P., Kasowski, M., Grubert, F., Seki, S., Garcia, M., Whirl-Carrillo, M., Gallardo, M., Blasco, M.A., Greenberg, P.L., Snyder, P., Klein, T.E., Altman, R.B., Butte, A.J., Ashley, E.A., Gerstein, M., Nadeau, K.C., Tang, H., Snyder, M.: Personal omics profiling reveals dynamic molecular and medical phenotypes. Cell **148**(6), 1293–307 (2012)
7. Dekking, F.M.: A Modern Introduction to Probability and Statistics: Understanding Why and How. Springer Science & Business Media (2005)
8. Denker, M., Woyczynski, W.: Introductory Statistics and Random Phenomena: Uncertainty, Complexity and Chaotic Behavior in Engineering and Science. Springer Science & Business Media (2012)
9. Edgar, R., Domrachev, M., Lash, A.E.: Gene expression omnibus: NCBI gene expression and hybridization array data repository. Nucleic Acids Res. **30**(1), 207–210 (2002)
10. Feller, W.: An Introduction to Probability Theory and Its Applications: Volume I, vol. 3. Wiley, New York (1968)
11. Feller, W.: Introduction to the Theory of Probability and Its Applications, vol. 2, 2nd edn. New York, Wiley (1971)
12. Golub, T.R., Slonim, D.K., Tamayo, P., Huard, C., Gaasenbeek, M., Mesirov, J.P., Coller, H., Loh, M.L., Downing, J.R., Caligiuri, M.A., Bloomfield, C.D., Lander, E.S.: Molecular classification of cancer: class discovery and class prediction by gene expression monitoring. Science **286**(5439), 531–537 (1999)
13. Heyries, K.A., Tropini, C., VanInsberghe, M., Doolin, C., Petriv, I., Singhal, A., Leung, K., Hughesman, C.B., Hansen, C.L.: Megapixel digital PCR. Nat. Methods **8**(8), 649–651 (2011)
14. Karolchik, D., Hinrichs, A.S., Kent, W.J.: The UCSC genome browser. Current Protocols in Bioinformatics Chapter 1, Unit1 4 (2012)
15. Kolesnikov, N., Hastings, E., Keays, M., Melnichuk, O., Tang, Y.A., Williams, E., Dylag, M., Kurbatova, N., Brandizi, M., Burdett, T., Megy, K., Pilicheva, E., Rustici, G., Tikhonov, A., Parkinson, H., Petryszak, R., Sarkans, U., Brazma, A.: Arrayexpress update–simplifying data submissions. Nucleic Acids Res. **43**(Database issue), D1113—D1116 (2015)
16. Kuhn, R.M., Haussler, D., Kent, W.J.: The UCSC genome browser and associated tools. Brief Bioinform **14**(2), 144–61 (2013)
17. Lawler, G.F., Coyle, L.N.: Lectures on Contemporary Probability, vol. 2. American Mathematical Society (1999)

18. Marcobal, A., Yusufaly, T., Higginbottom, S., Snyder, M., Sonnenburg, J.L., Mias, G.I.: Metabolome progression during early gut microbial colonization of gnotobiotic mice. Sci. Rep. **5**, 11,589 (2015)
19. Mias, G., Snyder, M.: Personal genomes, quantitative dynamic omics and personalized medicine. Quant. Biol. **1**(1), 71–90 (2013)
20. Mias, G.I., Snyder, M.: Multimodal dynamic profiling of healthy and diseased states for future personalized health care. Clin. Pharmacol. Ther. **93**(1), 29–32 (2013)
21. Mias, G.I., Yusufaly, T., Roushangar, R., Brooks, L.R., Singh, V.V., Christou, C.: MathIOmica: An integrative platform for dynamic omics. Sci. Rep. **6**, 37,237 (2016)
22. Pavlidis, P.: Using ANOVA for gene selection from microarray studies of the nervous system. Methods **31**(4), 282–289 (2003) (Candidate Genes from DNA Array Screens: application to neuroscience)
23. Pavlidis, P., Noble, W.S.: Matrix2png: a utility for visualizing matrix data. Bioinformatics **19**(2), 295–296 (2003)
24. Rees, D.: Essential Statistics. Springer, Berlin (2013)
25. Sandberg, R., Yasuda, R., Pankratz, D.G., Carter, T.A., Del Rio, J.A., Wodicka, L., Mayford, M., Lockhart, D.J., Barlow, C.: Regional and strain-specific gene expression mapping in the adult mouse brain. Proc. Natl. Acad. Sci. **97**(20), 11038–11043 (2000)
26. Vogelstein, B., Kinzler, K.W.: Digital PCR. Proc. Natl. Acad. Sci. **96**(16), 9236–9241 (1999)
27. Wasserman, L.: All of statistics: a concise course in statistical inference. Springer Science & Business Media (2013)
28. Wolfram Alpha LLC: Wolfram|Alpha (2017). Accessed November 2017
29. Wolfram Research, Inc.: Mathematica, Version 11.2. Champaign, IL, 2017
30. Yamagishi, M.E.B., Shimabukuro, A.I.: Nucleotide frequencies in human genome and Fibonacci numbers. Bull. Math. Biol. **70**(3), 643–653 (2008)

Chapter 4
Databases: E-Utilities and UCSC Genome Browser

4.1 Connecting to Databases

There are many different types of databases available online that serve the scientific community. These provide large set of information that has been compiled and organized for uniformity and ease of use.

Typically, access to a database is done through an API (Application Programming interface Interface) that can be used to communicate with online resources. A lot of modern web services rely on Representational State Transfer (REST) interfaces, that can return data through a use of a Uniform Resource Identifier (URI), such as a website service URL (Uniform Resource Locator) to request and retrieve information. The information is typically returned in some standard format, such as XML (Extensible Markup Language), HTML (Hypertext Markup Language), or JSON (JavaScript Object Notation). The JSON format has gained a lot of popularity for its simplicity of structure, low footprint in terms of structure and size and flexibility of information encoding. Other possibilities are to use well structured traditional databases based on the Structured Query Language (SQL, e.g. MySQL).

The Wolfram Language has many functions that allow interaction and data retrieval from both local and external databases. In this chapter we will look at some examples, particularly in detail how to access information from the National Center for Biotechnology Information (NCBI) databases [5–7] and the University of California at Santa Cruz Genome Browser databases [3, 4].

Electronic supplementary material The online version of this chapter (https://doi.org/10.1007/978-3-319-72377-8_4) contains supplementary material, which is available to authorized users.

4.2 NCBI Entrez Programming Utilities

The National Center for Biotechnology Information (NCBI) offers a set of programs that allow an interface for query and data retrieval from its database system. These are called the Entrez Programming Utilities (E-utilities). They can be accessed through a URL interface, following a certain syntax and can be a very powerful tool for processing bioinformatics data and information. We will show multiple of examples of how to use the E-utilities through the Wolfram Language programming style. Extensive information is available directly from NCBI [5, 7], and here we go over information on the website in the context of using the Wolfram Language directly as an interface, as well as a guide to the E-utilities.

The base URL for all the E-utilities is:
"https://eutils.ncbi.nlm.nih.gov/entrez/eutils/".

An extension to this URL is the "eutilname.fcgi", where the *eutilname* corresponds to the E-utilities described below. The utilities use unique identifiers (UIDs such as GI numbers, RefSeq accessions or PMIDs from PubMed) to identify information and retrieve records. There are multiple utilities, which we describe in detail how to access and work with using the Wolfram Language. Our goal is to provide a reference that we can use as needed on how to access the information, rather than explain in detail the obtained information.

4.2.1 EInfo

There are multiple NCBI databases available that we can access. These are retrieved by calling the base URL "https://eutils.ncbi.nlm.nih.gov/entrez/eutils/einfo.fcgi" and simply getting the values in JSON format, by setting the parameter "retmode" (retrieval mode) to "json". We can use **URLExecute** to "POST" the request with the parameter "RequestMethod", and specify that the returned result is parsed as "RawJSON", so we can get an association:

```
In[1]:= eutilsDatabases=URLExecute[
  "https://eutils.ncbi.nlm.nih.gov/entrez/eutils/einfo.fcgi",
      {"RequestMethod"->"POST",
       "retmode"->"json"},
      "RawJSON"]
Out[1]= <|header-><|type→einfo,version→0.3|>,
      einforesult→
        <|dblist→{pubmed,protein,nuccore,ipg,
          nucleotide,nucgss,nucest,structure,
          sparcle,genome,annotinfo,assembly,
          bioproject,biosample,blastdbinfo,
          books,cdd,clinvar,clone,gap,gapplus,
          grasp,dbvar,gene,gds,geoprofiles,
          homologene,medgen,mesh,ncbisearch,
          nlmcatalog,omim,orgtrack,pmc,popset,
          probe,proteinclusters,pcassay,
```

4.2 NCBI Entrez Programming Utilities

```
           biosystems, pccompound, pcsubstance,
           pubmedhealth, seqannot, snp, sra, taxonomy,
           biocollections, unigene, gencoll, gtr }|>|>
```

We see that there is a header key and the result:

In[2]:= **Keys** @ eutilsDatabases
Out[2]= {header, einforesult}

The result has a key "dblist" that gives us all the available databases:

In[3]:= databasesEutils=**Query**["einforesult"]@eutilsDatabases
Out[3]= <|dblist→{pubmed, protein, nuccore, ipg,
           nucleotide, nucgss, nucest, structure,
           sparcle, genome, annotinfo, assembly,
           bioproject, biosample, blastdbinfo, books,
           cdd, clinvar, clone, gap, gapplus, grasp,
           dbvar, gene, gds, geoprofiles, homologene,
           medgen, mesh, ncbisearch, nlmcatalog, omim,
           orgtrack, pmc, popset, probe, proteinclusters,
           pcassay, biosystems, pccompound, pcsubstance,
           pubmedhealth, seqannot, snp, sra, taxonomy,
           biocollections, unigene, gencoll, gtr }|>

For any database we can now extract additional information by setting the "db" parameter to the database of interest. For example, let us extract the database information for the snp database:

In[4]:= snpInfo=**URLExecute**[
    "https://eutils.ncbi.nlm.nih.gov/entrez/eutils/einfo.fcgi",
      {"RequestMethod"→"POST",
      "retmode"→"json",
      "db"→ "snp" },
      "RawJSON"];

The returned information is in an association. It has the same outer keys as before, so we look for the keys in the inner association:

In[5]:= **Query**[**All** , **Keys**]@snpInfo
Out[5]=<|header→{type, version}, einforesult→{dbinfo}|>

Now we see that information is embedded further in the association. There is one list with a **Key**. Let us probe further to get its keys:

In[6]:= **Query**["einforesult", 1, **Keys**]@snpInfo
Out[6]= {dbname, menuname, description, dbbuild,
           count, lastupdate, fieldlist, linklist}

Now we can extract more useful information. We can extract one key at a time:

In[7]:= **Query**["einforesult", 1, "dbname"]@snpInfo
Out[7]= snp

Or we can extract multiple keys:

In[8]:= **Query**["einforesult", 1, {"description", "dbbuild", "count"}]@
           snpInfo

136   4 Databases: E-Utilities and UCSC Genome Browser

Out[8]= <|description→Single Nucleotide Polymorphisms,
           dbbuild→Build170908−1425m.1,
           count→1070220676|>

The "fieldlist" information has multiple members, corresponding to a description of fields that can be used in queries. There are 39 members:

In[9]:= **Query**["einforesult",1,"fieldlist",**Length**]@snpInfo
Out[9]= 39

We can extract information for any of these:

In[10]:= **Query**["einforesult",1,"fieldlist",1]@snpInfo
Out[10]= <|name→ALL, fullname→All Fields, description→
           All terms from all searchable fields,
           termcount→1632565546, isdate→N,
           isnumerical→N, singletoken→N,
           hierarchy→N, ishidden→N|>

Or we may want to get all the names, which are useful if we are building searches:

In[11]:= snpFieldList=
           Query["einforesult",1,"fieldlist",All,
           {"name","fullname"}]@snpInfo
Out[11]= { <|name→ALL, fullname→All Fields|>,
           <|name→UID, fullname→UID|>,
           <|name→FILT, fullname→Filter|>,
           <|name→RS, fullname→Reference SNP ID|>,
           <|name→CHR, fullname→Chromosome|>,
           <|name→GENE, fullname→Gene Name|>,
           <|name→HAN, fullname→Submitter Handle|>,
           <|name→ACCN, fullname→Accession|>,
           <|name→LLID, fullname→LocusLink ID|>,
           <|name→ORGN, fullname→Organism|>,
           <|name→FXN, fullname→Function Class|>,
           <|name→GTYP, fullname→Genotype|>,
           <|name→NREF, fullname→non reference assembly|>,
           <|name→HETZ, fullname→Heterozygosity|>,
           <|name→MPWT, fullname→Map Weight|>,
           <|name→VALI, fullname→Validation Status|>,
           <|name→SRAT, fullname→Success Rate|>,
           <|name→CBID, fullname→Create Build ID|>,
           <|name→UBID, fullname→Update Build ID|>,
           <|name→PDAT, fullname→Publication Date|>,
           <|name→MDAT, fullname→Modification Date|>,
           <|name→PCLS, fullname→Population Class|>,
           <|name→MCLS, fullname→Method Class|>,
           <|name→SS, fullname→Submitter SNP ID|>,
           <|name→SID, fullname→Local SNP ID|>,
           <|name→VARI, fullname→Allele|>,
           <|name→SCLS, fullname→SNP Class|>,
           <|name→GDSC, fullname→Gene Description|>,
           <|name→CPOS, fullname→Base Position|>,
           <|name→GPOS, fullname→Contig Position|>,
           <|name→WORD, fullname→Text Word|>,

4.2 NCBI Entrez Programming Utilities

```
<|name→WTAA,fullname→ Reference Amino Acid|>,
<|name→MTAA,fullname→ Variant Amino Acid|>,
<|name→RSNP,fullname→ Reference SNP|>,
<|name→SIDX,fullname→SNP Index|>,
<|name→ALOR,fullname→SNP Allele Origin|>,
<|name→SUSP,fullname→Suspected false variation|>,
<|name→CLIN,fullname→Clinical Significance|>,
<|name→GMAF,fullname→Global Minor Allele Frequency|>}
```

Finally the link information can be used by Elink (see further below).

In[12]:= snpLinkList=**Query**["einforesult",1,"linklist"]@snpInfo;

We can look at a few of these, which are named links, and have a "dbto" association to the linked database.

In[13]:= snpLinkList[[1;;3]]
Out[13]= {<|name→ snp_bioproject,menu→BioProject Links,
         description→Related BioProject record,
         dbto→bioproject|>,
       <|name→ snp_biosample,menu→BioSample Links,
         description→Related BioSample record,
         dbto→biosample|>,
       <|name→ snp_clinvar,menu→ClinVar,description→
         Clinical reports for this variation,
         dbto→clinvar|>}

4.2.2 ESearch

The ESearch utility allows us to carry out searches to obtain lists of unique identifiers (UIDs). There are two required parameters, `"db"` referring to a database, and `"term"` which is a text query. For example, let us search the "snp" database for autism:

In[14]:= esearchSNPAutism=**URLExecute**[
    "https://eutils.ncbi.nlm.nih.gov/entrez/eutils/esearch.fcgi",
       {"RequestMethod"->"POST",
        "retmode"->"json",
        "db"-> "snp",
        "term"-> "autism"},"RawJSON"]
Out[14]= <|header→ <|type→esearch,version→0.3|>,
        esearchresult→
          <|count→30,retmax→20,retstart→0,idlist→
          {386572987,122468182,121912597,121908445,
           104894663,60754679,60350369,59359388,
           58778612,56567265,52837965,17851360,
           17826777,17551405,17350302,17295902,
           10373747,10348618,10314229,7794745},
          translationset→{},translationstack→
          {<|term→ autism[All Fields],
            field→All Fields,count→30,
            explode→N|>,GROUP},

            querytranslation→autism[All Fields]|>|>

The structure of the returned information is the same as with EInfo. There is a header, and a key for the results,"esearchresult". The results have additional **Keys** with values:

In[15]:= **Query**["esearchresult", Keys]@esearchSNPAutism
Out[15]= {count,retmax,retstart,idlist,translationset,
           translationstack,querytranslation}

We notice that there are 30 results, but only 20 have been returned ("retmax" parameter):

In[16]:= **Query**["esearchresult",{"count","retmax","retstart"}]@
          esearchSNPAutism
Out[16]= <|count→30,retmax→20,retstart→0|>

The UIDs can be obtained from the "idlist"

In[17]:= esearchSNPAutismIDs=**Query**["esearchresult","idlist"]@
          esearchSNPAutism
Out[17]= {386572987,122468182,121912597,121908445,
          104894663,60754679,60350369,59359388,
          58778612,56567265,52837965,17851360,
          17826777,17551405,17350302,17295902,
          10373747,10348618,10314229,7794745}

The "retmax" parameter that is set to 20 has a maximum value of 100,000 and can be requested in **URLExecute**. If your search has more than 100,000 results, you can change the position of returned IDs by using the "retstart" parameter, which indicates after which position to return. For example, let us return the results from any position in the list we want - here, we return 5 results, locations 3 through 7 by returning data *after* position 2:

In[18]:= **URLExecute**[
    "https://eutils.ncbi.nlm.nih.gov/entrez/eutils/esearch.fcgi",
       {"RequestMethod"->"POST",
        "retmode"->"json",
        "db"-> "snp",
        "term"-> "autism",
        "retmax"-> 5,
        "retstart"-> 2},"RawJSON"]
Out[18]= <|header→<|type→esearch,version→0.3|>,
          esearchresult→
            <|count→30,retmax→5,retstart→2,
            idlist→{121912597,121908445,104894663,
              60754679,60350369},translationset→{},
            translationstack→{<|term→autism[All Fields],
              field→All Fields,count→30,
              explode→N|>,GROUP},
            querytranslation→autism[All Fields]|>|>

We can see that these correspond to the correct list items (3–7) from the original search:

4.2 NCBI Entrez Programming Utilities

In[19]:= {**Query**["esearchresult","idlist"]@%,
        esearchSNPAutismIDs[[3;;7]]}
Out[19]= {{121912597,121908445,104894663,
        60754679,60350369},{121912597,
        121908445,104894663,60754679,60350369}}

An additional parameter can be passed, `"rettype"` which can take either its default value `"uilist"`, i.e. return a list of UIDs, or `"count"` to omit the UIDs and just return the `"count"` key without additional information:

In[20]:= **URLExecute**[
    "https://eutils.ncbi.nlm.nih.gov/entrez/eutils/esearch.fcgi",
        {"RequestMethod"->"POST",
        "retmode"->"json",
        "db"-> "snp",
        "term"-> "autism",
        "rettype"-> "count"},"RawJSON"]
Out[20]= <|header→ <|type→esearch,version→0.3|>,
        esearchresult→ <|count→30|>|>

4.2.2.1 Multiple Queries and Boolean Operators

We can also construct queries by combining terms in the form of `"term"` parameter strings `"term1[field1] Boolean Operator term2[field2] ... Boolean Operator termN[fieldN]"`. The possible operators are AND,OR,NOT. The fields are strings specific to the database, and we have seen how to extract them in the previous section using EInfo. As an example we construct a search looking for `"autism"` in the title field, AND location on chromosome 3:

In[21]:= autismANDChromosome3SNP=**URLExecute**[
    "https://eutils.ncbi.nlm.nih.gov/entrez/eutils/esearch.fcgi",
        {"RequestMethod"->"POST",
        "retmode"->"json",
        "db"-> "snp",
        "term"-> "autism[ALL]AND 3[CHR]",
        "retmax"-> 5},"RawJSON"]
Out[21]= <|header→ <|type→esearch,version→0.3|>,
        esearchresult→ <|count→1,retmax→1,
        retstart→0,idlist→{121912597},
        translationset→{},translationstack→
          {<|term→autism[ALL],field→ALL,count→30,
            explode→N|>,<|term→3[CHR],field→CHR,
            count→71071058,explode→N|>,AND},
        querytranslation→autism[ALL] AND 3[CHR]|>|>

Notice that the `"translationstack"` key gives a list of the result counts for the individual terms being searched, and the operator connecting them:

In[22]:= **Query**["esearchresult","translationstack"]@
        autismANDChromosome3SNP
Out[22]= {<|term→autism[ALL],field→ALL,count→30,

140 4 Databases: E-Utilities and UCSC Genome Browser

```
         explode→Nl>,<lterm→3[CHR],field→CHR,
         count→71071058,explode→Nl>,AND}
```

We can get both counts:

```
In[23]:= Query["esearchresult","translationstack",All,
            {"term","count"}]@
         autismANDChromosome3SNP
Out[23]= {<lterm→autism[ALL],count→30l>,
          <lterm→3[CHR],count→71071058l>,
          Missing[PartInvalid,{term,count}]}
```

We see that though there are 30 and 71071058 terms in total, the AND operator filters only the ones that are overlapping in the database. We can construct more general queries by grouping the logical operations in brackets. For example, we may want chromosome 3 or 6 terms:

```
In[24]:= URLExecute[
         "https://eutils.ncbi.nlm.nih.gov/entrez/eutils/esearch.fcgi",
            {"RequestMethod"->"POST",
            "retmode"->"json",
            "db"-> "snp",
            "term"-> "autism[ALL] AND (3[CHR] OR 6[CHR])",
            "retmax"-> 5},"RawJSON"]
Out[24]= <lheader→<ltype→esearch,version→0.3l>,
         esearchresult→
            <lcount→11,retmax→5,retstart→0,
            idlist→{386572987,121912597,52837965,
            17851360,17551405},translationset→{},
            translationstack→{<lterm→autism[ALL],
               field→ALL,count→30,explode→Nl>,
               <lterm→3[CHR],field→CHR,count→
               71071058,explode→Nl>,<lterm→6[CHR],
               field→CHR,count→56692539,explode→Nl>,
               OR,GROUP,AND},querytranslation→
            autism[ALL] AND (3[CHR] OR 6[CHR])l>l>
```

4.2.2.2 Other ESearch Parameters: Limit Results by Date and Sort

We can limit the results returned by date. The availability will depend on the database. There are three relevant parameters:

- `"datetype"` parameter, which can take values:
  - `"edat"` corresponding to Entrez Date
  - `"pdat"` corresponding to publication date
  - `"mhdat"` corresponding to the date a citation was indexed with MeSH

- `"reldate"` parameter takes as value n, and returns results limited to the last n days. The search will use the field specified by `"datetype"`

4.2 NCBI Entrez Programming Utilities 141

- `"mindate"` and `"maxdate"` parameters which *must* be used can specify a range of dates taking the start and end date respectively. The date is formatted as YYYY/MM/DD or YYYY/MM or YYYY

For an example, let us extract records from PubMED with field `"autism"` in the last 20 days using Entrez Date:

```
In[25]:= URLExecute[
    "https://eutils.ncbi.nlm.nih.gov/entrez/eutils/esearch.fcgi",
        {"RequestMethod"->"POST",
         "retmode"->"json",
         "db"-> "pubmed",
         "term"-> "autism[ALL]",
         "retmax"-> 5,
         "datetype"-> "edat",
         "reldate"-> 20},"RawJSON"]
Out[25]= <|header-> <|type->esearch,version->0.3|>,
         esearchresult-> <|count->258,retmax->5,
           retstart->0,idlist->{29195167,
             29195157,29194420,29193861,29193847},
           translationset->{<|from->autism[ALL],
             to->"autistic disorder"[MeSH Terms]
                 OR ("autistic"[All Fields]
                 AND "disorder"[All Fields])
                 OR "autistic disorder"[All
                 Fields] OR "autism"[All Fields]|>},
           translationstack->
             {<|term->"autistic disorder"[MeSH Terms],
               field->MeSH Terms,
               count->18382,explode->Y|>,
             <|term->"autistic"[All Fields],
               field->All Fields,
               count->22129,explode->N|>,
             <|term->"disorder"[All Fields],
               field->All Fields,count->566097,
               explode->N|>,AND,GROUP,OR,
             <|term->"autistic disorder"[All Fields],
               field->All Fields,
               count->18765,explode->N|>,
            OR,<|term->"autism"[All Fields],
               field->All Fields,count->35376,
               explode->N|>,OR,GROUP,
              <|term->2017/11/13[EDAT],field->EDAT,
               count->0,explode->N|>,
              <|term->2017/12/03[EDAT],field->EDAT,
               count->0,explode->N|>,
            RANGE,AND},querytranslation->
             ("autistic disorder"[MeSH Terms]
              OR ("autistic"[All Fields]
              AND "disorder"[All Fields])
              OR "autistic disorder"[All
              Fields] OR "autism"[All
              Fields]) AND 2017/11/13[EDAT]
              : 2017/12/03[EDAT]|>|>
```

We get 251 records for these dates (your result will of course vary depending on when you run the command):

In[26]:= **Query**["esearchresult",{"count","idlist"}]@%
Out[26]= <|count→258,idlist→{29195167,
29195157,29194420,29193861,29193847}|>

We can instead search by publication date:

In[27]:= **URLExecute**[
 "https://eutils.ncbi.nlm.nih.gov/entrez/eutils/esearch.fcgi",
  {"RequestMethod"->"POST",
  "retmode"->"json",
  "db"-> "pubmed",
  "term"-> "autism[ALL]",
  "retmax"-> 5,
  "datetype"-> "pdat",
  "reldate"-> 20},"RawJSON"]
Out[27]= <|header→ <|type→esearch,version→0.3|>,
 esearchresult→ <|count→310,retmax→5,
  retstart→0,idlist→{29195167,
   29195157,29194420,29193861,29193847},
  translationset→{<|from→autism[ALL],
   to→ "autistic disorder"[MeSH Terms]
     OR ("autistic"[All Fields]
     AND "disorder"[All Fields])
     OR "autistic disorder"[All
     Fields] OR "autism"[All
     Fields]|>},translationstack→
  {<|term→ "autistic disorder"[MeSH Terms],
   field→MeSH Terms,
   count→18382,explode→Y|>,
   <|term→ "autistic"[All Fields],
   field→All Fields,
   count→22129,explode→N|>,
   <|term→ "disorder"[All Fields],
   field→All Fields,count→566097,
   explode→N|>,AND,GROUP,OR,
   <|term→ "autistic disorder"[All Fields],
   field→All Fields,
   count→18765,explode→N|>,
   OR,<|term→ "autism"[All Fields],
   field→All Fields,count→35376,
   explode→N|>,OR,GROUP,
   <|term→2017/11/13[PDAT],field→PDAT,
   count→0,explode→N|>,
   <|term→2017/12/03[PDAT],field→PDAT,
   count→0,explode→N|>,
   RANGE,AND},querytranslation→
   ("autistic disorder"[MeSH Terms]
    OR ("autistic"[All Fields]
    AND "disorder"[All Fields])
    OR "autistic disorder"[All
    Fields] OR "autism"[All
    Fields]) AND 2017/11/13[PDAT]

4.2 NCBI Entrez Programming Utilities

: 2017/12/03[PDAT]|>|>

We get a different list in this case:

In[28]:= **Query**["esearchresult",{"count","idlist"}]@%
Out[28]= <|count→310,idlist→{29195167,
       29195157,29194420,29193861,29193847}|>

We can also specify the date range:

In[29]:= **URLExecute**[
    "https://eutils.ncbi.nlm.nih.gov/entrez/eutils/esearch.fcgi",
      {"RequestMethod"−>"POST",
      "retmode"−>"json",
      "db"−> "pubmed",
      "term"−> "autism[ALL]",
      "retmax"−> 5,
      "datetype"−> "pdat",
      "mindate"−> "2017/11/07",
      "maxdate"−> "2017/11/27"},"RawJSON"]
Out[29]= <|header→<|type→esearch,version→0.3|>,
       esearchresult→<|count→213,retmax→5,
        retstart→0,idlist→{29195167,
         29191694,29187318,29186576,29186566},
        translationset→{<|from→autism[ALL],
          to→ "autistic disorder"[MeSH Terms]
            OR ("autistic"[All Fields]
            AND "disorder"[All Fields])
            OR "autistic disorder"[All
            Fields] OR "autism"[All
            Fields]|>},translationstack→
         {<|term→ "autistic disorder"[MeSH Terms],
            field→MeSH Terms,
            count→18382,explode→Y|>,
          <|term→ "autistic"[All Fields],
            field→All Fields,
            count→22129,explode→N|>,
          <|term→ "disorder"[All Fields],
            field→All Fields,count→566097,
            explode→N|>,AND,GROUP,OR,
          <|term→ "autistic disorder"[All Fields],
            field→All Fields,
            count→18765,explode→N|>,
          OR,<|term→ "autism"[All Fields],
            field→All Fields,count→35376,
            explode→N|>,OR,GROUP,
          <|term→2017/11/07[PDAT],field→PDAT,
            count→0,explode→N|>,
          <|term→2017/11/27[PDAT],field→PDAT,
            count→0,explode→N|>,
          RANGE,AND},querytranslation→
         ("autistic disorder"[MeSH Terms]
           OR ("autistic"[All Fields]
           AND "disorder"[All Fields])
           OR "autistic disorder"[All

            Fields] OR "autism"[All
            Fields]) AND 2017/11/07[PDAT]
            : 2017/11/27[PDAT]|>|>

In[30]:= Query["esearchresult",{"count","idlist"}]@%
Out[30]= <|count→213,idlist→{29195167,
          29191694,29187318,29186576,29186566}|>

The ESearch utility will give out a list of UIDs. We can also specify how to sort the data by setting the "sort" parameter. The values to sort by are data base specific, and the easiest way is to retrieve them online the first time you need to use them. For example, if we search PubMed for identifiers and sort by title or Last Author:

In[31]:= **URLExecute**[
         "https://eutils.ncbi.nlm.nih.gov/entrez/eutils/esearch.fcgi",
               {"RequestMethod"->"POST",
                "retmode"->"json",
                "db"-> "pubmed",
                "term"-> "autism[ALL]",
                "retmax"-> 5,
                "sort"-> "title"},"RawJSON"]
         Query["esearchresult",{"count","idlist"}]@%
Out[31]= <|header→<|type→esearch,version→0.3|>,
          esearchresult→<|count→40052,retmax→5,
            retstart→0,idlist→{26071405,
             20832509,24767651,24613754,27892236},
            translationset→{<|from→autism[ALL],
              to→ "autistic disorder"[MeSH Terms]
                 OR ("autistic"[All Fields]
                 AND "disorder"[All Fields])
                 OR "autistic disorder"[All
                 Fields] OR "autism"[All
                 Fields]|>},translationstack→
             {<|term→ "autistic disorder"[MeSH Terms],
                field→MeSH Terms,
                count→18382,explode→Y|>,
              <|term→ "autistic"[All Fields],
                field→All Fields,
                count→22129,explode→N|>,
              <|term→ "disorder"[All Fields],
                field→All Fields,count→566097,
                explode→N|>,AND,GROUP,OR,
              <|term→ "autistic disorder"[All Fields],
                field→All Fields,
                count→18765,explode→N|>,
              OR,<|term→ "autism"[All Fields],
                field→All Fields,
                count→35376,explode→N|>,
              OR,GROUP},querytranslation→
             "autistic disorder"[MeSH Terms] OR
               ("autistic"[All Fields] AND
               "disorder"[All Fields]) OR
               "autistic disorder"[All Fields]
               OR "autism"[All Fields]|>|>

4.2 NCBI Entrez Programming Utilities   145

Out[32]= <|count→40052,idlist→{26071405,
             20832509,24767651,24613754,27892236}|>

In[33]:= **URLExecute**[
    "https://eutils.ncbi.nlm.nih.gov/entrez/eutils/esearch.fcgi",
        {"RequestMethod"–>"POST",
         "retmode"–>"json",
         "db"–> "pubmed",
         "term"–> "autism[ALL]",
         "retmax"–> 5,
         "sort"–> "Last Author"},"RawJSON"];
       Query["esearchresult",{"count","idlist"}]@%
Out[34]= <|count→40052,idlist→{28905926,
             28493759,28218600,28703053,28097989}|>

4.2.2.3 Advanced Searching Using the History Server

For advanced searches and combining the results of different searches it is possible to use the History server instead of the programmatic methods shown above. The E-utilities allow us to actually store data online and combine searches that are previously stored. In Esearch the relevant parameter is `usehistory`. This allows us to actually use the history server by setting its value to `y`. Let us do this for our dbSNP search:

In[35]:= historyExampleAutism=**URLExecute**[
    "https://eutils.ncbi.nlm.nih.gov/entrez/eutils/esearch.fcgi",
        {"RequestMethod"–>"POST",
         "retmode"–>"json",
         "db"–> "snp",
         "term"–> "autism",
         "retmax"–> 5,
         "usehistory"–> "y"},"RawJSON"]
Out[35]= <|header→ <|type→esearch,version→0.3|>,
           esearchresult→ <|count→30,retmax→5,
             retstart→0,querykey→1,webenv→
               NCID_1_44322648_130.14.18.34_9001_1512335349
                 _1326347342_0MetA0_S_MegaStore_F_1,
             idlist→{386572987,122468182,
               121912597,121908445,104894663},
             translationset→{},translationstack→
               {<|term→autism[All Fields],
                 field→All Fields,count→30,
                 explode→N|>,GROUP},
             querytranslation→autism[All Fields]|>|>

We notice that there are additional parameters in the results: `"webenv"` and `"querykey"`:

In[36]:= **Query**["esearchresult",{"querykey","webenv"}]@
           historyExampleAutism
Out[36]= <|querykey→1,webenv→

        NCID_1_44322648_130.14.18.34_9001_1512335349_
        1326347342_0MetA0_S_MegaStore_F_1|>

The `"webenv"` is a web environment string that is an identifier for other E-utilities to use to combine queries. The querykey is the query index/number in the Web environment.

In[37]:= **Query**[`"esearchresult"`,`"querykey"`]@historyExampleAutism
Out[37]= 1

These parameter values provide a reference for combining queries using Boolean operators.

We can carry out for example our previous search of looking for autism results on chromosome 3 in dbSNP by combining a search for chromosome 3 with the above search results that are stored in the server. We use the `"WebEnv"` parameters to specify where the results should be placed on the history server:

In[38]:= chromosome3SNP=**URLExecute**[
    "https://eutils.ncbi.nlm.nih.gov/entrez/eutils/esearch.fcgi",
        {"RequestMethod"->"POST",
        "retmode"->"json",
        "db"-> "snp",
        "term"-> "3[CHR]",
        "usehistory"-> "y",
        "WebEnv"-> **Query**["esearchresult",
            "webenv"]@historyExampleAutism,
        "retmax"-> 5},"RawJSON"]
Out[38]= <|header→<|type→esearch,version→0.3|>,
        esearchresult→<|count→71071058,retmax→5,
            retstart→0,querykey→2,webenv→
            NCID_1_44322648_130.14.18.34_9001_1512335349
                _1326347342_0MetA0_S_MegaStore_F_1,
            idlist→{1135402756,1135402755,
                1135402754,1135401948,1135401796},
            translationset→{},translationstack→
                {<|term→3[CHR],field→CHR,
                count→71071058,explode→N|>,GROUP},
            querytranslation→3[CHR]|>|>

We used **Query**[`"esearchresult"`,`"webenv"`]@historyExampleAutism to pass the correct string. From the results, we notice how the `"querykey"` value is now incremented to 2, but the web environment is the same as above.

In[39]:= **Query**["esearchresult",{"querykey","webenv"}]@
    chromosome3SNP
Out[39]= <|querykey→2,webenv→
    NCID_1_44322648_130.14.18.34
    _9001_1512335349_1326347342_0MetA0_S_MegaStore_F_1|>

To combine the results we can run a third query and combine the result sets using Boolean operators in the `"term"`. The queries combined are referred to by "#" followed by its `"querykey"` value:

In[40]:= autismSNPChromosome3Combined=**URLExecute**[

4.2 NCBI Entrez Programming Utilities

```
        "https://eutils.ncbi.nlm.nih.gov/entrez/eutils/esearch.fcgi",
         {"RequestMethod"->"POST",
          "retmode"->"json",
          "db"-> "snp",
          "term"-> "#1 AND #2",
          "usehistory"-> "y",
          "WebEnv"->  Query["esearchresult",
              "webenv"]@historyExampleAutism,
          "retmax"-> 5},"RawJSON"]
```
Out[40]= <|header→ <|type→esearch, version→0.3|>,
         esearchresult→<|count→1,retmax→1,
           retstart→0,querykey→3,webenv→
             NCID_1_44322648_130.14.18.34_9001_1512335349
               _1326347342_0MetA0_S_MegaStore_F_1,
           idlist→{121912597},
           translationset→{},translationstack→{AND},
           querytranslation→#1 AND #2|>|>

The search term is "#1 AND #2", i.e. combine those two searches' UIDs to get their intersection. We can combine multiple new operations and the results from the server. For example, we extend to searching for SNPs on chromosome 6 or 3.

In[41]:= autismSNPChromosome3Combined=URLExecute[
    "https://eutils.ncbi.nlm.nih.gov/entrez/eutils/esearch.fcgi",
         {"RequestMethod"->"POST",
          "retmode"->"json",
          "db"-> "snp",
          "term"-> "#1 AND (#2 OR 6[CHR])",
          "usehistory"-> "y",
          "WebEnv"-> Query["esearchresult",
              "webenv"]@historyExampleAutism,
          "retmax"-> 5},"RawJSON"]
Out[41]= <|header→ <|type→esearch, version→0.3|>,
         esearchresult→<|count→11,retmax→5,
           retstart→0,querykey→4,webenv→
             NCID_1_44322648_130.14.18.34_9001_1512335349
               _1326347342_0MetA0_S_MegaStore_F_1,
           idlist→{386572987,121912597,
             52837965,17851360,17551405},
           translationset→{},translationstack→
             {<|term→6[CHR],field→CHR,count→
               56692539,explode→N|>,OR,GROUP,AND},
           querytranslation→#1 AND (#2 OR 6[CHR])|>|>

4.2.3 EPost

The epost utility allows us to store a set UIDs (required parameter `"id"` which takes a list of comma separated UIDs) from a database (specified by the required `"db"` parameter) to the History Server. If an existing web environment will be used, the optional server parameter `"WebEnv"` must be provided, and if it is not a new Web environment will be created.

```
In[42]:= epostSNPs=URLExecute[
   "https://eutils.ncbi.nlm.nih.gov/entrez/eutils/epost.fcgi",
       {"RequestMethod"->"POST",
        "db"-> "snp",
        "id"-> "59359388,58778612,56567265,52837965,17851360",
        "WebEnv"-> Query["esearchresult",
             "webenv"]@historyExampleAutism }]
Out[42]= <?xml version="1.0" encoding="UTF-8" ?>
       <!DOCTYPE ePostResult PUBLIC
        "-//NLM//DTD epost 20090526//EN"
        "https://eutils.ncbi.nlm.nih.gov/eutils/dtd/
        20090526/epost.dtd"><ePostResult>
          <QueryKey>5</QueryKey>


          <WebEnv>NCID_1_44322648_130.14.18.34_9001_
          1512335349_1326347342_0MetA0_S_MegaStore_F_1</
          WebEnv>
       </ePostResult>
```

Please note that as of this writing the result returned is in XML, and not JSON. We can easily parse the string to extract an association by using string patterns. First let us get the querykey:

```
In[43]:= StringCases[epostSNPs,
          "<QueryKey>"~~x:__~~"</QueryKey>"-> x]
Out[43]= {5}
```

Here, we have have extracted the named string pattern x, that matches a string __ that is enclosed by <QueryKey> and </QueryKey> and use a rule −> x to return the matching string. Similarly, we can extract the string contained within the <WebEnv> </WebEnv> strings for the Web environment:

```
In[44]:= StringCases[epostSNPs,
          "<WebEnv>"~~y:__~~"</WebEnv>"-> y]
Out[44]= {NCID_1_44322648_130.14.18.34_9001_1512335349_
          1326347342_0MetA0_S_MegaStore_F_1}
```

We can put it all together in an association

```
In[45]:= extractedEpostInfo=
       Association@StringCases[epostSNPs,
          {("<QueryKey>"~~x:__~~"</QueryKey>")->
           "querykey"-> x,
           ("<WebEnv>"~~y:__~~"</WebEnv>")->
           "webenv"-> y}]
Out[45]= <|querykey→5,webenv→
          NCID_1_44322648_130.14.18.34_9001_1512335349_
          1326347342_0MetA0_S_MegaStore_F_1|>
```

In this code, the string patterns x and y are extracted in a subpattern list and each rule makes an assignment of the keys `"querykey"` and `"webenv"` to x and y respectively. Finally, the **Association** @ (prefix) at the start converts the list to an association.

4.2.4 ESummary

ESummary allows us to retrieve document summaries corresponding to UIDs from a given database. There are two mutually exclusive ways to carry this out:

- We can provide a set of UIDs directly as `"id"` parameter values, in the form of a list of comma separated UIDs.

```
In[46]:= esummarySNPs=URLExecute[
    "https://eutils.ncbi.nlm.nih.gov/entrez/eutils/esummary.fcgi",
        {"RequestMethod"->"POST",
         "retmode"->"json",
         "db"-> "snp",
         "id"-> "59359388,58778612,56567265,
                52837965,17851360"},
    "RawJSON"];
```

```
In[47]:= esummarySNPs//Short
Out[47]//Short=  <|header→ <|type→esummary,→ 0.3|>, ... →<<1>>|>
```

- Alternatively, we can use UIDs retrieved from the History Server by specifying the `"query_key"` parameter identifying the query location and the corresponding `"WebEnv"` parameter identifying the Web environment the query is from. For our example we use the search we had posted using EPost in the previous section

```
In[48]:= esummarySNPsHistoryServer=URLExecute[
    "https://eutils.ncbi.nlm.nih.gov/entrez/eutils/esummary.fcgi",
        {"RequestMethod"->"POST",
         "retmode"->"json",
         "query_key"-> Query["querykey"]@extractedEpostInfo,
         "webenv"-> Query["webenv"]@extractedEpostInfo},
    "RawJSON"];
```

As with ESearch notice that we have set the retrieval type to JSON for easy parsing into Associations using `"RawJSON"` as the input format in **URLExecute**.

Also, we can define other retrieval parameters to use with ESummary that are the same as before : `"retmax"` sets a limit to the number of document summaries to be retrieved. The maximum is 10,000. If your search has more than 10,000 results, you can change the position of returned IDs by using the `"retstart"` variable, which indicates which position to return.

Note that the outputs are suppressed since they are fairly large:

```
In[49]:= esummarySNPs//Short
Out[49]//Short=  <|header→ <|type→esummary, version→0.3|>,
                  ... →<<1>>|>
```

The keys are a header and the result:

```
In[50]:= Query[Keys]@esummarySNPs
Out[50]= {header,result}
```

Within the result there are multiple **Keys**:

```
In[51]:= Query["result",Keys]@esummarySNPs
Out[51]= {uids,59359388,58778612,
         56567265,52837965,17851360}
```

The uids is a list of the UIDs we have provided. Each of the other keys is the result for the particular UID. Let us look at the first one.

```
In[52]:= Query["result","56567265",Keys]@esummarySNPs
Out[52]= {uid,snp_id,organism,allele_origin,global_maf,
         global_population,global_samplesize,suspected,
         clinical_significance,genes,acc,chr,
         weight,handle,fxn_class,validated,gtype,
         nonref,docsum,het,srate,tax_id,chrrpt,
         orig_build,upd_build,createdate,updatedate,
         pop_class,method_class,snp3d,linkout,ss,
         locsnpid,allele,snp_class,chrpos,contigpos,
         text,lookup,sort_priority,snp_id_sort,
         clinical_sort,human_sort,cited_sort,
         weight_sort,chrpos_sort,merged_sort}
```

```
In[53]:= Query["result","56567265",Length]@esummarySNPs
Out[53]= 47
```

There are many terms of interest from the 47 keys and we can return any ones we want:

```
In[54]:= Query["result","56567265",
         {"snp_id","fxn_class","genes","updatedate","docsum"}]@
         esummarySNPs
Out[54]= <|snp_id→7794745,fxn_class→intron-variant,
         genes→{<|name→CNTNAP2,gene_id→26047|>},
         updatedate→2017/02/15 16:05,docsum→
           CLINICAL=Y|HGVS=NC_000007.13:g.146489606A&gt;T,
           NC_000007.14:g.146792514A&gt;T,NG_007092.2:
           g.681154A&gt;T,NM_014141.5:c.208+18133A&gt;
           T,XM_017011950.1:c.208+18133A&gt;T|SEQ=
           CAGGTCAGGACCTGGAAAGGCCTAA[A/T]
           TGATAAGACTAAGTGTCAAAATCAG|GENE=CNTNAP2:
           26047|>
```

We may want to get more information out of the `"docsum"` above. We can extract it and split the string along the separator |.

```
In[55]:= StringSplit[#,"|"]&@
         (Query["result","56567265","docsum"]@esummarySNPs)
Out[55]= {CLINICAL=Y,
         HGVS=NC_000007.13:g.146489606A&gt;T,NC_000007.14
           :g.146792514A&gt;T,NG_007092.2:g.681154A&gt;T
           ,NM_014141.5:c.208+18133A&gt;T,XM_017011950.1
           :c.208+18133A&gt;T,
         SEQ=CAGGTCAGGACCTGGAAAGGCCTAA[A/T]
           TGATAAGACTAAGTGTCAAAATCAG,
         GENE=CNTNAP2:26047}
```

4.2 NCBI Entrez Programming Utilities

For another example we can retrieve various records from the `pubmed` database, for two PubMED ids that we are interested in, `"4913914"` and `"2231712"`:

In[56]:= idPUBMED=**URLExecute**[
  "http://eutils.ncbi.nlm.nih.gov/entrez/eutils/esummary.fcgi?",
       {"RequestMethod"−>"POST",
        "retmode"−>"json",
        "db"−> "pubmed",
        "id"−> "4913914,2231712"},
       "RawJSON"];

Like other results, by setting the output as RawJSON we get a hierarchy of keys and values that we can navigate:

In[57]:= Keys@idPUBMED
Out[57]= {header, result}

In[58]:= **Query**["header"]@idPUBMED
Out[58]= <|type→esummary, version→0.3|>

In[59]:= **Query**["result", **Keys**]@idPUBMED
Out[59]= {uids, 4913914, 2231712}

We can extract the full results for the first ID:

In[60]:= **Query**["result","4913914"]@idPUBMED
Out[60]= <|uid→4913914,pubdate→1970 Aug 8,
      epubdate→, source→Nature,
      authors→{<|name→Crick F, authtype→Author,
          clusterid→|>}, lastauthor→Crick F,
      title→Central dogma of molecular biology.,
      sorttitle→central dogma of molecular biology,
      volume→227, issue→5258, pages→561−3,
      lang→{eng}, nlmuniqueid→0410462,
      issn→0028−0836, essn→1476−4687,
      pubtype→{Historical Article, Journal Article},
      recordstatus→PubMed − indexed for MEDLINE,
      pubstatus→4, articleids→{<|idtype→pubmed,
         idtypen→1, value→4913914|>,
        <|idtype→rid, idtypen→8, value→4913914|>,
        <|idtype→eid, idtypen→8, value→4913914|>},
      history→{<|pubstatus→pubmed, date→
         1970/08/08 00:00|>, <|pubstatus→medline,
         date→1970/08/08 00:01|>, <|pubstatus→
         entrez, date→1970/08/08 00:00|>},
      references→{}, attributes→{},
      pmcrefcount→212, fulljournalname→Nature,
      elocationid→, doctype→citation,
      srccontriblist→{}, booktitle→,
      medium→, edition→, publisherlocation→,
      publishername→, srcdate→,
      reportnumber→, availablefromurl→,
      locationlabel→, doccontriblist→{},
      docdate→, bookname→, chapter→,
      sortpubdate→1970/08/08 00:00,
      sortfirstauthor→Crick F, vernaculartitle→|>

We can see that there is a lot of information, again as **Keys**. For the second record, let us first get all the **Keys**:

In[61]:= Query["result","2231712",**Keys**]@idPUBMED
Out[61]= {uid,pubdate,epubdate,source,authors,
    lastauthor,title,sorttitle,volume,
    issue,pages,lang,nlmuniqueid,issn,essn,
    pubtype,recordstatus,pubstatus,articleids,
    history,references,attributes,pmcrefcount,
    fulljournalname,elocationid,doctype,
    srccontriblist,booktitle,medium,edition,
    publisherlocation,publishername,srcdate,
    reportnumber,availablefromurl,locationlabel,
    doccontriblist,docdate,bookname,chapter,
    sortpubdate,sortfirstauthor,vernaculartitle}

Let us get some bibliographical information for the second example:

In[62]:= Query["result","2231712",
    {"authors","title","source","volume","issue","pages"}]@
    idPUBMED
Out[62]= <|authors→{<|name→Altschul SF,
    authtype→Author,clusterid→|>,
    <|name→Gish W,authtype→Author,
    clusterid→|>,
    <|name→Miller W,authtype→Author,
    clusterid→|>,
    <|name→Myers EW,authtype→Author,
    clusterid→|>,<|name→Lipman DJ,
    authtype→Author,clusterid→|>},
    title→Basic local alignment search tool.,
    source→J Mol Biol,
    volume→215,
    issue→3,
    pages→403−10|>

4.2.5 EFetch

EFetch allows us to retrieve actual data records for a list of UIDs in a given database. Like e-summary, we have required parameters for providing the UIDs. Either:

- We can provide a set of UIDs directly as "id" parameter values, in the form of a list of comma separated UIDs.
- We can use UIDs retrieved from the History Server by specifying the known "query_key" parameter identifying the query location and the corresponding "WebEnv" parameter identifying the Web environment the query is from.

Some behavior is similar to the utilities considered above: The retrieval options "retmax" set the maximum retrieval of records (10,000) and "retstart" is the

4.2 NCBI Entrez Programming Utilities

parameter indicating at which position to start returning (the first record is at position 0 for EFetch). The retrieval parameters `"retmode"` (format of records returned) and `"rettype"` (record view returned) are not uniform across databases and need to be adjusted depending on what you want to download. Please consult the database of interest or Entrez Programming Utilities Help [Internet]. Bethesda (MD): National Center for Biotechnology Information (US); 2010-. Available from: `"https://www.ncbi.nlm.nih.gov/books/NBK25499/"` page for information on allowable values for each database. Depending on the combination you will also need to adjust the input format for the **URLExecute** command (last argument, that we have been setting to RawJSON mostly).

Continuing our example with the SNP data we retrieved before, we can get FASTA files:

```
In[63]:= efetchSNPs=URLExecute[
    "https://eutils.ncbi.nlm.nih.gov/entrez/eutils/efetch.fcgi",
        {"RequestMethod"->"POST",
        "retmode"->"text",
        "rettype"->"fasta",
        "db"-> "snp",
        "id"-> "59359388,56567265"},{"FASTA","LabeledData"}];
```

The above query will return a FASTA format. We have set the retrieval mode `"retmode"` to `"text"`, and the `"rettype"` to `"fasta"`. Additionally, as we will see next chapter when processing `"FASTA"` files, we have the option to import the information as `"labeled"` data to maintain an identifier to sequence rule.

```
In[64]:= efetchSNPs[[1]]//Short
Out[64]//Short= gnl|dbSNP|rs7794745 ... :0.4946|clinsig=
                  other→<<1103>>

In[65]:= Keys @ efetchSNPs
Out[65]= {gnl|dbSNP|rs7794745
            rs=7794745|pos=501|len=100|taxid=9606|mol="
            genomic"|class=snp|alleles="A/T"|build=150|
            suspect=?|GMAF=A:2477:0.4946|clinsig=other,
          gnl|dbSNP|rs7794745
            rs=7794745|pos=501|len=100|taxid=9606|mol="
            genomic"|class=snp|alleles="A/T"|build=150|
            suspect=?|GMAF=A:2477:0.4946|clinsig=other}
```

Some additional considerations regarding parameters available for EFetch:

- There are additional optional parameters for sequence databases. Some are self-explanatory: `"strand"` (DNA strand to retrieve, `"1"`:plus strand, `"2"`: negative strand), `"seq_start"` (integer coordinate of first position to retrieve) and `"seq_stop"` corresponds to the last position to return.
- Finally, there is a `"complexity"` parameter, which takes values that return different amounts of data, for entries that are part of a larger entry `"blob"`. This can take parameters that are listed here just for reference:

0 return entire info
1 return bioseq
2 return minimal bioseq-set
3 return minimal nuc-prot
4 return minimal pub-set

As another example, we extract the feature table for NM_003988:

```
In[66]:= genBankExample=URLExecute[
    "https://eutils.ncbi.nlm.nih.gov/entrez/eutils/efetch.fcgi",
        {"RequestMethod"->"POST",
         "retmode"->"text",
         "rettype"->"ft",
         "db"-> "nuccore",
         "id"-> "NM_003998"},"Text"];
In[67]:= genBankExample//Short
Out[67]//Short= >Feature  ref|NM_00399\...
            4085         4085              polyA_site
```

4.2.6 ELink (Entrez Links)

The ELink utility allows us to submit a list of UIDs and returns related UIDs in either the same or a related database. Additionally it can return a set of URLs to related resources, including from external providers. The approach offers advanced functionality in connecting information across databases.

4.2.6.1 ELink Main Parameters

In addition to the optional "retmode" parameter (retrieval) which we set to "json" for convenience, there are a number of required parameters, first the "id" and "dbfrom":

- "id": a list of UIDs provided in a string list, separated by commas.
- "dbfrom": which is the database that the provided UIDs came from.

If the History Server will be used, the above two parameters should not be used. Instead you can retrieve UIDs from the History Server by specifying the known/saved "query_key" parameter identifying the query location (integer) and the corresponding "WebEnv" parameter identifying the Web environment the query is from (see the previous sections for more details).

Additional parameters should include:

- "db": is the database we want to extract linked UIDs from. The parameter does not need to be provided if similar records from the same originating database or URL links are required.

4.2 NCBI Entrez Programming Utilities

- `"cmd"`: is the command mode to use. The parameter can take values as listed in the paragraphs below:

1. `"neighbor"`: this is the default value and will return neighboring/related UIDs. As an example we see if we can connect some gene UIDs to proteins.

   In[68]:= neighborGeneProtein=**URLExecute**[
       "https://eutils.ncbi.nlm.nih.gov/entrez/eutils/elink.fcgi",
           {"RequestMethod"->"POST",
           "retmode"->"json",
           "dbfrom"-> "gene",
           "db"-> "protein",
           "id"-> "4790,7157",
           "cmd"-> "neighbor"},"RawJSON"];

   There are 2 outer keys:

   In[69]:= **Query**[**Keys**]@neighborGeneProtein
   Out[69]= {header, linksets}

   We are interested in the `"linksets"`, which also have 3 keys:

   In[70]:= **Query**[**All**, **Keys**]@neighborGeneProtein
   Out[70]= <|header→{type, version},
           linksets→{{dbfrom, ids, linksetdbs}}|>

   Some of the information is verification of the input, from `"db"` `"gene"` and the two ids:

   In[71]:= **Query**["linksets",1,{"dbfrom","ids"}]@
           neighborGeneProtein
   Out[71]= <|dbfrom→gene, ids→{4790,7157}|>

   There are two linked sets:

   In[72]:= **Query**["linksets",**All**,"linksetdbs",**Keys**]@
           neighborGeneProtein
   Out[72]= {{{dbto, linkname, links},{dbto, linkname, links}}}

   We can see where these are to:

   In[73]:= **Query**["linksets",**All**,
           "linksetdbs",**All**,{"dbto","linkname"}]@
           neighborGeneProtein
   Out[73]= {{<|dbto→protein, linkname→gene_protein|>,
           <|dbto→protein, linkname→gene_protein_refseq|>}}

   These `"dbname_dbnames"` specify the computational neighbor info (a full list is maintained at:

   In[74]:= **SystemOpen**["https://eutils.ncbi.nlm.nih.gov/entrez/query/
       static/entrezlinks.html"]

   The links are contained in the `"links"` key:

   In[75]:= (**Query**["linksets",**All**,"linksetdbs",**All**,"links"]
       @neighborGeneProtein);

2. `"neighbor_score"`: which will return similar or related UIDs (from the same db) and a similarity score for each

   In[76]:= neighborScoreGeneProtein=**URLExecute**[
      "https://eutils.ncbi.nlm.nih.gov/entrez/eutils/elink.fcgi",
      {"RequestMethod"–>"POST",
       "retmode"–>"json",
       "dbfrom"–> "gene",
       "db"–> "gene",
       "id"–> "4790,7157",
       "cmd"–> "neighbor_score"},"RawJSON"];

   As before information is embedded within lists:

   In[77]:= **Query**["linksets",**All**,"linksetdbs",**Keys**]@
            neighborScoreGeneProtein
   Out[77]= {{{dbto,linkname,links},
             {dbto,linkname,links}}}

   We extract the first entry in the first list:

   In[78]:= **Query**["linksets",**All**,"linksetdbs",1,**All**]@
            neighborScoreGeneProtein//**Short**
   Out[78]//**Short**= {<|dbto→gene,<<1>>,
             links→{<|<<1>>|>,<<59>>,<|<<1>>|>}|>}

   We see the similarity score now associated with each id, and the linkname explaining the similarity.

3. `"neighbor_history"`: can store the result to the History Server, and provides the Web environment query values to use them as indexes for other searches using the History Server.

   In[79]:= neighborScoreGeneProteinHistoryServer=**URLExecute**[
      "https://eutils.ncbi.nlm.nih.gov/entrez/eutils/elink.fcgi",
      {"RequestMethod"–>"POST",
       "retmode"–>"json",
       "dbfrom"–> "gene",
       "db"–> "gene",
       "id"–> "4790,7157",
       "cmd"–> "neighbor_history"},"RawJSON"]
   Out[79]= <|header→<|type→elink, version→0.3|>,
             linksets→{<|dbfrom→gene,
               ids→{4790,7157},linksetdbhistories→
                 {<|dbto→gene,linkname→gene_gene_h3k4me3,
                   querykey→1|>,<|dbto→gene,
                   linkname→gene_gene_neighbors,
                   querykey→2|>},webenv→
                 NCID_1_44323195_130.14.18.34_9001_1512335353
                 _2031828891_0MetA0_S_MegaStore_F_1|>
             }|>

4. `"acheck"`: can return all links for the input UIDs to the `"db"` provided (or to all databases if the `"db"` was not provided)

4.2 NCBI Entrez Programming Utilities

```
In[80]:= allLinks=URLExecute[
    "https://eutils.ncbi.nlm.nih.gov/entrez/eutils/elink.fcgi",
        {"RequestMethod"->"POST",
         "retmode"->"json",
         "dbfrom"-> "gene",
         "id"-> "4790,7157",
         "cmd"-> "acheck"},"RawJSON"];
```

There are 2 keys, one corresponding to the input database, `"dbfrom"`, and the `"idchecklist"`:

```
In[81]:= Query["linksets",All,Keys]@allLinks
Out[81]= {{dbfrom,idchecklist}}
```

The checklist has a set `"idlinksets"`:

```
In[82]:= Query["linksets",All,"idchecklist",Keys]@
           allLinks
Out[82]= {{idlinksets}}
```

There are two sets within `"idchecklist"`, one for each input id :

```
In[83]:= Query["linksets",All,"idchecklist",
           "idlinksets",Keys]@allLinks
Out[83]= {{{id,linkinfos},{id,linkinfos}}}

In[84]:= Query["linksets",All,"idchecklist",
           "idlinksets",All,{"id"}]@allLinks
Out[84]= {{<|id→4790|>,<|id→7157|>}}
```

We can look at the first entry, so for id 4790, we have link name and tags, and we can look at the first 3 as an example:

```
In[85]:= Query["linksets",All,"idchecklist",
           1,1,"linkinfos",1;;3]@allLinks
Out[85]= {{<|dbto→bioconcepts,linkname→
             gene_bioconcepts,menutag→BioConcepts,
             htmltag→BioConcepts,priority→128|>,
           <|dbto→bioproject,linkname→gene_bioproject,
             menutag→BioProjects,
             htmltag→BioProjects,priority→128|>,
           <|dbto→biosystems,linkname→gene_biosystems,
             menutag→BioSystem Links,
             htmltag→BioSystems,priority→128|>}}
```

5. `"ncheck"`: can check if there are any links between input UIDs and other UIDs in the database.

```
In[86]:= nCheck=URLExecute[
    "https://eutils.ncbi.nlm.nih.gov/entrez/eutils/elink.fcgi",
        {"RequestMethod"->"POST",
         "retmode"->"json",
         "dbfrom"-> "gene",
         "id"-> "4790,7157",
         "cmd"-> "ncheck"},"RawJSON"]
Out[86]= <|header→<|type→elink,version→0.3|>,
```

```
linksets→{<|dbfrom→gene,idchecklist→
    <|ids→{<|value→4790,hasneighbor→N|>,
        <|value→7157,hasneighbor→N|>}|>|>}|>
```

For each id we can get information if it has neighbors or not:

In[87]:= Query["linksets",**All**,"idchecklist","ids"]@
nCheck
Out[87]= {{<|value—>4790,hasneighbor—>N|>,
<|value—>7157,hasneighbor—>N|>}}

6. "lcheck": can check if external links exist for the input UIDs - these are from so-called *LinkOut* providers:

In[88]:= lCheck=**URLExecute**[
   "https://eutils.ncbi.nlm.nih.gov/entrez/eutils/elink.fcgi",
       {"RequestMethod"—>"POST",
        "retmode"—>"json",
        "dbfrom"—> "gene",
        "id"—> "4790,7157",
        "cmd"—> "lcheck"},"RawJSON"]
Out[88]= <|header→ <|type→elink, version→0.3|>,
    linksets→{<|dbfrom→gene,idchecklist→
        <|ids→{<|value→4790,haslinkout→Y|>,
            <|value→7157,haslinkout→Y|>}|>|>}|>

For each id we can get information if it has linkouts or not:

In[89]:= Query["linksets",**All**,"idchecklist","ids"]@lCheck
Out[89]= {{<|value→4790,haslinkout→Y|>,
<|value→7157,haslinkout→Y|>}}

7. "llinks": can check if external links exist and return the URLs (non-library providers)

In[90]:= lLinks=**URLExecute**[
   "https://eutils.ncbi.nlm.nih.gov/entrez/eutils/elink.fcgi",
       {"RequestMethod"—>"POST",
        "retmode"—>"json",
        "dbfrom"—> "gene",
        "id"—> "4790",
        "cmd"—> "llinks"},"RawJSON"];

There is a list of URLs that is returned:

In[91]:= Query["linksets",**Keys**]@lLinks
Out[91]= {{dbfrom,idurllist}}

This contains URLs in the "objurls" that can be processed as you require:

In[92]:= Query["linksets",1,"idurllist"]@lLinks//**Short**
Out[92]//Short= {<|id→4790,
   objurls→{<|<<1>>|>,<<54>>,<|<<1>>|>}|>}

Here is an example:

4.2 NCBI Entrez Programming Utilities

In[93]:= Query["linksets",1,"idurllist",1,"objurls",4]@
lLinks

Out[93]= <|url→ <|value→
http://natural.salk.edu/cgi-bin/creb?DB=human
&TABLE=Human_CRE_prediction&
FIELD=LocusLink&QUERY=4790|>,
iconurl→ <|lng→EN, value→
//www.ncbi.nlm.nih.gov/corehtml/query/egifs/
http:--natural.salk.edu-CREB-Icon.gif
|>,
subjecttypes→{gene/protein/disease-specific},
categories→{Molecular Biology Databases},
attributes→{free resource},
provider→ <|name→CREB Target Gene Database,
nameabbr→CREBDB, id→5380, url→ <|lng→EN,
value→http://natural.salk.edu/CREB/|>,
iconurl→ <|lng→EN, value→
http://natural.salk.edu/CREB/Icon.gif|>|>|>

8. `"llinkslib"`: can check if external links exist for the input UIDs, and returns URLs (all providers). For this XML return mode works fairly reliably:

In[94]:= lLinkslib=**URLExecute**[
    "https://eutils.ncbi.nlm.nih.gov/entrez/eutils/elink.fcgi",
        {"RequestMethod"→"POST",
        "retmode"→"XML",
        "dbfrom"→ "gene",
        "id"→ "4790",
        "cmd"→ "llinkslib" },"text"];
        (*output is text that needs string parsing*)

9. `"prlinks"`: can return the primary provided URL for input UIDs.

In[95]:= prLinks=**URLExecute**[
    "https://eutils.ncbi.nlm.nih.gov/entrez/eutils/elink.fcgi",
        {"RequestMethod"→"POST",
        "retmode"→"JSON",
        "dbfrom"→ "pubmed",
        "id"→ "27883025",
        "cmd"→ "prlinks" },"RawJSON"]

Out[95]= <|header→ <|type→elink, version→0.3|>,
linksets→{<|dbfrom→pubmed, idurllist→
{<|id→27883025, objurls→{<|url→ <|value→
http://dx.doi.org/10.1038srep37237
|>,
iconurl→ <|lng→EN, value→
//www.ncbi.nlm.nih.gov/corehtml/
query/egifs/http:--www.
nature.com-images-lo_npg.
gif|>,
subjecttypes→{publishers/providers},
categories→{**Full Text** Sources},
attributes→
{free resource, full-text online,
publisher of information in url},

```
                    provider→ <|name→
                        Nature Publishing Group,
                        nameabbr→NPG, id→3094,
                        url→ <|lng→EN, value→
                        http://www.nature.com|>|>|>}|>}|>
        }|>
```

4.2.6.2 Other ELink Parameters: Limiting Output

Certain parameters can be used to limit the output:

- "linkname": for cmd values of neighbor or neighbor_history will return the specified linkame (dbfrom_db_subset).
- "term": takes as value a query that can limit the input
- "holding": for cmd values of llinks or llinkslib only, will limit URLs returned to those of the LinkOut provider specified (see nameabbr key in outputs).

We can limit the results returned by date as with e-search. The availability will depend on the database, as discussed above - the parameters are the same – see the previous sections for usage.

- "datetype" parameter, which can take values:
  - "edat" corresponding to Entrez Date
  - "pdat" corresponding to publication date
  - "mhdat" corresponding to the date a citation was indexed with MeSH.
- "reldate" parameter takes as value n, and returns results limited to the last n days. The search will use the field specified by "datetype".
- "mindate" and "maxdate" parameters, which must be used together can specify a range of dates taking the start and end date respectively. The date is formatted as YYYY/MM/DD or YYYY/MM or YYYY

4.2.7 EGQuery

EGQuery returns the number of records in All databases. The required parameter is "term" and it is a text.

In[96]:= autismEGQuery=**URLExecute**[
   "https://eutils.ncbi.nlm.nih.gov/entrez/eutils/egquery.fcgi",
      {"RequestMethod"–>"POST","term"–> "autism"}];

The result is an XML format that can be parsed as needed using string patterns:

In[97]:= **StringTake**[autismEGQuery,1;;450]
Out[97]= <?xml version="1.0" encoding="UTF-8"?>
   <!DOCTYPE Result PUBLIC "-//NLM//DTD

4.2 NCBI Entrez Programming Utilities

```
eSearchResult, January 2004//EN"
"https://www.ncbi.nlm.nih.gov/entrez/query/DTD/
egquery.dtd">
<Result>

    <Term>autism</Term>

    <eGQueryResult>

    <ResultItem><DbName>pubmed</DbName><MenuName>
    PubMed</MenuName><Count>40052</Count><Status>Ok
    </Status></ResultItem>

    <ResultItem><DbName>pmc</DbName><MenuName>
    PubMed
    Central</MenuName><Count>50534</Count>
```

In the example XML above we see how the DbName is given, e.g. pubmed, and a count for the items in response to our search, 40052. Similarly for pmc, the count is 50534.

4.2.8 ESpell

The ESpell utility provides spelling suggestions. The parameters required are `"db"` the database to search, and the query value for parameter `"term"`. The default output is XML:

```
In[98]:= URLExecute[
    "https://eutils.ncbi.nlm.nih.gov/entrez/eutils/espell.fcgi",
        {"RequestMethod"->"POST",
         "db"-> "pubmed",
         "term"-> "iinflueenza"},"Plaintext"]
Out[98]= <?xml version="1.0"?>
        <!DOCTYPE eSpellResult PUBLIC "-//NLM//DTD
        eSpellResult, 23 November 2004//EN"
        "http://www.ncbi.nlm.nih.gov/entrez/query/DTD/
        eSpell.dtd">
        <eSpellResult>
            <Database>pubmed</Database>
            <Query>iinflueenza</Query>
            <CorrectedQuery>influenza</CorrectedQuery>

            <SpelledQuery><Original></Original><Replaced>
            influenza</Replaced></SpelledQuery>
                <ERROR/>
            </eSpellResult>
```

As we can see from the strings above, our Query for iinflueenza is corrected to influenza.

4.2.9 ECitMatch

The final E-utility we are considering is ECitMatch. This can retrieve PubMed IDs (PMIDs) from citation strings. The required parameter is `"db"` and actually only supports `"pubmed"` as its values, while the `"rettype"` parameter can only retrieve XML. The input citation format is provided using `"bdata"` as a parameter. The citation string has the format:

journal_title|year|volume|first_page|author_name|your_key|

If you want to enter more than one citation string you must enter them line by line. Finally, the last format entry in the citation string, `"your_key"` is a label that the user provides (for example could be a citation key in your reference software manager). We can look at an example:

```
In[99]:= bdataExample=
    "Nature|1970|227|561|Crick F|citation1|
    j mol biol|215|1990|Altschul SF|citation2|"

In[100]:= URLExecute[
    "https://eutils.ncbi.nlm.nih.gov/entrez/eutils/ecitmatch.cgi",
        {"RequestMethod"->"POST",
        "db"-> "pubmed",
        "retmode"-> "xml",
        "bdata"-> bdataExample}]
Out[100]= Nature|1970|227|561|Crick F|citation1|4913914
    j mol biol|215|1990|Altschul SF|||NOT_FOUND
```

As we can see, one of our citation searches works, and we got a match, whereas the second one was not found. Note also how the user labels are also returned with all the information.

4.2.10 API Keys and Usage Guidelines

You should avoid posting more than 3 requests per second without an API key. An API key can be obtained though the Settings page of your NCBI account (if you do not have one, you can check out `"www.ncbi.nlm.nih.gov/account/"`). The API can then be included as a parameter `"api_key"` -> `"API KEY"`

If you use the E-utilities or create software that does utilize them please check the NCBI's Disclaimer and Copyright notice,

`"https:// www.ncbi.nlm.nih.gov/About/disclaimer.html"`

and make sure it is evident to users of your product. Please check the website for additional information:

```
In[101]:= SystemOpen[
    "https://www.ncbi.nlm.nih.gov/books/NBK25497/#_chapter2_\
    Usage_Guidelines_and_Requiremen_"]
```

4.2 NCBI Entrez Programming Utilities 163

Finally, as a reminder, you should provide your e-mail with every search, using an additional `"email"` parameter, with a string value of your e-mail address, with no spaces.

4.3 UCSC Genome Browser

4.3.1 Sequence Retrieval

The UCSC Genome Browser [3, 4] houses one of the most comprehensive sequence information databases available. Here we see how we can retrieve sequences from the server, and then how we can write a simple function example to parse the information retrieved. We discuss the details in the first part, but you can skip to the end product in the second part of the subsection if you wish.

4.3.2 Sequence Retrieval Details

Specific sequence segments can be obtained through a simple URL execution:

```
In[102]:= ucscExampleSequence=Import[
"http://genome.ucsc.edu/cgi-bin/das/hg38/dna?segment=chr2:10000,10100"]
Out[102]= <?xml version="1.0" standalone="no"?>
         <!DOCTYPE DASDNA SYSTEM
           "http://www.biodas.org/dtd/dasdna.dtd">
         <DASDNA>
         <SEQUENCE id="chr2" start="10000"
           stop="10100" version="1.00">
         <DNA length="101">
         ncgtatcccacacaccacacccacacaccacacccacacacacccacacc
         cacacccacacacaccacacccacacaccacacccacacccacacaccac
         a
         </DNA>
         </SEQUENCE>
         </DASDNA>
```

The URL utilizes the UCSC DAS [1, 2], and the query is of the form:

```
http://genome.ucsc.edu/cgi-bin/das/[db_name]/entry_points
http://genome.ucsc.edu/cgi-bin/das/[db_name]/types
```

In our example above, we queried the hg38 annotation, and retrieve dna segment from chromosome 2, positions 10000–10100. Per the UCSC Genome Browser website, this approach should be restricted to a maximum of 1 hit per 15 s and less than 5000 per day, so please do not use it for mass downloading. See also

`"http://genome.ucsc.edu/FAQ/FAQdownloads.html#download23"`

We can extract the sequence information using a string pattern:

In[103]:= sequenceInformation=
           StringCases[ucscExampleSequence,
             "<SEQUENCE "~~**Shortest**[x:__]~~">"-> x]
Out[103]= {id="chr2" start="10000"
           stop="10100" version="1.00"}

Above we have extracted the shortest named string pattern x (using **Shortest**), that follows the string "SEQUENCE " up until the closing ">", and then use a rule to return x instead of the whole matching string.

We can easily parse this further if we wanted. We split the string wherever there are spaces:

In[104]:= splitSequenceInformation=
           StringSplit[sequenceInformation [[1]]]
Out[104]= {id="chr2",start="10000",
           stop="10100",version="1.00"}

We then create a rule by splitting the substrings in the list on the "=" sign, and use a pure function to assign a rule from the first sublist element to the second, and finally **Apply** the **Head** to be an **Association** instead of a **List**:

In[105]:= **Association** @@
              ((#[[1]]-> #[[2]]& @ StringSplit[#,"="]) &/@
              splitSequenceInformation)
Out[105]= <|id->"chr2",start->"10000",
           stop->"10100",version->"1.00"|>

Similarly we can parse the DNA part of the sequence, to get the shortest sequence enclosed within the "<DNA " and closest ">" strings:

In[106]:= dnaInformation=
           StringCases[ucscExampleSequence,
             "<DNA "~~**Shortest**[x:__]~~">"-> x]
Out[106]= {length="101"}

We can again make this an **Association**:

In[107]:= **Association** @@
              ((#[[1]]-> #[[2]]& @ **StringSplit**[#,"="])&/@
              dnaInformation)
Out[107]= <|length->"101"|>

We can then get the actual sequence by finding what is enclosed between the <DNA ...> and </DNA> elements:

In[108]:= dnaSequence=**StringCases**[ucscExampleSequence,
             "<DNA "~~__~~">"~~dna:__~~"</DNA>"-> dna]
Out[108]= {
           ncgtatcccacacaccacacccacacaccacacccacacacacccacacc
           cacacccacacacaccacacccacacaccacacccacacccacacaccac
           a
           }

We notice that the string is segmented in lines, so we **StringSplit** to get a list and then **StringJoin** the segments to get a continuous line:

4.3 UCSC Genome Browser   165

```
In[109]:= StringJoin[StringSplit[dnaSequence[[1]]]]
Out[109]= ncgtatcccacacaccacacccacacaccacacccacacacacccaca
         cccacacccacacacaccacacccacacaccacacccacacccac
         acaccaca
```

Now let us put this all together in a single pattern match using **StringCases**:

```
In[110]:= StringCases[ucscExampleSequence,
          "<SEQUENCE "~~x:__~~">"~~__~~"<DNA "~~
              y:__~~">"~~seq:__~~"</DNA>"->
          {x,y,seq}]
Out[110]= {{id="chr2" start="10000" stop="10100"
              version="1.00",length="101",
           ncgtatcccacacaccacacccacacaccacacccacacacacccacacc
           cacacccacacacaccacacccacacaccacacccacacccacacaccac
           a
           }}
```

We can compile a final function to generalize this approach in the next section.

4.3.2.1 Sequence Retrieval Function

As an aside, note that we could of course write all the code in the previous section in one line (just for reference and for a different style). Our function takes as input the information from the UCSC DAS server::

```
In[111]:= ucscDASSequenceParser[input_]:=
          Module[{str=input,stringsAll,info1,
              dnaLength,sequenceInfo},
              stringsAll=
              Association@
                Flatten@StringCases[input,
                  "<SEQUENCE "~~x__~~">"~~__~~
                  "<DNA "~~y__~~">"~~seq:__~~
                  "</DNA>":>
                  {#[[1]]-> #[[2]]&/@
                    (StringSplit[#,"="]&/@
                     StringSplit[x]),
                    (#[[1]]-> #[[2]]&@
                      StringSplit[#,"="])&@y,
                    "sequence"->StringJoin[
                     StringSplit[seq]]}];
              Return[stringsAll]]
```

Note that in the function we extract the string patterns as we had done in the previous section. We instead use a delayed rule :>, which allows us to perform action on the extracted named pattern.

We can now use our simple parser directly, say to retrieve from hg38 the dna sequence from chromosome 1 coordinates 10200 to 10278:

```
In[112]:= ucscExampleSequence2=
          ucscDASSequenceParser[
```

Import[
"http://genome.ucsc.edu/cgi-bin/das/hg38/dna?segment=chr1:10200,10278
"]]

Out[112]= <|id->"chr1",start->"10200",stop->"10278",
version->"1.00",length->"79",sequence->
accctaaccctaaccctaaccctaaccctaaccctaaccctaac
cctaaaccctaaaccctaaccctaaccctaaccc|>

Additionally, we could automate the import as well:

In[113]:= UCSCSequence[genomeVersion_,chromosome_,
start_,end_]:=
Module[{ gen=genomeVersion,chr=chromosome,
st=start,en=end,queryOut,stringsAll,
info1,dnaLength,sequenceInfo },
queryOut=
Import["http://genome.ucsc.edu/cgi-bin/das/"<>
gen<>"/dna?segment="<>chr<>":"<>
st<>","<>en];
stringsAll=
Association@
Flatten@StringCases[queryOut,
"<SEQUENCE "~~x__~~">"~~__~~
"<DNA "~~y__~~">"~~seq:__~~
"</DNA>":>
{#[[1]]-> #[[2]]&/@
(**StringSplit**[#,"="]&/@
StringSplit[x]),
(#[[1]]->#[[2]]&@
StringSplit[#,"="])&@y,
"sequence"->StringJoin@
StringSplit@seq }];
Return[stringsAll]]

Let us now use our function, for chromosome 17, from 11,1234 to 11,2112

In[114]:= UCSCSequence["hg38","chr17","111234","112112"]
Out[114]= <|id->"chr17",start->"111234",stop->"112112",
version->"1.00",length->"879",sequence->
acaaaggctgctgtgacaaaaaagcagggaaagggaatttttttt
ttaaaagcaaacaacaacaacaaaaaccccacagaaaagc
aaacaacaaacaaacaaaaaacagaggaagaagttgaaca
ccccgggctgtgactacttccaggaaggggctacaagagg
cagttggaaattctatttgttttgcaactgtgggttttcc
ggcctgcttcctttctaaagtatattactctgcttttggt
tcatgaagttatccatttctgttttctggaacagctatgt
attttctttatctatcatctatctatctatttaccatcta
tcttttctacctttgctatcaagagcttgtgtcaagcag
gatagaattccagtgtatgttcactctaccgtttaaaaca
agagctcttgtgggcattctccatcacatcataaacctga
gctttctaaaacagagtgtggcaaactaccatgcatggac
catgtctgacacagtctgcgtttgtaagtaaagttgtaat
gggacacaaccaatacatgtgttacataatgtctctggct
actttcatggtataatggaagagctgagtcattgagagag
agaccatatggcttggaaaacttaaaatatttaacattta

4.3 UCSC Genome Browser   167

```
                  gccccttgcagaaaatacttgctgactcttgttttaaaag
                  atctctgtttagaatgctacctattgcgttctggatagaa
                  tcacaactctttaccacaattgacacagcttcagccctgc
                  ttctatatccagcctcatctatttctgctcctcctcctta
                  ttttccttctggccatgctgatggattgtcagcttcccag
                  atgtgcaagaatctctcctcccttcccaacattc|>
```

Let us get the same information from hg19:

In[115]:= **UCSCSequence**["hg19","chr17","111234","112112"]
Out[115]= <|id->"chr17",start->"111234",stop->"112112",
           version->"1.00",length->"879",sequence->
```
              tctggatcagccagcccccttgatcttggacttccctgtccccaga
                  gctctgagaaacaaattcccactgtttacaccactcagtc
                  tatggtatgttgttatggtggctcaagttgactaccatgg
                  aaaagataggccagctataaggaaaccagacccagaatcg
                  cattagacttatcactggtaacagtggccgcccaaagact
                  ctggatctagaattctatcagccaaactatcaacacaatg
                  ggagggcagaacacagatgttttttaaactcataaggacgc
                  tgagtttcccttacacacactcttttttaagaacttaggac
                  gggttccagaaaaacacaagtgaaaaacaagaaagggaaa
                  acacaggatcctggaaacagtctccaactcaaaaggtgag
                  ctgagggaagttcccggaaggcagcgctcaggaactact
                  ctggggtctggagggcgtctcacaggaaaacagaaggctc
                  cacacgatagatggtaaaataaaaaagctgcataaaagag
                  gatgtaaagtttcattattctgttgtcaataagaaacaca
                  gtcaataaaaactccaggaaaaacgaaaagttacacaaga
                  acatcatggtccaaatattgggtaaattaaaatatgacag
                  ggatgaagaaattcagagttaaaaaggaatgtttgctcag
                  aatggtgggaacattctccttggggtggtctagggggtcaa
                  ggtcgttataaaaatgtgatcctcacatacgtttgagcct
                  aagacgtttcatgtttacagttttttttttagatggtgtc
                  tcgctctgtcacccaggctggagtgcagcggcacaatctc
                  ggctcactgcaacctccgcctcccgggttcaagc|>
```

4.3.2.2 Tables Programmatic Access - MySQL

We can also access the UCSC Genome Browser database tables directly. This involves performing MySQL queries directly. Some of the functionality is already available in MathIOmica through the UCSCBrowserSQL function.

Please note that SQL functionality might be problematic in connecting to a database, in case of firewalls, or other connectivity issues that may prevent connection to the UCSC Genome Browser.

We assume here some familiarity with MySQL commands. For table naming please consult the UCSC Genome Browser tables directly.

We need to load the DatabaseLink package first.

In[116]:= **Needs**["DatabaseLink`"]

Next, we open a SQL connection, Fig. 4.1. We use **SQLConnections**[] to see if any connections are open.

In[117]:= `ucscDatabase = OpenSQLConnection[`
`    JDBC["MySQL(Connector/J)",`
`    "genome-mysql.cse.ucsc.edu"],`
`    "Username" -> "genomep",`
`    "Password" -> "password"];`

In[118]:= `SQLConnections[]`

Out[118]= {SQLConnection[⊞ ⇌ Name: **None** ID: 1   ]}
                                Status: **Open** Catalog:

Fig. 4.1 Out[118]= Opening a connection to the UCSC MySQL databases

Now we can use the connection to probe the databases.

In[119]:= ucscDatabases=**SQLExecute**[ucscDatabase,
        "SHOW DATABASES"];
In[120]:= **Length**[ucscDatabases]
Out[120]= 229

Let us look at the first few entries:

In[121]:= ucscDatabases[[1;;10]]
Out[121]= {{information_schema},{ailMel1},
        {allMis1},{anoCar1},{anoCar2},{anoGam1},
        {apiMel1},{apiMel2},{aplCal1},{aptMan1}}

Now let us use specifically the hg38 database:

In[122]:= **SQLExecute**[ucscDatabase,"USE hg38"]
Out[122]= 0

We can get the hg38 Tables:

In[123]:= hg38Tables=**SQLExecute**[ucscDatabase,"SHOW TABLES"];

Let us look at the first few entries:

In[124]:= hg38Tables[[1;;10]]
Out[124]= {{affyGnf1h},{affyU133},{affyU95},
        {all_est},{all_mrna},{all_sts_primer},
        {all_sts_seq},{altLocations},
        {altSeqLiftOverPsl},{altSeqLiftOverPslP11}}

There are 849 tables as of this writing:

In[125]:= hg38Tables//**Short**
Out[125]//**Short**= {{affyGnf1h},<<847>>,{xenoRefSeqAli}}

For this example we will look at the kgXref Table, that gives us information linking together a known Gene ID (kgID) and a gene alias. Let us see what the schema looks like:

4.3 UCSC Genome Browser

In[126]:= **SQLExecute**[ucscDatabase,"DESCRIBE kgXref"]
Out[126]= {{kgID,varchar(255),NO,MUL,**Null**,},
  {mRNA,varchar(255),NO,MUL,**Null**,},
  {spID,varchar(255),NO,MUL,**Null**,},
  {spDisplayID,varchar(255),NO,MUL,**Null**,},
  {geneSymbol,varchar(255),NO,MUL,**Null**,},
  {refseq,varchar(255),NO,MUL,**Null**,},
  {protAcc,varchar(255),NO,MUL,**Null**,},
  {description,longblob,NO,,**Null**,},
  {rfamAcc,varchar(255),NO,MUL,**Null**,},
  {tRnaName,varchar(255),NO,MUL,**Null**,}}

We can select all the columns where the mRNA matches the accession NM_003998:

In[127]:= **SQLExecute**[ucscDatabase,
  "SELECT * FROM kgXref WHERE mRNA='NM_003998'"]
Out[127]= {{uc011cep.2,NM_003998,P19838,
  NFKB1_HUMAN,NFKB1,NM_003998,NM_003998,
  SQLBinary[{72,111,109,111,32,115,97,112,
    105,101,110,115,32,110,117,99,108,
    101,97,114,32,102,97,99,116,111,114,
    32,111,102,32,107,97,112,112,97,32,
    108,105,103,104,116,32,112,111,108,
    121,112,101,112,116,105,100,101,32,
    103,101,110,101,32,101,110,104,97,
    110,99,101,114,32,105,110,32,66,45,
    99,101,108,108,115,32,49,32,40,78,
    70,75,66,49,41,44,32,116,114,97,
    110,115,99,114,105,112,116,32,118,
    97,114,105,97,110,116,32,49,44,32,
    109,82,78,65,46,32,40,102,114,111,
    109,32,82,101,102,83,101,113,32,78,7
    7,95,48,48,51,57,57,56,41}],,}}

You will notice that the data is in binary form. We can convert to characters using **FromCharacterCode**:

In[128]:= FromCharacterCode@@
  SQLBinary[{72,111,109,111,32,115,97,112,
    105,101,110,115,32,110,117,99,108,
    101,97,114,32,102,97,99,116,111,114,
    32,111,102,32,107,97,112,112,97,32,
    108,105,103,104,116,32,112,111,108,
    121,112,101,112,116,105,100,101,32,
    103,101,110,101,32,101,110,104,97,
    110,99,101,114,32,105,110,32,66,45,
    99,101,108,108,115,32,49,32,40,78,
    70,75,66,49,41,44,32,116,114,97,
    110,115,99,114,105,112,116,32,118,
    97,114,105,97,110,116,32,49,44,32,
    109,82,78,65,46,32,40,102,114,111,
    109,32,82,101,102,83,101,113,32,78,7
    7,95,48,48,51,57,57,56,41}]
Out[128]= Homo sapiens nuclear factor of

kappa light polypeptide gene enhancer
in B-cells 1 (NFKB1), transcript
variant 1, mRNA. (from RefSeq NM_003998)

Finally we can close our connection to the database:

In[129]:= CloseSQLConnection[ucscDatabase];

In practice if you plan to use the MySQL capabilities for the UCSC Genome Browser, you need some familiarity also with the table schemas. The Table Browser is a great starting point to get information about the schemas. Simply make your selections and press the describe table schema button at:

In[130]:= **SystemOpen**["https://genome.ucsc.edu/cgi-bin/hgTables?"]
 (*Accessed 11/23/2017*)

References

1. Dowell, R.D., Stein, L., Eddy, S.: Distributed sequence annotation system (das)
2. Dowell, R.D., Jokerst, R.M., Day, A., Eddy, S.R., Stein, L.: The distributed annotation system. BMC Bioinform. **2**, 7–7 (2001)
3. Karolchik, D., Hinrichs, A.S., Kent, W.J.: The UCSC genome browser. Curr. Protoc. Bioinform. **Chapter 1**, Unit1 4 (2012)
4. Kuhn, R.M., Haussler, D., Kent, W.J.: The UCSC genome browser and associated tools. Brief Bioinform. **14**(2), 144–161 (2013)
5. National Center for BIotechnology Information (US): Entrez programming utilities help [internet]
6. NCBI Resource Coordinators: Database resources of the national center for biotechnology information. Nucleic Acids Res. **45**(D1), D12–D17 (2017)
7. Sayers, E.: A general introduction to the e-utilities. In: Entrez programming utilities help [internet]

Chapter 5
Genomic Sequence Data and BLAST

5.1 Sequences of Genomic Information

Deoxyribonucleic Acid (DNA) contains the genetic information in living organisms. Based on the central dogma of molecular biology, the genomic sequence information from DNA used through transcription to Ribonucleic Acid (RNA) and translation to proteins, provides the necessary code for systems level signaling in our cells. For a brief review, can look at an example of a short DNA segment by the following PDB (Protein Data Bank) example file [4], Fig. 5.1.

As you can see the DNA is arranged in a helix, with a sugar-phosphate outer backbone and pairs of nucleotides in the inner region. These nucleotides are called adenine (A), guanine (G), thymine (T) and cytosine (C). The two DNA strands comprising a double helix, are complementary, and held together by hydrogen bonds between complementary nucleotide pairs. These pairs are A-T and G-C and vice versa. We can actually get the sequence information from our PDB file, that has many `Elements`:

```
In[2]:= Import["1bna.pdb",{"PDB",{"Elements"}}]
Out[2]= {AdditionalAtoms, AdditionalCoordinates,
    AdditionalCoordinatesList, AdditionalIndex,
    AdditionalResidues, Authors, Comments,
    DepositionDate, Graphics3D, Organism,
    PDBClassification, PDBID, References, ResidueAtoms,
    ResidueChainLabels, ResidueCoordinates,
    ResidueCoordinatesList, ResidueIndex, ResidueRoles,
    Residues, Resolution, SecondaryStructure, Sequence,
    Title, VertexCoordinates, VertexCoordinatesList,
    VertexTypes}
```

For example, we can import the sequences and classification:

Electronic supplementary material The online version of this chapter (https://doi.org/10.1007/978-3-319-72377-8_5) contains supplementary material, which is available to authorized users.

In[1]:= `SetDirectory[NotebookDirectory[]]; Import["1bna.pdb"]`

Out[1]=

Fig. 5.1 Out[1]= Example DNA sequence from a Protein Data Bank (PDB) file [4]

In[3]:= `Import["1bna.pdb",`
`{"PDB",{"Sequence","PDBClassification"}}]`
Out[3]= `{{CGCGAATTCGCG,CGCGAATTCGCG}, DNA}`

The nucleotides belong to two families: A and G are purines, and C and Ts are pyrimidines. We can look up these DNA bases from a `"Chemical"` entity in the Wolfram Language [7, 8].

In[4]:= `dnaBases=EntityClass["Chemical","DNABases"];`
 `(* or type in: ctrl= DNA Bases *)`

We can extract values from the **Entity**. for example, we can get the structure. We return the results in an `"EntityAssociation"` which is easy to both visualize and for information access, Fig. 5.2.

Notice how the purines have a two-ring structure, and the pyrimidines have a single ring structure. The sequence of nucleotides in the two complementary strands is represented by a string of letters, an alphabet of {A,C,T,G}. As the DNA strands have a directionality (corresponding to the backbone pentose ring bond connections), this is reflected by convention in writing the sequence from the 5' to the 3' direction.

5.1 Sequences of Genomic Information

```
In[5]:= EntityValue[dnaBases, "BlackStructureDiagram",
        "EntityAssociation"]
```

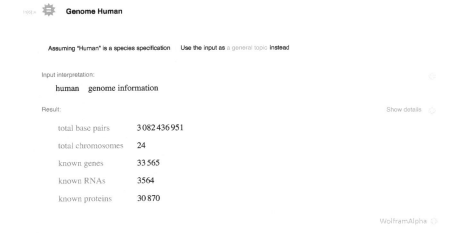

Fig. 5.2 Out[5]= Association of **Entity** to value, for adenine, guanine, thymine and cytosine

Fig. 5.3 Out[6]= We can get genome information from Wolfram|Alpha

The complementary strand is actually antiparallel and oriented in exactly the opposite way. The entirety of the sequence of DNA is called the Genome. In humans, it is approximately 3.1 Gbp (giga base pairs), as can be seen from a Wolfram|Alpha search, Fig. 5.3.

Sequences are extensively analyzed, compared, shared and stored. We will see how to parse certain sequence files and how to do alignments of sequences in the following sections.

5.2 Parsing Files

We have seen in the previous chapter how we can obtain sequencing data from online databases. Standard traditional files in sharing sequence information include FASTA and GenBank [3] file formats. The Wolfram Language has parsers for both formats, which we review in this section.

5.2.1 FASTA Sequences

The FASTA file format is a plain text file format where identifiers and sequences are listed in order, with identifier lines starting with ">" and sequences listed right below it in multiple lines. The files are typically given the extension .fasta or .fa and are often zipped (gzip) so they are more compact for file transfer (.gz). The Wolfram Language can recognize zipped FASTA sequences and parses them automatically.

We can import various elements from the sequence. Let us look at an example that contains three sequences. The file we need is called "example_sequence.fasta":

```
In[7]:= SetDirectory[NotebookDirectory[]];
In[8]:= fastaExample=Import["example_sequence.fasta"];
```

If we look at the result it is a list only, without the identifier:

```
In[9]:= fastaExample // Short
Out[9]//Short= {GTGAGAGAGTGAGCGAGACA...TTGATGACCTCAAAAAAAAA,
                ..., ... }
```

We can import a list of which "Elements" can be obtained from the file:

```
In[10]:= fastaExampleElements=Import["example_sequence.fasta",
         "Elements"]
Out[10]= {Accession,Data,Description,GenBankID,Header,
          LabeledData,Length,Plaintext,Sequence}
```

For example we can get the "Accession":

```
In[11]:= Import["example_sequence.fasta","Accession"]
Out[11]= {NM_001165412.1,NM_003998.3,NM_001319226.1}
```

We can get multiple "Elements" at the same time:

```
In[12]:= Import["example_sequence.fasta",
         {{"Header","GenBankID","Description"}}]
Out[12]= {{gi|259155301|ref|NM_001165412.1| Homo
```

5.2 Parsing Files

```
              sapiens nuclear factor kappa B subunit
                1 (NFKB1), transcript variant 2, mRNA,
              gi|259155300|ref|NM_003998.3| Homo sapiens
                nuclear factor kappa B subunit
                1 (NFKB1), transcript variant 1, mRNA,
              gi|984880772|ref|NM_001319226.1| Homo sapiens
                nuclear factor kappa B subunit
                1 (NFKB1), transcript variant 3, mRNA},
              {259155301,259155300,984880772},
              {Homo sapiens nuclear factor kappa B subunit
                1 (NFKB1), transcript variant 2, mRNA,
              Homo sapiens nuclear factor kappa B subunit
                1 (NFKB1), transcript variant 1, mRNA,
              Homo sapiens nuclear factor kappa B subunit
                1 (NFKB1), transcript variant 3, mRNA}}
```

Perhaps it is most useful to get the `"LabeledData"`, which is a list of rules from headers to sequences:

In[13]:= fastaExampleLabeled=
   Import["example_sequence.fasta","LabeledData"];

We can see that the **Keys** are the headers:

In[14]:= **Keys** @ fastaExampleLabeled
Out[14]= { gi|259155301|ref|NM_001165412.1| Homo
   sapiens nuclear factor kappa B subunit
   1 (NFKB1), transcript variant 2, mRNA,
  gi|259155300|ref|NM_003998.3| Homo sapiens
   nuclear factor kappa B subunit 1
   (NFKB1), transcript variant 1, mRNA,
  gi|984880772|ref|NM_001319226.1| Homo
   sapiens nuclear factor kappa B subunit
   1 (NFKB1), transcript variant 3, mRNA}

And the **Values** are the corresponding sequences:

In[15]:= **Values** @ fastaExampleLabeled // **Short**
Out[15]//Short= {GTGAGAGAGTGAGCGAGACA ... TTGATGACCTCAAAAAAAAA
 ,... ,... }

Please note that the parser assumes a standard `"HeaderFormat"` option,

`"HeaderFormat"`->{`"gi|"`,`"DatabaseIndex"`,`"|gb|"`,`"Accession"`,`"|"`, `"Description"` }.

Export can also export strings to FASTA format. FASTA file export options include a `"LineWidth"`, and `"ToUpperCase"`, which allow fine control of the exported strings. We can also instead of using files, export and import FASTA Strings to output, using **ExportString** with the format:

ExportString[{{ Header List }, { Sequences List }}]

Here is a short example:

```
In[16]:= exampleFastaString=ExportString[{{"Seq1","Seq2","Seq3
    "},
        {"GATCACAGGTCTATCA","ACCCTATTAACC",
        "AGCCCACACG"}},"FASTA"]
Out[16]= >Seq1
        GATCACAGGTCTATCA
        >Seq2
        ACCCTATTAACC
        >Seq3
        AGCCCACACG
```

If a "Header" list is not provided the Wolfram Language will generate one automatically :

```
In[17]:= ExportString[{"GATCACAGGTCTATCA","ACCCTATTAACC",
        "AGCCCACACG"},"FASTA"]
Out[17]= >Sequence 1 | Created with the
            Wolfram Language : www.wolfram.com
        GATCACAGGTCTATCA
        >Sequence 2 | Created with the
            Wolfram Language : www.wolfram.com
        ACCCTATTAACC
        >Sequence 3 | Created with the
            Wolfram Language : www.wolfram.com
        AGCCCACACG
```

We can easily make an association of the imported information:

```
In[18]:= Association @@ ImportString[exampleFastaString,
            {"FASTA","LabeledData"}]
Out[18]= <|Seq1→GATCACAGGTCTATCA,
            Seq2→->ACCCTATTAACC, Seq3→AGCCCACACG|>.
```

5.2.2 GenBank Records

We can also parse GenBank [3] files, .gbk/.gb extensions, which are the format of records for NCBI (National Center for Biotechnology Information). Similarly to the parsing of FASTA files above, we can import and export GenBank files. There are multiple elements in a recognized GenBank file:

```
In[19]:= Import["NM_003998.gb", "Elements"]
Out[19]= {Comments, Definition, Features,
        GenBankID, Keywords, Locus, NCBIAccession,
        NCBIAccessionVersion, Organism, Origin,
        Plaintext, PRIMARY, Reference, Sequence}
```

The comments include important information about the record:

```
In[20]:= commentsExample=Import["NM_003998.gb", "Comments"];
```

5.2 Parsing Files

We use a postfix //**Short**[#,n]& as a function to get about n lines of the output:

In[21]:= commentsExample//**Short**[#,15]&
Out[21]//Short= REVIEWED REFSEQ: This record has been curated by
    NCBI staff. The reference sequence was
    derived from BC051765.1, AF213884.2 and
    AI076882.1. This sequence is a reference
    standard in the RefSeqGene project.
    On Sep 24, 2009 this sequence version
    replaced gi:34577121. Summary: This gene
 ... ecord to access additional publications.
    ##Evidence−Data−START## Transcript
    exon combination :: BC051765.1,
    M55643.1 [ECO:0000332]
    RNAseq introns :: single sample supports all
    introns SAMEA1965299, SAMEA1966682
    [ECO:0000348] ##Evidence−Data−END##
    COMPLETENESS: complete on the 3' end.

We may also want to get the definition:

In[22]:= **Import**["NM_003998.gb", "Definition"]
Out[22]= Homo sapiens nuclear factor kappa B subunit
    1 (NFKB1), transcript variant 1, mRNA.

The "Features" has various keys itself:

In[23]:= **Import**["NM_003998.gb", "Features"]//**Keys**
Out[23]= {CodingSequence, Exon, Gene, MiscellaneousFeature,
    PolyASite, Regulatory, SequenceTaggedSite, Source}

For example we can get the "Gene" Feature:

In[24]:= **Import**["NM_003998.gb",{ "Features","Gene"}]
Out[24]= {Location→{{1,4093},TopStrand},
    GeneSymbol→NFKB1,
    GeneSynonym→CVID12; EBP−1; KBF1;
      NF−kappa−B; NF−kappaB; NF−kB1;
      NFkappaB; NFKB−p105; NFKB−p50; p105; p50,
    Note→nuclear factor kappa B subunit 1,
      CrossReferences −>{GeneID→4790},
      CrossReferences −>{HGNC:HGNC→7794},
      CrossReferences→MIM→164011}}

Or the "Source" information:

In[25]:= **Import**["NM_003998.gb",{ "Features","Source"}]
Out[25]= {Location→{{1,4093},TopStrand},
    Organism→Homo sapiens, MoleculeType→mRNA,
    CrossReferences→{TaxonomyID→9606},
    Chromosome→4,Map→4q24}

If we want the sequence, the "Plaintext" element contains a version (here we again look at about 10 lines of output using **Short**):

In[26]:= **Import**["NM_003998.gb","Plaintext"]//**Short**[#,10]&
Out[26]//Short=   1   1 gtgagagagt gagcgagaca

```
            gaaagagaga gaagtgcacc agcgagccgg ggcaggaaga
        61  ggaggtttcg ccaccggagc ggcccggcga cgcgctgaca
gcttcccctg cccttcccgt
       121  cggtcgggcc gccag.... ttccccctttt
      3961  tctgcatttt gctattgtaa atatgttttt tagatcaaat
actttaaagg aaaaaatgtt
      4021  ggatttataa atgctatttt ttattttact tttataataa
aaggaaaagc aaattgatga
      4081  cctcaaaaaa aaa
```

"Reference" imports the biographical information:

In[27]:= exampleGenBankReferences=**Import**["NM_003998.gb",
        {"Reference"}];

Let us look at a couple of examples of the references included:

In[28]:= exampleGenBankReferences[[1;;2]]
Out[28]= {{Authors→{Lapid **D**,Lahav−Baratz S,Cohen S.},
        Title→A20 inhibits both the degradation
            and limited processing of the
            NF−kappaB p105 precursor: A novel
            additional layer to its regulator role,
        Journal→Biochem. Biophys. Res. Commun.
            493 (1), 52−57 (2017),
        PubMedID→28923245,Remark→
        GeneRIF: Our data propose an additional novel
            mechanism to explain the known NF−kappaB
            inhibitory effects of A20: by affecting
            p105 ubiquitination and subsequently its
            degradation and limited processing.},
    {Authors→{Azizan **N**,Suter MA,Liu Y,
            Logsdon CD.},
        Title→RAGE maintains high levels of
            NFkappaB and oncogenic Kras
            activity in pancreatic cancer,
        Journal→Biochem. Biophys. Res. Commun.
            493 (1), 592−597 (2017),
        PubMedID→28867179,Remark→
        GeneRIF: These data indicate that RAGE
            plays a central role in maintaining
            inflammatory signaling in PDAC
            that benefits tumor growth.}}

We can also import the sequence as a string:

In[29]:= exampleGenBankSequence=
        Import["NM_003998.gb","Sequence"];

In[30]:= exampleGenBankSequence // **Short**
Out[30]//**Short**= GTGAGAGAGTGAGCGAGACAGA...AAATTGATGACCTCAAAAAAAAA

We can also get the length of the sequence using **StringLength**:

In[31]:= **StringLength**[exampleGenBankSequence]
Out[31]= 4093

5.3 Sequence Alignment

Sequences allow us to compare both within a species and across species in nature to look for similarities. Homologous sequences arise from common ancestors in the course of evolution. Additionally, sequence variation compared to a reference (a corresponding healthy baseline) can give us information that may be related to disease if there are alterations in the DNA. Alterations can arise in the form of point variation, substitutions, insertions and deletions. Additionally, structural variation is possible in terms of big rearrangements in chromosomes, hundreds to thousands of base pairs long.

We consider the following short sequence. We have put in a substitution in the last position for the second sequence for a T- to a G:

In[32]:= referenceExample1="ACGT";
          substitutionExample1="ACGG";

In the Wolfram Language alignment of strings is possible using the **SequenceAlignment** function:

In[34]:= ?**SequenceAlignment**
         SequenceAlignment[s_1, s_2] finds an optimal
           alignment of sequences of elements in the
           strings or lists s_1 and s_2, and yields a list of
           successive matching and differing sequences. >>

In[35]:= alignmentExample12=
         SequenceAlignment[referenceExample1,
           substitutionExample1]
Out[35]= {ACG,{T,G}}

The overall list contains the alignment results. The list within the result list {T,G} in the above result corresponds to positions that do not align for the two strings, with the two possibilities listed. The aligned part ACG is common, and is not enclosed as a list. Now consider the following simple case where we have a deletion:

In[36]:= deletionExample1="AGT";
         SequenceAlignment[referenceExample1,
           deletionExample1]
Out[37]= {A,{C,},GT}

Now we see that the second sequence as aligned differs from the first, as shown by the middle list {C, }. Here the missing part, gap in the second sequence, is denoted as an empty string. Now let us consider another example, a bit longer:

In[38]:= dnaA="TAATAAAAGGAAAAGCAAATTGATGACCT";
         dnaB="TACAAGAGGAAAAGTTAGATAGATGAGCGGAACT";

In[40]:= alignmentShortAB=**SequenceAlignment**[dnaA,dnaB]
Out[40]= {T,{A,},A,{T,C},AA,{A,G},AGGAAAAG,{C,TT},
         A,{A,G},AT,{T,A},GATGA,{,G},C,{,GGAA},CT}

The embedded lists in the above result again correspond to positions that do not align for stringA and string B respectively. Notice the {C,TT} and { ,GGAA} entries indicating further possible variation, and longer deletions, insertions. While the format is easy to use for short strings, it may be simpler to write the alignment out to see it by using a simple function:

In[41]:= alignmentView[alignment_]:=
          Transpose[
          (If[!ListQ[#],ConstantArray[#,2],#]&/@
          alignment)]//MatrixForm

The function checks the list to see if the head of the element is not **List**, meaning the string is the same for both input sequences. If so it generates two copies, one for each sequence by using **ConstantArray**[#,2]. Finally, we put the results in a matrix form. Here is an implementation on our alignment result from above:

In[42]:= alignmentView[alignmentShortAB]
Out[42]//MatrixForm=
$$\begin{pmatrix} T\ A\ A\ T\ AA\ A\ AGGAAAAG\ C\ A\ A\ AT\ T\ GATGA\ \ \ C\ \ \ \ \ \ CT \\ T\ \ \ \ A\ C\ AA\ G\ AGGAAAAG\ TT\ A\ G\ AT\ A\ GATGA\ G\ C\ GGAA\ CT \end{pmatrix}$$

The function **SequenceAlignment** can align strings both globally, or locally, through the option "Method", with "Global" being the default value, which tries to produce an overall optimal global alignment. For a longer discussion, including dynamic programming, see for example Isaev, 2016 [5]. Local alignment tries to find pairs of subsequences that have the highest alignment scores. We can carry out local alignment on our example from before and compare:

In[43]:= alignmentLocalAB=
         SequenceAlignment[dnaA,dnaB,"Method"-> "Local"]
Out[43]= {{TAA,},TA,{ ,C},AA,{ ,G},AGGAAAAG,{C,TT},A,
         {A,G},AT,{T,A},GATGA,{C,G},C,{T,GGAACT}}

In[44]:= alignmentView[%]
Out[44]//MatrixForm=
$$\begin{pmatrix} TAA\ TA\ \ \ \ AA\ \ \ \ \ AGGAAAAG\ C\ A\ A\ AT\ T\ GATGA\ C\ C\ \ \ \ \ T \\ \ \ \ \ \ TA\ C\ AA\ G\ AGGAAAAG\ TT\ A\ G\ AT\ A\ GATGA\ G\ C\ GGAACT \end{pmatrix}$$

There are various scoring schemes for similarity that can be used to decide on a string alignment. The option "GapPenalty" can assign a penalty to a gap in trying to find an optimal alignment. additionally "SimilarityRules" can take different values that correspond to a scoring rule for similarities. The form for "SimilarityRules" is a rule {string/pattern 1, string/pattern 2}—>similarity score .

The default value of SimilarityRules is {{{a_, a_} –> 1, {a_, b_} –> −1}} , meaning matches get a +1 score and mismatches a -1 score. {" ",string/pattern}} stands for insertions and {string/pattern," "} represents deletions.

We can import sequences directly from NCBI as we saw in the previous chapter to align them:

In[45]:= sequencesAB=
         URLExecute[
         "https://eutils.ncbi.nlm.nih.gov/entrez/eutils/

5.3 Sequence Alignment

```
            efetch.fcgi?",
         {"RequestMethod"->"POST",
          "db"->"nucleotide",
          "id"->"NM_001165412,NM_008689",
          "retmode"->"text",
          "rettype"-> "fasta"},
         {"FASTA","LabeledData"}];
```

These correspond to a human and a mouse sequence related to NFKB1:

In[46]:= **Keys** @ sequencesAB
Out[46]= {NM_001165412.1 Homo sapiens
    nuclear factor kappa B subunit 1
    (NFKB1), transcript variant 2, mRNA,
    NM_008689.2 Mus musculus nuclear factor of
    kappa light polypeptide gene enhancer
    in B cells 1, p105 (Nfkb1), mRNA}

We can try to align the sequences to find the longest common subsequence (we apply the function **LongestCommonSubsequence** to the list of **Values** of the sequence):

In[47]:= **LongestCommonSubsequence** @@ **Values** @ sequencesAB
Out[47]= GTGCAGGATGAGAATGGGGACAGTGTCTTACACTTAGC

The longest common sequence is a concatenation of the common sequences:

In[48]:= LongestCommonSequence@@Values@sequencesAB // **Short**
Out[48]//Short= GTGAGAGAGTGAGGAGACAGAA...AGCAATTGTGACCTAAAAAAAAA

The longest common sequence is 3444 bases:

In[49]:= **StringLength**[%]
Out[49]= 3444

Let us look at another alignment with "BLAST" as the similarity rule:

In[50]:= alignmentSequencesAB=
    SequenceAlignment[Sequence@@Values@sequencesAB,
    "SimilarityRules"-> "BLAST"];

Code Note: We are using **Sequence** @@ within the function to take the elements of the list of **Values** and pass them as arguments (without the **List** head) to the **SequenceAlignment** function. Here are the first few elements of the alignment

In[51]:= alignmentSequencesAB[[1;;10]]
Out[51]= {GT,{GA,CC},G,{AGA,TCT},GT,{GA,CT},GC,
    {GAGACAGAA, TCTCTCTCG},A,{GAGAGAGAA,CGTCAGTGG}}

And we can use our simple visualization tool for these:

In[52]:= alignmentView[alignmentSequencesAB[[1;;10]]]
Out[52]//MatrixForm=
$$\begin{pmatrix} GT & GA & G & AGA & GT & GA & GC & GAGACAGAA & A & GAGAGAGAA \\ GT & CC & G & TCT & GT & CT & GC & TCTCTCTCG & A & CGTCAGTGG \end{pmatrix}$$

We can calculate various alignment similarity scores using known algorithms:

In[53]:= **Apply**[NeedlemanWunschSimilarity, Values@sequencesAB]
Out[53]= 2597.

```
In[54]:= Apply[SmithWatermanSimilarity, Values@sequencesAB]
Out[54]= 2645.
```

5.4 BLAST(n)

One of the most used tools in bioinformatics and sequence analysis is the Basic Local Alignment Search Tool (BLAST) [1, 2]. BLAST can search for local similarity between sequences and extends matches to obtain longer alignments. BLAST works both with nucleotides and proteins, and can compare sequences across genes, proteins and organisms. In this section we will see how we can use the online BLAST Web API (application program interface) to carry out alignments using BLAST.

5.4.1 BLAST API Parameter List

The BLAST API uses the URL https://blast.ncbi.nlm.nih.gov/Blast.cgi. The community guidelines are that this should only be used once every ten seconds for submission, and polling result completion should not be more often than once a minute. An EMAIL parameter should be provided as well. There are multiple parameters that can be used with the RESTful BLAST API.

```
In[55]:= SystemOpen["https://ncbi.github.io/blast-cloud/dev/
   api.html"] (*Accessed 11/20/17*)
```

The API can be accessed using the Wolfram Language (as we had done for NCBI E-utilities in the previous chapter). Specifically, we will use **URLExecute** to access the API and POST requests. There are many parameters that can be used, but we will only review important ones in this chapter. The required parameters are:

- "QUERY": This is the actual search. As with the web version, this can be any of: {Accession, GI, FASTA sequence}.
- "DATABASE": This specifies the database to use. The databases are listed in appendix 2 of the BLAST guide, or can be found from the online database.
- "PROGRAM": This specifies which BLAST program to use, and can be one of: {"blastn", "megablast", "blastp", "blastx", "tblastn", "tblast"}
- "CMD" The command mode can be one of: {"Put", "Get", "Delete"}, depending on whether you are submitting a search, retrieving the results, or deleting a previous search respectively.

Other BLAST API parameters are listed in Table 5.1

5.4 BLAST(n)

Table 5.1 Out[56]= Other BLAST API parameters[a]

Parameter	Description
RID	The request ID, which is a string returned with search submitted
ALIGNMENTS	Number of alignments to print for HTML or Text
DESCRIPTIONS	Sets the number of descriptions to print, as a positive Integer
FILTER	Can be set to F to disable filtering. T or L enable filtering. We can prepend m for mask, e.g. mL
FORMAT_TYPE	Indicates how results should be returned in {HTML,Text, XML, JSON2 or Tabular}
EXPECT	Expect value (number>0)
NUCL_REWARD	Matching base reward, as a positive integer (to be used for blastn and megablast)
NUCL_PENALTY	FMismatch penalty cost, as a negative integer
GAPCOSTS	For setting the cost of gap existence and extensions (pairs of integers separated by space)
MATRIX	Scoring matrix to use for alignments, can be one of {BLOSUM45, BLOSUM50, BLOSUM62, BLOSUM80,BLOSUM90 - PAM250, PAM30 or PAM70}. BLOSUM 62 default
HitLIST_SIZE	Indicates how many sequences to keep as a positive integer
NCBI_GI	Sets whether to show NCBI GIs in the report (T or F string)
THRESHOLD	Neighboring score for initial words, a positve integer (not to be used with blastn or megablast)
WORD_SIZE	A positive integer indicating the size of word for initial matches
COMPOSITION_BASED_STATISTIC	Sets whether to use composition-based statistics:
	0 No composition-based statistics
	1 As in Schaffer et al. [6]
	2 Score adjustment as in Yu et al. [9], conditioned on sequence properties
	3 As in Yu et al. [9], unconditionally
FORMAT_OBJECT	Object type when getting results/status. Can take values of SearchingInfo (status) or Alignment (report formatting)

[a]Infomation based on parameter table at https://ncbi.github.io/blast-cloud/dev/api.html (accessed 11/30/17)

5.4.2 Running Web Based BLAST Example 1

Here we run an example of blastn, running a query on entry 4507564 to see what we get. We return the result as `"Text"` so we can parse it later:

In[57]:= blastExample=
          URLExecute["https://blast.ncbi.nlm.nih.gov/Blast.cgi?"
      ,
                    {"QUERY"-> "4507564","DATABASE"-> "nt",
                     "PROGRAM"-> "blastn","CMD"-> "Put"},"Text"
          ];

The above query will generate HTML output which we parse as text. The relevant info we need is in a "QBlastInfo" block, which we can now extract the Request ID (RID) from:

In[58]:= qBlastInfo=**StringCases**[blastExample,
              "QBlastInfoBegin"~~__~~"QBlastInfoEnd"]
Out[58]= {QBlastInfoBegin
               RID = 2BUKVWAW015
               RTOE = 14
           QBlastInfoEnd}

We can extract the RID for use directly.

In[59]:= rid=**StringSplit**[qBlastInfo][[1,4]]
Out[59]= 2BUKVWAW015

Once we wait a bit, we can check to get the result back. Please note that you will get a different RID when you perform the search, as this is a unique identifier. That is why we will use the variable rid set above in the examples below as the "RID" parameter value. The request can be used to obtain information in text form:

In[60]:= blastReturnStatus=
          URLExecute[
            "https://blast.ncbi.nlm.nih.gov/Blast.cgi?",
            {"CMD"-> "Get","RID"-> rid,
             "FORMAT_OBJECT"-> "status"},"Text"];

In[61]:= **StringCases**[#,"QBlastInfoBegin"~~__~~
                   "QBlastInfoEnd"]&@blastReturnStatus
Out[61]= {QBlastInfoBegin
               Status=READY
           QBlastInfoEnd}

From the "Status" above we can see whether our result is either "READY" or in "WAITING". Once the status is "READY" we can get the results. If you get a status of "WAITING" please wait and repeat the commands a bit later.

Again "CMD" is set to "Get", the "RID" to what we extracted above (rid). We use the Tabular format, and set/limit the Descriptions to 1, to see what this looks like:

In[62]:= blastReturnTabular=**URLExecute**[
            "https://blast.ncbi.nlm.nih.gov/Blast.cgi?",
            {"CMD"-> "Get","RID"-> rid,
             "FORMAT_TYPE"->"Tabular",
             "FORMAT_OBJECT"-> "Alignment",
             "DESCRIPTIONS"-> "1"}]

5.4 BLAST(n)

```
Out[62]= <p><!--
         QBlastInfoBegin
                  Status=READY
         QBlastInfoEnd
         --></p>
         <PRE>
         # blastn
         # Iteration: 0
         # Query: NM_003839.1 Homo sapiens tumor
           necrosis factor receptor superfamily, member
           11a, activator of NFKB (TNFRSF11A), mRNA
         # RID: 2BUKVWAW015
         # Database: nt
         # Fields: query id, subject
           ids, query acc.ver, subject acc.ver, %
           identity, alignment length, mismatches,
           gap opens, q. start, q. end, s.
           start, s. end, evalue, bit score
         # 1 hits found
         gi|4507564|ref|NM_003839.1|
            gi|2612917|gb|AF018253.1|AF018253
            NM_003839.1              AF018253.1
            100.000              3136              0           0          1
            3136          1         3136          0.0        5656
         </PRE>
```

We can extract the column information:

```
In[63]:=  columnsBLAST=
          StringSplit[StringCases[blastReturnTabular,
            "# Fields: "~~x:__~~"#"-> x],{","}]
Out[63]= {{query id, subject ids,
          query acc.ver, subject acc.ver,
          % identity, alignment length,
          mismatches, gap opens, q. start, q. end,
          s. start, s. end, evalue, bit score
         }}
```

We can use the CSV file and `"ALIGNMENT_VIEW" -> "Tabular"` to return the set of this information. Again we look at only a few alignment examples, and the first few results of the output:

```
In[64]:=  blastReturnCSV=
          URLExecute[
           "https://blast.ncbi.nlm.nih.gov/Blast.cgi?",
           {"CMD"-> "Get","RID"-> rid,
            "FORMAT_TYPE"->"CSV",
            "FORMAT_OBJECT"-> "Alignment",
            "ALIGNMENT_VIEW"-> "Tabular",
            "DESCRIPTIONS"-> "10",
            "ALIGNMENTS"-> "10"},"CSV"];

In[65]:=  blastReturnCSV[[1;;3]]
Out[65]= {{gi|4507564|ref|NM_003839.1|,
           gi|2612917|gb|AF018253.1|AF018253,
           NM_003839.1,AF018253.1,100.,3136,
```

186   5 Genomic Sequence Data and BLAST

```
           0,0,1,3136,1,3136,0.,5656},
         {gi|4507564|ref|NM_003839.1|,
           gi|401015114|ref|NM_003839.3|,
           NM_003839.1,NM_003839.3,99.808,3129,
           3,2,1,3129,29,3154,0.,5609},
         {gi|4507564|ref|NM_003839.1|,
           gi|1034604957|ref|XM_011526244.2|,
           NM_003839.1,XM_011526244.2,99.327,
           3121,3,3,24,3129,1,3118,0.,5537}}
```

To get more information we can use the text return format, which returns multiple lists. In this case we can restrict "ALIGNMENTS" as well as "DESCRIPTIONS" to the number we desire, e.g. 2:

```
In[66]:=  blastReturnText=
          URLExecute[
            "https://blast.ncbi.nlm.nih.gov/Blast.cgi?",
            {"CMD"-> "Get","RID"-> rid,
             "FORMAT_TYPE"->"Text",
             "FORMAT_OBJECT"-> "Alignment",
             "ALIGNMENTS"-> 2,"DESCRIPTIONS"-> "2"}];
```

Here is a short view of the information (about 20 lines using a **Short** postfix):

```
In[67]:=  blastReturnText//Short[#,20]&
Out[67]//Short=  {{<p><!-- },{QBlastInfoBegin},
           {Status=READY},{QBlastInfoEnd},
           {--><p>},{<PRE>},{BLASTN 2.7.1+},
           {Reference: Stephen F. Altschul,
           Thomas L. Madden, Alejandro},
            {A. Schaffer, Jinghui Zhang, Zheng Zhang,
             Webb Miller, and},{David J. Lipman (1997),
             Gapped BLAST and PSI-BLAST: a new
           generation of protein database search programs,
             Nucleic},{Acids Res. 25:3389-3402.},
            {},{},{RID: 2BUKVWAW015},{},<<481>>,
            {Effective length of query: 3096},
            {Effective length of database: 161695478452},
            {Effective search space: 500609201287392},
            {Effective search space used: 500609201287392},
            {A: 0},{X1: 22 (20.1 bits)},
            {X2: 33 (29.8 bits)},
            {X3: 110 (99.2 bits)},{S1: 28 (26.5 bits)},
            {S2: 50 (46.4 bits)},{},{},{},{}}
```

The following allows us to obtain the alignments in JSON format in a single file, by using the "FORMAT_TYPE" -> "JSON2_S":

```
In[68]:=  blastReturnJSONSingleFile=
          URLExecute[
            "https://blast.ncbi.nlm.nih.gov/Blast.cgi?",
            {"CMD"-> "Get","RID"-> rid,
             "FORMAT_TYPE"->"JSON2_S",
             "FORMAT_OBJECT"-> "Alignment",
             "DESCRIPTIONS"-> "10"},"RawJSON"];
```

5.4 BLAST(n)

And again, we look at 15 lines from the data:

```
In[69]:=  blastReturnJSONSingleFile // Short[#,15]&
Out[69]//Short=
          <|BlastOutput2 ->
          {<|report -><|program ->blastn,<<4>>,
              results -><|search ->
              <|query_id ->NM_003839.1,query_title ->
                Homo sapiens tumor necrosis
                    factor receptor superfamily, member
                    11a, activator of
                    NFKB (TNFRSF11A), mRNA,
              query_len ->3136,query_masking ->
                {<|from ->7,to ->110|>,
                 <|from ->2198,to ->2207|>,
                 <|from ->2317,to ->2355|>,
                 <|from ->2507,to -><<4>>|>,
                 <|from ->2830,to ->2849|>,
                 <|from ->3127,to ->3135|>},hits ->
                {<<1>>},stat -><|db_num ->45448023,
                 db_len ->304642124,hsp_len ->40,
                 eff_space ->500609201287392,
                 kappa ->0.41,lambda ->0.625,
                 entropy ->0.78|>|>|>|>|>|>|>
```

5.4.2.1 Some Format Type Variations

There are multiple `"FORMAT_TYPE"` values that can be passed and here we look at a few variations. We note that by setting the parameter value for `"FORMAT_TYPE"` to `"JSONSA"` we can get a JSON alignment file:

```
In[70]:=  blastReturnJSONSingleAlignment=
          URLExecute[
            "https://blast.ncbi.nlm.nih.gov/Blast.cgi?",
            {"CMD"-> "Get","RID"-> rid,
             "FORMAT_TYPE"->"JSONSA",
             "RESULTS_FILE"-> "ON",
             "FORMAT_OBJECT"-> "Alignment",
             "DESCRIPTIONS"-> "10","ALIGNMENTS"-> "10"},
            "RawJSON"];
```

If we instead set the `"FORMAT_TYPE"` parameter to a `"JSON2"` format, we actually get a zip format, which we will instead have to process a file at a time:

```
In[71]:=  blastReturnJSONZipFileNames=
          URLExecute[
            "https://blast.ncbi.nlm.nih.gov/Blast.cgi?",
            {"CMD"-> "Get","RID"-> rid,
             "FORMAT_TYPE"->"JSON2",
             "FORMAT_OBJECT"-> "Alignment",
             "DESCRIPTIONS"-> "10","ALIGNMENTS"-> "10"},
            {"ZIP","FileNames"}]
```

Out[71]= {2BUKVWAW015.json,2BUKVWAW015_1.json}

We also notice that the name of the two files retrieved above match the RID number. Now that we have the file names, we can parse these ZIP files one at a time:

```
In[72]:= blastReturnJSONZipFile1=
         URLExecute[
          "https://blast.ncbi.nlm.nih.gov/Blast.cgi?",
          {"CMD"-> "Get","RID"-> rid,
           "FORMAT_TYPE"->"JSON2",
           "FORMAT_OBJECT"-> "Alignment",
           "DESCRIPTIONS"-> "10","ALIGNMENTS"-> "10"},
          {"ZIP",{#}}]&/@blastReturnJSONZipFileNames;
```

We can then convert the information into an association by using a trick to export the string to JSON and then import as RawJSON. We can do this on both files, but we only use the first short file here as an example:

```
In[73]:= blastReturnAssociation=
         ImportString[
          ExportString[blastReturnJSONZipFile1[[1]],
           "JSON"],"RawJSON"]
Out[73]= <|BlastJSON→{<|File→2BUKVWAW015_1.json|>}|>
```

This file is simply an association of the contained filename in the archive. In the example in the next section we will use instead a file format "JSON2_S", which is a single JSON file for multiple searches.

5.4.3 BLAST Example 2

Now let us suppose we have a list of sequence strings we are interested in. We can actually submit a FASTA format with multiple queries. First, we create the FASTA formatted string:

```
In[74]:= sequenceString2=
         ExportString[{{"test1","test2","test3",},
           {"TTTTTCCCGCTACAGGGGGGGGCCTGAGGCACTGCAGAAAGTG
             GGCCTGAGCCTCGAGGAAAA",
            "NNNNNCCCGCTACAGGGGGGGNNCTGAAACACTNNAGAAAGTG
             GGCCTGAGCCTCGAGGTTTT",
            "GTGAGAGAGTGAGGAGACAGAAAGAGAGAGAAGTACCAGCGAG
             CCGGGCAGGAAGAGGAGGTTTTATAAATGCTATTTTT
             TATTTTACTTTTATAATAAAAGAAAGCAATTGTGACC
             TAAAAAAAAA"}},"FASTA"]
Out[74]= >test1
         TTTTTCCCGCTACAGGGGGGGGCCTGAGGCACTGCAGAAAGTGGGCCT
           GAGCCTCGAGGAAAA
         >test2
         NNNNNCCCGCTACAGGGGGGGNNCTGAAACACTNNAGAAAGTGGGCCT
           GAGCCTCGAGGTTTT
         >test3
```

5.4 BLAST(n)

```
         GTGAGAGAGTGAGGAGACAGAAAGAGAGAGAAGTACCAGCGAGCCGGG
          CAGGAAGAGGAGGTTTTATAAATGCTATTTTTTATTTTACTTTTA
      TAATAAAAGAAAGCAATTGTGACCTAAAAAAAAA
```

The above FASTA string variable can be passed as the "QUERY" in the BLAST search directly:

```
In[75]:= blastExample2=
         URLExecute[
          "https://blast.ncbi.nlm.nih.gov/Blast.cgi?",
          {"QUERY"-> sequenceString2 ,
           "PROGRAM"-> "blastn",
           "DATABASE"-> "nt",
           "CMD"-> "Put"},"Text"];
```

The information we need is in a QBlastInfo block, which we can now extract the Request ID (RID) from:

```
In[76]:= qBlastInfo2=StringCases[blastExample2 ,
             "QBlastInfoBegin"~~__~~"QBlastInfoEnd"]
Out[76]= {QBlastInfoBegin
             RID = 2BUZ80S001R
             RTOE = 15
          QBlastInfoEnd }
```

We take the fourth entry after we split the string as the RID:

```
In[77]:= rid2=StringSplit[qBlastInfo2][[1,4]]
Out[77]= 2BUZ80S001R
```

We can run a "Get" command request to check the status information:

```
In[78]:= blastReturnStatus2=
         URLExecute[
          "https://blast.ncbi.nlm.nih.gov/Blast.cgi?",
          {"CMD"-> "Get",
           "RID"-> rid2 ,
           "FORMAT_TYPE"->"Tabular",
           "FORMAT_OBJECT"-> "status"},"Text"];
```

Using **StringCases** we can extract the information on the status from the text:

```
In[79]:= StringCases[#,"QBlastInfoBegin"~~__~~"QBlastInfoEnd"
     ]& @blastReturnStatus2
Out[79]= {QBlastInfoBegin
             Status=READY
          QBlastInfoEnd }
```

We can use the CSV file and "Alignment_View" as "Tabular" to get the data information, corresponding to the columns extracted before:

```
In[80]:= blastReturnCSV2=
         URLExecute[
          "https://blast.ncbi.nlm.nih.gov/Blast.cgi?",
          {"CMD"-> "Get","RID"-> rid2 ,
           "FORMAT_TYPE"->"CSV",
           "FORMAT_OBJECT"-> "Alignment",
```

Out[80]= {{test1, gi|405832|gb|U00001.1|HSCDC27,
```
              "ALIGNMENT_VIEW" -> "Tabular",
              "DESCRIPTIONS" -> "2"}]
```
test1, U00001.1, 98.246, 57, 0, 1,
4, 60, 4, 59, 1.14×10^{-16}, 95.1},
{test1, gi|1244127952|ref|XM_022519950.1|,
test1, XM_022519950.1, 96.296, 54, 2,
0, 7, 60, 101, 154, 4.83×10^{-15}, 89.7},
{test2, gi|405832|gb|U00001.1|HSCDC27,
test2, U00001.1, 88.889, 54, 5,
1, 6, 59, 6, 58, 4.53×10^{-9}, 69.8},
{test2, gi|1211850968|ref|NR_148340.1|,
test2, NR_148340.1, 88.679, 53, 5,
1, 7, 59, 116, 167, 1.58×10^{-8}, 68.},
{test3, gi|1028908930|ref|NG_050628.1|,
test3, NG_050628.1, 94.118, 68, 0, 3,
1, 64, 5001, 5068, 1.14×10^{-16}, 95.1},
{test3, gi|1028908930|ref|NG_050628.1|,
test3, NG_050628.1, 92.063, 63, 1, 3, 60,
118, 120910, 120972, 7.17×10^{-13}, 82.4},
{test3, gi|259155301|ref|NM_001165412.1|,
test3, NM_001165412.1, 94.118, 68,
0, 3, 1, 64, 1, 68, 1.14×10^{-16}, 95.1},
{test3, gi|259155301|ref|NM_001165412.1|,
test3, NM_001165412.1, 91.667, 72, 2, 3, 60, 127,
4018, 4089, 3.97×10^{-16}, 93.3}, {}, {}, {}, {}}

The columns correspond to what we extracted before:

In[81]:= columnsBLAST
Out[81]= {{query id, subject ids,
          query acc.ver, subject acc.ver,
          % identity, alignment length,
          mismatches, gap opens, q. start, q. end,
          s. start, s. end, evalue, bit score}}

We see that our accessions are aligned with the subject IDs, the alignment length is given and an e-value. We can extract any of the information as necessary.

We can also obtain JSON format in a single file, by using the "FORMAT_TYPE" -> "JSON2_S":

In[82]:=   blastReturnJSONSingleFile2=
            URLExecute[
                "https://blast.ncbi.nlm.nih.gov/Blast.cgi?",
                {"CMD" -> "Get", "RID" -> rid2,
                 "FORMAT_TYPE" -> "JSON2_S",
                 "FORMAT_OBJECT" -> "Alignment",
                 "DESCRIPTIONS" -> "10",
                 "ALIGNMENTS" -> "10"}, "RawJSON"];

Instead, by using "JSONSA" we can get a JSON alignment file:

In[83]:=   blastReturnJSONSingleAlignment2=
            URLExecute[

5.4 BLAST(n)

```
              "https://blast.ncbi.nlm.nih.gov/Blast.cgi?",
              {"CMD"-> "Get","RID"-> rid2 ,
               "FORMAT_TYPE"->"JSONSA",
               "RESULTS_FILE"-> "ON",
               "FORMAT_OBJECT"-> "Alignment",
               "DESCRIPTIONS"-> "10",
               "ALIGNMENTS"-> "10"},"RawJSON"];
```

Let us look at the single JSON file first. The Output is denoted by a key named `"BlastOutput2"`:

In[84]:= Keys@blastReturnJSONSingleFile2
Out[84]= {BlastOutput2}

There are 3 reports, one each for our 3 sequences

In[85]:= Query[All, Keys]@blastReturnJSONSingleFile2
Out[85]= <|BlastOutput2→{{report},{report},{report}}|>

We can look at the **Keys** of the first one:

In[86]:= Query[All,1,"report",Keys]@blastReturnJSONSingleFile2
Out[86]= <|BlastOutput2→{program, version,
         reference, search_target, params, results}|>

The **Keys** give us information about which BLAST program was ran, and the target database and parameters:

In[87]:= Query[All,1,"report",
         {"program","version","search_target",
         "params"}]@blastReturnJSONSingleFile2
Out[87]= <|BlastOutput2→
         <|program→blastn, version→BLASTN 2.7.1+,
         search_target→<|db→nt|>,
         params→<|expect→10,sc_match→2,
         sc_mismatch→−3,gap_open→5,
         gap_extend→2,filter→L;m;|>|>|>

We can keep searching to extract any further information we are interested in. The sequence alignment is actually further down the hierarchy:

In[88]:= Query[1,2,"report","results","search",
         "hits",4,"hsps",1,
         {"qseq","midline","hseq"}]@
         blastReturnJSONSingleFile2
Out[88]= <|qseq→
         CCGCTACAggggggggNNCTGAAACACTNNAGAAAGTGGGCCTGAGC
             CTCGAGG,
         midline→ ||||||||||||||| ||||
             |||| |||||||||||||||||||||||| ,
         hseq→CCGCTACAGGGGGGGC−
             CTGAGGCACTGCAGAAAGTGGGCCTGAGCCTCGAGG|>

We can actually use the strings and the midline to visualize this alignment with our simple function below (see also Fig. 5.4):

```
⎛C C G C T A C A G G G G G G G N N C T G A A A C A C T N N A G A A A G T G G G C C T G A G C C T C G A G G⎞
⎜| | | | | | | |   | | | | | |   | | | | | |     | | | |     | | | | | | | | | | | | | | | | | | | | | |⎟
⎝C C G C T A C A G G G G G G G C - C T G A G G C A C T G C A G A A A G T G G G C C T G A G C C T C G A G G⎠
```

Fig. 5.4 Out[89]= Aligning sequences with BLAST

```
In[89]:= MatrixForm @
           Characters @
             Values @
               Query[1,2,"report","results","search",
                 "hits",4,"hsps",1,
                 {"qseq","midline","hseq"}]@
               blastReturnJSONSingleFile2
Out[89]//MatrixForm= (*See Fig. 5.4*)
```

You can probe any of the information from the BLAST output similarly, and include it in your workflow as necessary for a bioinformatics solution.

References

1. Altschul, S.F., Gish, W., Miller, W., Myers, E.W., Lipman, D.J.: Basic local alignment search tool. J. Mol. Biol. **215**(3), 403–410 (1990)
2. Altschul, S.F., Madden, T.L., Schaffer, A.A., Zhang, J., Zhang, Z., Miller, W., Lipman, D.J.: Gapped blast and psi-blast: a new generation of protein database search programs. Nucleic Acids Res. **25**(17), 3389–402 (1997)
3. Clark, K., Karsch-Mizrachi, I., Lipman, D.J., Ostell, J., Sayers, E.W.: Genbank. Nucleic Acids Res. **44**(D1), D67–D72 (2016)
4. Drew, H.R., Wing, R.M., Takano, T., Broka, C., Tanaka, S., Itakura, K., Dickerson, R.E.: Structure of a b-dna dodecamer: conformation and dynamics. Proc. Natl. Acad. Sci. U S A **78**(4), 2179–2183 (1981)
5. Isaev, A.: Introduction to Mathematical Methods in Bioinformatics. Springer, Berlin (2006)
6. Schäffer, A.A., Aravind, L., Madden, T.L., Shavirin, S., Spouge, J.L., Wolf, Y.I., Koonin, E.V., Altschul, S.F.: Improving the accuracy of psi-blast protein database searches with composition-based statistics and other refinements. Nucleic Acids Res. **29**(14), 2994–3005 (2001)
7. Wolfram Alpha LLC: Wolfram|Alpha (Access November 2017) (2017)
8. Wolfram Research, Inc.: Mathematica, Version 11.2. Champaign, IL (2017)
9. Yu, Y.K., Altschul, S.F.: The construction of amino acid substitution matrices for the comparison of proteins with non-standard compositions. Bioinformatics **21**(7), 902–911 (2005)

Chapter 6
Transcriptomics Examples

6.1 Transcriptomic Analysis

Transcriptomics refers to the set of all possible transcripts in a cell (e.g. mRNA, small RNA, non-coding RNA). In contrast to the DNA which is fairly constant across cells (besides cellular variation), gene expression varies widely in different kind of cells, and in relation to developmental stages and health status. Many methods are used to study transcriptomes, ranging from microarrays [2–5, 21] to state-of-the art RNA-sequencing [8, 9, 11, 12, 16, 17]. We often have to analyze high-throughput datasets that need preprocessing and then additional post processing appropriate to large numbers of tests, genes and conditions. In this chapter we will analyze some well known gene expression datasets to show how the Wolfram Language [18–20] can be used in general working examples of such analysis. We will see how differential gene expression can be assessed, including using an analysis of variance approach that is extensible to other big datasets.

6.2 Golub ALL AML Training Set

Let us consider the Golub dataset [6, 7], introduced in Sect. 3.7.4. The dataset is distributed with the Mathematica notebooks accompanying this book. First we set the directory to the notebook directory:

In[1]:= SetDirectory[NotebookDirectory[]];

Let us import the original training dataset:

In[2]:= golubALLAMLTrain=
    Import["Golub_data_set_ALL_AML_train.tsv","TSV"];

Electronic supplementary material The online version of this chapter (https://doi.org/10.1007/978-3-319-72377-8_6) contains supplementary material, which is available to authorized users.

© Springer International Publishing AG 2018
G. Mias, *Mathematica for Bioinformatics*,
https://doi.org/10.1007/978-3-319-72377-8_6

We can look at a few lines:

In[3]:= golubALLAMLTrain//**Short**[#,10]&
Out[3]//Short= {{Gene Description,Gene Accession Number,1,call,2,
call,3,call,4,call,5,call,6,call,7,call,
8,call,9,call,10,call,11,call,12,call,
13,call,14,call,15,call,16,call,17,call,
18,<<5>>,21,call,22,call,23,call,24,call,
25,call,26,call,27,call,34,call,35,call,
36,call,37,call,38,call,28,call,29,call,
30,call,31,call,32,call,33,call},<<7129>>}

Certain columns contain non-numeric data referring to a call as to whether a baseline is present. We can choose the positions that do not include those:

In[4]:= positionsTrain=,
Flatten@Position[golubALLAMLTrain[[1]],x_/;x!="call"]
Out[4]= {1,2,3,5,7,9,11,13,15,17,19,21,23,25,
27,29,31,33,35,37,39,41,43,45,47,49,51,
53,55,57,59,61,63,65,67,69,71,73,75,77}

We can the use the positions as indices to select the data that we want from the original list:

In[5]:= golubALLAMLTrainData=
golubALLAMLTrain[[**All**,positionsTrain]];

We can now take a look at the header and the first couple of rows of our selection:

In[6]:= golubALLAMLTrainData[[1;;3]]
Out[6]= {{Gene Description,Gene Accession Number,1,
2,3,4,5,6,7,8,9,10,11,12,13,14,15,
16,17,18,19,20,21,22,23,24,25,26,27,
34,35,36,37,38,28,29,30,31,32,33},
{AFFX–BioB–5_at (endogenous control),AFFX–BioB–5_at,
−214,−139,−76,−135,−106,−138,−72,−413,5,−88,
−165,−67,−92,−113,−107,−117,−476,−81,−44,
17,−144,−247,−74,−120,−81,−112,−273,−20,
7,−213,−25,−72,−4,15,−318,−32,−124,−135},
{AFFX–BioB–M_at (endogenous control),
AFFX–BioB–M_at,−153,−73,−49,−114,−125,−85,
−144,−260,−127,−105,−155,−93,−119,−147,
−72,−219,−213,−150,−51,−229,−199,−90,
−321,−263,−150,−233,−327,−207,−100,−252,
−20,−139,−116,−114,−192,−49,−79,−186}}

We can also see what the distributions of the data look like, in Fig. 6.1. These are rather broad distributions. Notice how the boxes are not distinguishable as the data are very packed in the lower intensities. We want to normalize the data to make the distributions comparable (and to account for some array to array variability), and also to bring them closer to being normal distributions, which we will do in the next section.

6.2 Golub ALL AML Training Set

```
In[7]:= BoxWhiskerChart[
    Transpose[N[golubALLAMLTrainData[[2;;, 3;;]]]],
    PlotLabel → "Golub Train Data Box plots",
    FrameLabel → {"Subject", "Intensity"}]
```

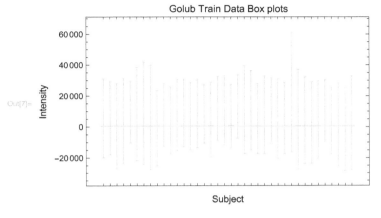

Fig. 6.1 Out[7]=Golub train data Box plots

6.2.1 Golub Original Set Normalization

Here we will follow the normalization approach by Dudoit et al. [6]. First we replace values matching certain criteria. Specifically, we want to replace low intensities, less than 100, and high intensities, more than 16000 to account for noise and saturation. We skip the header row 1, as well as the first two columns which contain gene information.

```
In[8]:= golubALLAMLTrainDataReplaced=
    (golubALLAMLTrainData[[2;;, 3;;]]/.x_/; x<100 -> 100)/.
      x_/; x>16000-> 16000;
```

Note that we have used two substitutions using patterns, the first one matching any variable x less than 100 and assigning it to 100, and the second one matching any x > 16000 and replacing it with 16000. We can verify by checking the data:

```
In[9]:= golubALLAMLTrainDataReplaced[[1;;3]]
Out[9]= {{100,100,100,100,100,100,100,100,100,100,
         100,100,100,100,100,100,100,100,100,100,
         100,100,100,100,100,100,100,100,100,100,
         100,100,100,100,100,100,100,100},
        {100,100,100,100,100,100,100,100,100,100,
         100,100,100,100,100,100,100,100,100,100,
         100,100,100,100,100,100,100,100,100,100,
         100,100,100,100,100,100,100,100},
        {100,100,100,265,100,215,238,100,106,100,
         100,100,100,100,100,100,100,100,100,100,
```

100,100,100,100,100,100,100,100,100,136,
124,100,100,100,100,100,100,100}}

Let us extract also the information regarding the measurements:

In[10]:= golubTrainIDs=golubALLAMLTrainData[[2;;,1;;2]];

In[11]:= golubTrainIDs//**Short**
Out[11]//Short= {{AFFX–BioB–5_at (endogenous control),... },
<<7128>>}

Next, we filter out data, keeping the genes for which the the absolute difference between the maximum and minimum expression is greater than 500, while the maximum expression across samples divided by the minimum expression across samples is greater than 5. We include a pattern:

x_/;((**Max**[x]−**Min**[x])>500 && (**Max**[x]/**Min**[x]>5))}

with the operator form of **Position**, that allows us to find the positions of a given pattern in the data:

In[12]:= locationsForSignalsTrain=
Flatten@
Position[
x_/;((**Max**[x]−**Min**[x])>500 &&
(**Max**[x]/**Min**[x]>5))]@
golubALLAMLTrainDataReplaced;

Based on this filtering we can see that 3051 signals remain.

In[13]:= **Length**[locationsForSignalsTrain]
Out[13]= 3051

We can extract the filtered signals by retrieving the data indexed at these positions:

In[14]:= golubTrainFiltered=
golubALLAMLTrainDataReplaced[[
locationsForSignalsTrain]];

Finally, we take the Log 10 value and standardize across each column: We first transpose the signal, then take Logarithm base 10 and standardize by subtracting the mean and dividing by the standard deviation. Finally, we transpose back to get the original matrix format. This can be done in a single step:

In[15]:= golubTrainNormed=
Transpose[**Standardize**[**Log10**[**N**[#]]]&/@
Transpose[golubTrainFiltered]];

We can plot again the distributions, Fig. 6.2. The distributions are fairly uniform across the data, and the range now is between −2 and 4 since we took a logarithm. We can now get the corresponding identity for each signal by using the positions for the signals we extracted in the ID list we created earlier:

In[17]:= golubTrainNormedIds=
golubTrainIDs[[locationsForSignalsTrain]];

6.2 Golub ALL AML Training Set

```
In[16]:= BoxWhiskerChart[Transpose[golubTrainNormed],
        PlotLabel → "Golub Train Data Boxplots",
        FrameLabel → {"Subject", "Intensity"},
        PlotTheme -> "Business"]
```

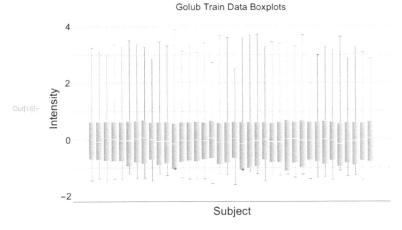

Out[16]=

Fig. 6.2 Out[16]= Normalized Golub train data box plots

From the data file we also have the information regarding the sample index number (which corresponds to the original study sample identifier).

In[18]:= golubALLAMLTrainData[[1,3;;]]
Out[18]= {1,2,3,4,5,6,7,8,9,10,11,12,13,14,
         15,16,17,18,19,20,21,22,23,24,25,26,
         27,34,35,36,37,38,28,29,30,31,32,33}

Additionally, we have the phenotype information for each sample that is from the file "Golub_table_ALL_AML_samples.doc". We have put this information in an association called golubPhenotypeData, which we have saved for convenience, and we import next:

In[19]:= golubPhenotypeData=<<"golubPhenotypeData";

In[20]:= golubPhenotypeData[[1;;3]]
Out[20]= <|Samples→{ALL/AML,BM/PB,
         T/B-cell (if ALL),FAB (if AML),Date,Gender,
         %Blasts,Treatment Response,PS,Source},
         1→{ALL,BM,B-cell,**Missing**[],9/4/1996,
         M,**Missing**[],**Missing**[],1.,DFCI},
         2→{ALL,BM,T-cell,**Missing**[],**Missing**[],
         M,**Missing**[],**Missing**[],0.41,DFCI}|>

In the association we have included a "Samples" **Key** that gives the information contained in each value list for each sample.

In[21]:= golubPhenotypeData[["Samples"]]

198   6 Transcriptomics Examples

```
Out[21]= {ALL/AML,BM/PB,T/B-cell (if ALL),
         FAB (if AML),Date,Gender,%Blasts,
         Treatment Response,PS,Source}
```

The disease condition for the train set can be obtained:

```
In[22]:= golubTrainClass=Query[1;;39,1]@golubPhenotypeData
Out[22]= <|Samples→ALL/AML,1→ALL,2→ALL,3→ALL,
         4→ALL,5→ALL,6→ALL,7→ALL,8→ALL,
         9→ALL,10→ALL,11→ALL,12→ALL,13→ALL,
         14→ALL,15→ALL,16→ALL,17→ALL,18→ALL,
         19→ALL,20→ALL,21→ALL,22→ALL,23→ALL,
         24→ALL,25→ALL,26→ALL,27→ALL,28→AML,
         29→AML,30→AML,31→AML,32→AML,33→AML,
         34→AML,35→AML,36→AML,37→AML,38→AML|>
```

We have 38 samples, and in addition a label key for a total of 39 entries.

6.2.2 Golub ALL AML Combined Dataset

We can also repeat all the process for the testing set from the same study. There are 34 additional samples that were originally used as independent testing data. The subjects are all again ALL and AML patients.

```
In[23]:= golubTestingClass=Query[40;;,1]@golubPhenotypeData
Out[23]= <|39->ALL,40->ALL,41->ALL,42->ALL,
         43->ALL,44->ALL,45->ALL,46->ALL,47->ALL,
         48->ALL,49->ALL,50->AML,51->AML,52->AML,
         53->AML,54->AML,55->ALL,56->ALL,57->AML,
         58->AML,59->ALL,60->AML,61->AML,62->AML,
         63->AML,64->AML,65->AML,66->AML,67->ALL,
         68->ALL,69->ALL,70->ALL,71->ALL,72->ALL|>
```

We can import the file and also create the cutoffs as before. We gather the calculations in one evaluation since they are the same as the previous section:

```
In[24]:= golubALLAMLIndependent=
           Import["Golub_data_set_ALL_AML_independent.tsv",
            "TSV"];
         positionsIndependent=
           Flatten@Position[golubALLAMLIndependent[[1]],
            x_/;x!="call"];
         golubALLAMLIndependentData=
           golubALLAMLIndependent[[All,
            positionsIndependent]];
         golubALLAMLIndependentDataReplaced=
           (golubALLAMLIndependentData[[2;;,3;;]]/.
            x_/;x<100 -> 100)/.x_/;x>16000-> 16000;
         golubIndependentIDs=
           golubALLAMLIndependentData[[2;;,1;;2]];
```

6.2 Golub ALL AML Training Set

Let us check if the gene IDs are in the same order, which is the case if all the genes are in the same order and identical:

In[29]:= golubIndependentIDs==golubTrainIDs
Out[29]= True

We can also compare the dimensions of the datasets:

In[30]:= Dimensions@golubALLAMLIndependentDataReplaced
Out[30]= {7129,34}

In[31]:= Dimensions@golubALLAMLTrainDataReplaced
Out[31]= {7129,38}

We see that the independent set has 34 samples, and the total is 72. We want to join these like matrices, and enlarge the rows. This will be in level 2 of each list included, and we use the **Join** function:

In[32]:= ?Join
         Join[list$_1$, list$_2$,...] concatenates lists or
              other expressions that share the same head.
         Join[list$_1$, list$_2$,...,n] joins the objects at level n in each
              of the list$_i$.?>>

In[33]:= golubALLAMLDataJoined=
             Join[golubALLAMLTrainDataReplaced,
             golubALLAMLIndependentDataReplaced,2];

This step concatenates the two datasets together. Overall we have 7129 genes from 72 samples, as we can verify by checking the dimensions of our new matrix:

In[34]:= **Dimensions**[golubALLAMLDataJoined]
Out[34]= {7129,72}

We again carry out filtering for large difference in the signals, and extract the filtered signals indexed at the positions identified:

In[35]:= locationsForSignalsJoined=
             Flatten@
             Position[
                x_/;((**Max**[x]−**Min**[x])>500 &&
                (**Max**[x]/**Min**[x]>5))]@golubALLAMLDataJoined;

Because of the additional samples, the locations will not be the same as they were for the training Golub dataset that was considered first above. We can extract the filtered signals by retrieving the data indexed at these positions:

In[36]:= golubFiltered=golubALLAMLDataJoined[[
             locationsForSignalsJoined]];

Finally, we take the Log 10 value and standardize across each column.

In[37]:= golubNormed=
             Transpose[**Standardize**[**Log10**[**N**[#]]]&/@
             Transpose[golubFiltered]];

After the filtering, our final extended dataset has 3571 genes in 72 samples:

In[38]:= **Dimensions**[golubNormed]
Out[38]= {3571,72}

We can also extract the corresponding identities.

In[39]:= golubNormedIDs=
golubIndependentIDs[[locationsForSignalsJoined]];

We also have the combined subject ids for both sets:

In[40]:= golubSampleIDs=
Join[golubALLAMLTrainData[[1,3;;]],
golubALLAMLIndependentData[[1,3;;]]]
Out[40]= {1,2,3,4,5,6,7,8,9,10,11,12,13,
14,15,16,17,18,19,20,21,22,23,24,
25,26,27,34,35,36,37,38,28,29,30,31,
32,33,39,40,42,47,48,49,41,43,44,45,
46,70,71,72,68,69,67,55,56,59,52,53,
51,50,54,57,58,60,61,65,66,63,64,62}

Using the IDs as **Keys** for the golubPhenotypeData association, we can get the corresponding phenotype for each set. We have to wrap the IDs with **Key** to distinguish that these are keys and not index positions:

In[41]:= golubClass=**Query**[**Key**[#]&/@golubSampleIDs,1]@
golubPhenotypeData
Out[41]= <|1→ALL,2→ALL,3→ALL,4→ALL,5→ALL,6→ALL,
7→ALL,8→ALL,9→ALL,10→ALL,11→ALL,12→ALL,
13→ALL,14→ALL,15→ALL,16→ALL,17→ALL,
18→ALL,19→ALL,20→ALL,21→ALL,22→ALL,
23→ALL,24→ALL,25→ALL,26→ALL,27→ALL,
34→AML,35→AML,36→AML,37→AML,38→AML,
28→AML,29→AML,30→AML,31→AML,32→AML,
33→AML,39→ALL,40→ALL,42→ALL,47→ALL,
48→ALL,49→ALL,41→ALL,43→ALL,44→ALL,
45→ALL,46→ALL,70→ALL,71→ALL,72→ALL,
68→ALL,69→ALL,67→ALL,55→ALL,56→ALL,
59→ALL,52→AML,53→AML,51→AML,50→AML,
54→AML,57→AML,58→AML,60→AML,61→AML,
65→AML,66→AML,63→AML,64→AML,62→AML|>

We can see which positions belong to either the "AML" or "ALL" class using **Select** in operator form:

In[42]:= ?**Select**
Select[*list*,*crit*] picks out
all elements e_i of list for which $crit[e_i]$ is **True**.
Select[*list*,*crit*,*n*] picks out the first *n* elements
for which $crit[e_i]$ is **True**.
Select[*crit*] represents an operator form of **Select**
that can be applied to an expression. >>

In[43]:= **Query**[**Select**[#=="AML"&]/***Keys**]@golubClass
Out[43]= {34,35,36,37,38,28,29,30,31,32,33,52,
53,51,50,54,57,58,60,61,65,66,63,64,62}

6.2 Golub ALL AML Training Set

```
In[44]:=  Query[Select[#=="ALL"&]/*Keys]@golubClass
Out[44]=  {1,2,3,4,5,6,7,8,9,10,11,12,13,
           14,15,16,17,18,19,20,21,22,23,24,
           25,26,27,39,40,42,47,48,49,41,43,
           44,45,46,70,71,72,68,69,67,55,56,59}
```

And we can get a count of the samples in each category. Note that the dataset is not balanced in terms of samples for each class.

```
In[45]:=  Tally[golubClass]
Out[45]=  {{ALL,47},{AML,25}}
```

We can also extract the positions of specific genes. For example, in the original paper by Golub et al. [7], they list C-myb(probe ID U22376) on the top of their heatmap in Figure 3B therein. Let us see what our information for the same gene indicates:

```
In[46]:=  Position[golubNormedIDs,
            x_/;StringContainsQ[x,"U22376"],{2},
            Heads-> False]
Out[46]=  {{2911,2}}
```

The gene is in position 2911, with the identifier in the second component, as we can verify:

```
In[47]:=  golubNormedIDs[[2911]]
Out[47]=  {C-myb gene extracted from Human (c-myb) gene,
            complete primary cds, and five complete
              alternatively spliced cds,U22376_cds2_s_at}
```

We can then look at the normalized expression:

```
In[48]:=  golubNormed[[2911]]
Out[48]=  {1.71002,0.773188,1.88837,2.32075,1.87385,
           2.02271,2.04337,2.13613,1.68993,1.37129,
           1.74838,0.630459,2.08134,1.13712,2.12751,
           2.1689,2.05944,1.17697,2.14585,1.86482,3.06356,
           2.52905,2.08468,2.32105,1.47215,1.95209,1.
           90282,0.87523,0.875575,-0.075546,-0.781754,
           0.419801,0.471177,1.32368,0.0314873,0.0159155,
           0.675457,-0.406858,1.51835,1.32106,1.49333,
           1.98039,2.24898,1.56569,2.37146,2.23193,
           2.23863,2.0087,2.08052,1.89405,-0.1543,
           1.10945,2.41711,2.01153,1.1975,1.24516,1.66756,
           2.51198,0.225917,1.40227,-0.00831519,0.92578,
           1.65879,2.07047,0.36254,1.98684,0.793666,
           0.194616,0.987969,0.0266237,1.54799,-0.0505007}
```

If we separate out the intensities for AML and ALL we can have a box plot. In Fig. 6.3 we have the **BoxWhiskerChart** as a pure function on the go and act on the data using the **Prefix** form, i.e. using @ after our function definition.

We can finally now create the association that was used in Chap. 3 and in the book to access the Golub et. al full data for examples. We create an association of probe ID to the list of normalized values for the filtered dataset:

In[50]:= golubAssociation=<|
         "AML"-> AssociationThread[golubNormedIDs[[All,2]],
           golubNormed[[All,
             Flatten@Position["AML"]@Values@golubClass]]],
         "ALL"-> AssociationThread[golubNormedIDs[[All,2]],
           golubNormed[[All,
             Flatten@Position["ALL"]@Values@golubClass]]]|>;

In[51]:= golubAssociation // Short
Out[51]//Short= <|AML→ <|AFFX–BioDn–3_at→{<<1>>},
         <<3569>>,...at→<<1>>|>,<<1>>|>

The data is separated into lists of normalized intensities for each disease per gene accession:

In[52]:= Query[All,"AFFX-BioDn-3_at"]@golubAssociation
Out[52]= <|AML→
         {−1.31426,−1.21439,−1.36559,−1.31182,−1.29915,

```
In[49]:= BoxWhiskerChart[#, PlotTheme → "Scientific",
         ChartStyle → "Pastel",
         FrameLabel → { "Category", "Normalized Intensity"},
         PlotLabel →
           "Expression profile of C-myb in Golub et. al study",
         ChartLabels → {"AML", "ALL"},
         ChartLegends → {"AML", "ALL"}] &@
  {golubNormed[[2911]][[Query[Select[# == "AML" &] /* Keys]@golubClass]],
   golubNormed[[2911]][[Query[Select[# == "ALL" &] /* Keys]@golubClass]]}
```

Out[49]=

Fig. 6.3 Out[49]= C-myb Normalized Expression in Golub et al. study

6.2 Golub ALL AML Training Set

$\quad\quad\quad -1.38573, -1.06345, -0.301638, -1.22849, -1.40419,$
$\quad\quad\quad -1.43473, 0.230166, -1.44362, -1.22702, -1.28071,$
$\quad\quad\quad -1.32727, -1.05801, -1.2241, -0.266702, -0.443805,$
$\quad\quad\quad -1.14316, -1.01868, -1.42036, -1.48163, -1.3111\},$
$\quad\quad\text{ALL} \to \{-0.78835, -1.33516, -1.4235, -0.941616,$
$\quad\quad\quad -1.37342, -1.19283, -1.08886, -1.33541, -1.42439,$
$\quad\quad\quad -0.104142, -1.24374, -1.01377, -1.31832, -1.35606,$
$\quad\quad\quad -1.33969, -0.585923, -1.51022, -1.17998, -1.20761,$
$\quad\quad\quad -0.58472, -1.04577, -1.1116, -1.18712, -1.43465,$
$\quad\quad\quad -1.21717, -1.24631, 0.11392, -1.22521, -0.0170479,$
$\quad\quad\quad -1.24638, -1.27091, -1.43155, 0.391399, -1.14543,$
$\quad\quad\quad -1.19663, -1.07245, -1.06248, -1.12033, -1.0413,$
$\quad\quad\quad -1.21065, -1.29024, -1.06932, -1.42247,$
$\quad\quad\quad -1.18331, 0.089294, -0.762898, 0.0807061\}|>$

We can also create an OmicsObject, though a bit redundant at this point as we do not have a set of Metadata specific to each probe for each sample. We **Transpose** the normed data, then for each member we create an association with the probe ID to each value, and an external association with the sample IDs to its probes:

In[53]:= golubOmicsObject=
   AssociationThread[
     golubSampleIDs –>
       (**AssociationThread**[golubNormedIDs[[All,2]] ,
          {{#},{"NA"}}&/@#]&/@
          Transpose[golubNormed])];

Again, we must be careful with **Keys** that are numbers in associations. We have to call **Key**[62] to get the corresponding entry for that sample **Key**:

In[54]:= **Query**[**Key**[62]]@golubOmicsObject // **Short**
Out[54]//Short= <|AFFX–BioDn–3_at –>{{-1.3111},{NA}},
            <<3569>>,M71243_f_at –><<1>>|>

Instead we have to call the number 62 to get the 62nd entry, which is actually different:

In[55]:= **Query**[62]@golubOmicsObject // **Short**
Out[55]//Short= <|AFFX–BioDn–3_at –>{{-1.28071},{NA}},
            <<3569>>,M71243_f_at –><<1>>|>

6.2.3 Golub Annotations

The identities of the probes on this Hu6800 array have been updated from Affymetrix, and we can use them to obtain information, from the Affymetrix website (as before, **SystemOpen** will use the default program for your operating system to open the link, i.e. probably your default browser).

In[56]:= **SystemOpen**[
"https://www.affymetrix.com/analysis/index.affx"]

The annotations are distributed with the Mathematica notebooks for this monograph. We import the gene annotation, which contains tab separated values ("TSV"). We set "Numeric" to False as we want to import strings:

In[57]:= hu6800Genes=**Import**["Hu6800_Genes.tsv","TSV",
"Numeric"->**False**];
In[58]:= hu6800Genes[[1;;3]]
Out[58]= {{Probe Set ID, Gene Symbol, Gene Title,
Entrez Gene ID, Cytoband}, {AB000409_at, MKNK1,
MAP kinase interacting serine/threonine kinase 1,
8569, 1p33}, {AB000410_s_at, OGG1,
8-oxoguanine DNA glycosylase, 4968, 3p26.2}}

The first line is a header, that indicates that the columns contain identifiers for Gene Symbol, Title, Entrez Gene IDs and Cytoband information. We can create a dictionary from Probe Set ID to Gene Symbol and the rest of the gene information:

In[59]:= hu6800IDtoAnnotation=
AssociationThread[hu6800Genes[[**All**,1]],
hu6800Genes[[**All**,2;;]]];

Let us look at the first 10 in the list. We can use **Query**:

In[60]:= **Query**[1;;10] @hu6800IDtoAnnotation
Out[60]= <|Probe Set ID→
{Gene Symbol, Gene Title, Entrez Gene ID, Cytoband},
AB000409_at→{MKNK1,
MAP kinase interacting serine/threonine kinase 1,
8569, 1p33}, AB000410_s_at→
{OGG1, 8-oxoguanine DNA glycosylase, 4968, 3p26.2},
AB000450_at→
{VRK2, vaccinia related kinase 2, 7444, 2p16.1},
AB000816_s_at→{ARNTL, aryl hydrocarbon receptor
nuclear translocator-like, 406, 11p15},
AB002365_at→{PRUNE2, prune homolog 2 (Drosophila),
158471, 9q21.2},
AB002382_at→{CTNND1 /// TMX2-CTNND1,
catenin (cadherin-associated protein), delta 1
/// TMX2-CTNND1 readthrough (NMD candidate),
1500 /// 100528016, 11q11}, AB003177_at→
{PSMD9, proteasome 26S subunit, non-ATPase 9,
5715, 12q24.31-q24.32}, AB003698_at→
{CDC7, cell division cycle 7, 8317, 1p22},
AB006190_at→{AQP7, aquaporin 7, 364, 9p13}|>

6.2 Golub ALL AML Training Set

We notice that some entries might have multiple identifiers, separated by "///", for example: AB002382_at. We can separate these further using a **StringSplit** through **Query** as necessary:

In[61]:= **Query**[1;;10,StringSplit[#," /// "]&]@hu6800IDtoAnnotation
Out[61]= <|Probe Set ID→{{Gene Symbol},
          {Gene Title},{Entrez Gene ID},{Cytoband}},
       AB000409_at→{{MKNK1},{MAP kinase interacting
          serine/threonine kinase 1},
       {8569},{1p33}},AB000410_s_at→{{OGG1},
          {8−oxoguanine DNA glycosylase},{4968},{3p26.2}},
       AB000450_at→{{VRK2},{vaccinia related kinase 2},
          {7444},{2p16.1}},
       AB000816_s_at→{{ARNTL},{aryl hydrocarbon
          receptor nuclear translocator−like},
       {406},{11p15}},AB002365_at→{{PRUNE2},
          {prune homolog 2 (Drosophila)},
          {158471},{9q21.2}},
       AB002382_at→{{CTNND1,TMX2–CTNND1},
          {catenin (cadherin−associated protein), delta 1,
          TMX2–CTNND1 readthrough (NMD candidate)},
          {1500,100528016},{11q11}},AB003177_at→
       {{PSMD9},{proteasome 26S subunit, non−ATPase 9},
          {5715},{12q24.31−q24.32}},
       AB003698_at→{{CDC7},{cell division cycle 7},
          {8317},{1p22}},AB006190_at→
       {{AQP7},{aquaporin 7},{364},{9p13}}|>

6.2.4 Leukemia Differential Expression Analysis

We will carry out a differential gene expression analysis on these data, comparing ALL to AML. Above we had extracted all the data from ALL and AML and created an association for the data. Here we pair together the values for each gene per disease so we can create a list of {AML gene data, ALL gene data} per gene that we can use for location testing (see Chap. 3 for Statistics).

We use **Merge** with **Identity** as the function to merge matching keys, which will create paired lists:

In[62]:= golubDifferentialSet=**Merge**[{golubAssociation["AML"],
       golubAssociation["ALL"]},**Identity**];

206   6   Transcriptomics Examples

We can verify how these data look for each accession:

In[63]:= **Query**["AFFX-BioDn-3_at"] @ golubDifferentialSet
Out[63]= {{−1.31426, −1.21439, −1.36559, −1.31182, −1.29915,
             −1.38573, −1.06345, −0.301638, −1.22849, −1.40419,
             −1.43473, 0.230166, −1.44362, −1.22702, −1.28071,
             −1.32727, −1.05801, −1.2241, −0.266702, −0.443805,
             −1.14316, −1.01868, −1.42036, −1.48163, −1.3111},
             {−0.78835, −1.33516, −1.4235, −0.941616, −1.37342,
             −1.19283, −1.08886, −1.33541, −1.42439, −0.104142,
             −1.24374, −1.01377, −1.31832, −1.35606, −1.33969,
             −0.585923, −1.51022, −1.17998, −1.20761,
             −0.58472, −1.04577, −1.1116, −1.18712, −1.43465,
             −1.21717, −1.24631, 0.11392, −1.22521, −0.0170479,
             −1.24638, −1.27091, −1.43155, 0.391399, −1.14543,
             −1.19663, −1.07245, −1.06248, −1.12033, −1.0413,
             −1.21065, −1.29024, −1.06932, −1.42247,
             −1.18331, 0.089294, −0.762898, 0.0807061}}

We have a total of 3571 accessions for which we want to carry out testing.

In[64]:= **Length**[golubDifferentialSet]
Out[64]= 3571

We will apply a location test for each gene using **LocationTest**. For example :

In[65]:= **LocationTest**[**Query**["AFFX-BioDn-3_at"]@
             golubDifferentialSet, **Automatic**,
             {"TestDataTable", **All** }]

Out[65]= $\dfrac{|\text{Statistic P-Value}}{\text{Mann–Whitney}|467.\quad 0.155795}$

We see that the Mann–Whitney test was selected automatically, and we have a p value of 0.16. We can get the "TestConclusion":

In[66]:= **LocationTest**[**Query**["AFFX-BioDn-3_at"]@
             golubDifferentialSet, **Automatic**, "TestConclusion"]
Out[66]= The null hypothesis that the mean difference is 0
             is not rejected at the 5 percent level
             based on the Mann–Whitney test.

We will calculate the test p value and conclusion for all our sets. We return the "PValue" and the "ShortTestConclusion" so we can check the numbers of hypotheses rejected or not rejected:

In[67]:= testGolubConclusions=
             Query[**All** ,
               LocationTest[#, **Automatic**,
                 {"PValue", "ShortTestConclusion"}]&]@
             golubDifferentialSet;

As an example we look at the first few results:

In[68]:= testGolubConclusions[[1;;3]]
Out[68]= <|AFFX–BioDn–3_at→{0.155795, Do not reject},
             AFFX–BioB–5_st→{0.705063, Do not reject},
             AFFX–HUMISGF3A/M97935_MA_at→
               {0.139275, Do not reject}|>

6.2 Golub ALL AML Training Set

We can see how many times we rejected or did not reject the null hypothesis. We group the results by **Last**, which is the last element that takes values "Reject" or "Do not reject" and we check the **Length** of each list:

In[69]:= Length[#]&/@GroupBy[testGolubConclusions, Last]
Out[69]= <|Do not reject→2151, Reject→1420|>

Based on the results we would reject the null hypothesis that there is no difference in the means of ALL and AML data in 1420 cases. However, we have to account for multiple hypothesis testing. Here we have checked 3571 tests, so at 0.05 we expect to get 179 results just by chance, and the likelihood of Type I error is high. One approach would be to use a Bonferroni correction, which is to divide the significance cutoff α by the number of hypotheses tested. We are checking 3571 sets, so we can set the p value cutoff at 0.05/3571:

In[70]:= testGolubConclusionsBFCorrected=
    Query[All,
      LocationTest[#, Automatic,
        {"PValue", "ShortTestConclusion"},
        SignificanceLevel ->(0.05/3571)]&]@
    golubDifferentialSet;

We see that our rejection/no rejection tally has now shifted, with now 259 rejections of the null hypothesis:

In[71]:= Length[#]&/@
    GroupBy[testGolubConclusionsBFCorrected, Last]
Out[71]= <|Do not reject→3312, Reject→259|>

A Bonferroni correction might be too strict. We can check the distribution of p values in Fig. 6.4.

Another approach is to use a Benjamini-Hochberg [1] False Discovery Rate(FDR) correction (see also our ANOVA example in the next section). The **BenjaminiHochbergFDR** function is available in MathIOmica [10]:

In[73]:= <<MathIOmica`

In[74]:= ?BenjaminiHochbergFDR
    BenjaminiHochbergFDR[pValues] calculates for a list of
      pValues the Benjamini Hochberg approach false discovery
      rates (FDR).

We set the SignificianceLevel to 0.05 as a statistical significance cutoff.

In[75]:= testGolubFDRResults=
    BenjaminiHochbergFDR[
      Values@testGolubConclusionsBFCorrected[[All,1]],
      SignificanceLevel -> 0.05];

The output has **Keys**:

In[76]:= Keys@testGolubFDRResults
Out[76]= {Results, p−Value Cutoff, q−Value Cutoff}

```
In[72]:= Histogram[
    testGolubConclusionsBFCorrected[[All, 1]],
    PlotTheme → "Scientific", PlotLabel →
    "P Value Distribution from 3571 Location Tests",
    FrameLabel → {"P Value", "Count"}]
```

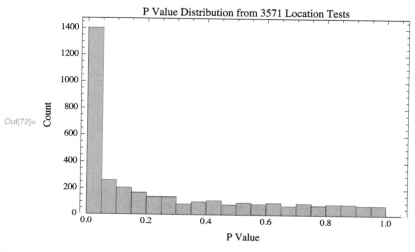

Fig. 6.4 Out[72]= Distribution of p values

The results contain list of values of the form: {original p−value, corrected p−value, **True**/**False**}. The last Boolean value in the list is an assertion on the significance of the p value based on the chosen SignificanceLevel. We take a look at the first few results:

```
In[77]:= Query["Results", 1;;5]@testGolubFDRResults
Out[77]= {{0.155795,0.293585,False},
         {0.705063,0.818524,False},
         {0.139275,0.270638,False},
         {0.524368,0.678947,False},
         {0.870155,0.925901,False}}
```

We can also obtain the "p-Value Cutoff" value for the p value corresponding to the significance level we decided on, and the "q-Value Cutoff" for the corresponding corrected cutoff value:

```
In[78]:= testGolubFDRResults["p-Value Cutoff"]
Out[78]= 0.0152203
```

```
In[79]:= testGolubFDRResults["q-Value Cutoff"]
Out[79]= 0.0499556
```

We next make an association for the features to the p value results :

6.2 Golub ALL AML Training Set

```
In[80]:= pValuesGolubFDR=
          Query["Results",
            (AssociationThread[
              Keys[testGolubConclusionsBFCorrected]-> #]&)]@
          testGolubFDRResults;
```

We can now select the results that give "True", and are significant based on the FDR:

```
In[81]:= Query[Select[#[[3]]==True&]]@pValuesGolubFDR;
```

There are in total 1088 features passing the FDR at 5 % (corresponding to a possible 55 false positive calls – the exact number being 54.4 rounded up):

```
In[82]:= Length[%]
Out[82]= 1088
```

We can extract the values and also calculate a fold change/effect size for all these values. Our data is already logarithmically transformed, so we look the the mean differences (which correspond to a fold) of the AML to ALL gene expressions. We can also see that the distribution of the mean changes is centered about zero, Fig. 6.5.

```
In[83]:= foldsGolubData=
          Query[All,Mean[#[[2]]] -Mean[#[[1]]]&]@
          golubDifferentialSet;
```

We can additionally calculate the negative Log base 10 of the relevant corrected p values:

```
In[85]:= pValueLog10GolubData=
          -Log10@Query[All,2]@pValuesGolubFDR;
```

We next use **Merge** to combine together the fold data and the p values:

```
In[86]:= volcanoPlotGolubData=
          Merge[{foldsGolubData,pValueLog10GolubData},
            Identity];
```

We can now construct a basic volcano plot in Fig. 6.6. We can also add lines to indicate our cutoffs for significance. We can add reference lines corresponding to cutoffs, at ± 1 for effect change, and also at - Log10[0.05] (negative base 10 log of 0.05) for significance. We use **GridLines** to add these, with the {{x grid lines },{y grid lines}} specification, where, in each grid line list, a line is indicated by {position, **Style**}, as shown in Fig. 6.7.

Alternatively we can use the **Epilog** option to insert a set of 2 dimensional graphics primitives, a set of lines at the right location provided a list as **Style**[**Line**[line positions], style information] , as shown in Fig. 6.8.

We can select features that have changes in the log mean greater than 1 in absolute values, and corrected p values smaller than 0.05:

```
In[90]:= selectedGenesGolub=
          Query[
            Select[
              Abs[#[[1]]]>1&& #[[2]] > (-Log10[0.05])&]]@
            volcanoPlotGolubData;
```

In[84]:= `Histogram[foldsGolubData,`
   `PlotTheme → "Scientific",`
   `PlotLabel → "Mean Log Change between AML-ALL",`
   `FrameLabel →`
     `{"Mean Log Change", "Count"}]`

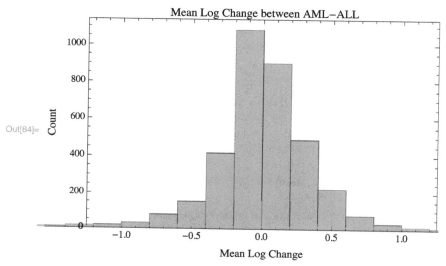

Fig. 6.5 Out[84]= Mean Log Change

The result is 92 genes:

In[91]:= `Length[selectedGenesGolub]`
Out[91]= 92

Here are the first few entries:

In[92]:= `selectedGenesGolub[[1;;10]]`
Out[92]= <|D10495_at→{−1.06097, 6.60748},
  D87433_at→{−1.14956, 3.48118},
  D88270_at→{2.29393, 5.11952},
  D88422_at→{−2.03815, 13.2644},
  HG1612−HT1612_at→{1.043, 13.2644},
  HG3494−HT3688_at→{−1.29344, 7.035},
  J03909_at→{−1.151, 3.63877},
  J04456_at→{−1.34649, 3.41742},
  J04615_at→{1.25548, 4.49444},
  J04990_at→{−1.05471, 4.46194}|>

We can use the accessions as keys for the array annotation we created earlier to obtain the gene names:

6.2 Golub ALL AML Training Set

```
In[87]:= ListPlot[Values@volcanoPlotGolubData,
    PlotTheme → "Scientific",
    PlotRange → Full,
    PlotLabel →
      "p Value Versus Mean Log Change Volcano Plot",
    FrameLabel → {"Mean Log Change", "-Log10 p Value"}]
```

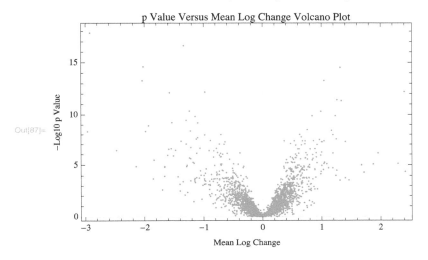

Fig. 6.6 Out[87]= Volcano plot

In[93]:= golubSelectedGeneNames=
    Query[Keys@selectedGenesGolub,1]@
       hu6800IDtoAnnotation;

Some gene information is missing, " --- ", from the annotation and there are multiple names for some others. For the multiple genes, we can create sublists of the multiple genes by using **StringSplit**, as discussed above:

In[94]:= **StringSplit**[#," /// "]&/@
       Values@golubSelectedGeneNames
Out[94]= {{PRKCD},{STAB1},{VPREB1},{CSTA},{---},
    {---},{IFI30},{LGALS1},{SNRPN,SNURF},{CTSG},
    {SPTAN1},{SERPINA1},{NPY},{CYFIP2},{DNTT},
    {LYN},{MPO},{S100A8},{CTSA},{CD33},{CST3},
    {IL7R},{FCER1G},{CD81},{CD9},{LGALS3},{CSF3R},
    {S100A4},{CFD},{CD79B},{SPINK2},{CCND3},
    {CTGF},{SERPINB1},{AZU1},{NAMPT},{CD79A},
    {LRMP},{VCAN},{MGST1},{FHL1},{PLD3},{DPYSL2},
    {ANXA1},{PLEK},{SRGN},{SPI1},{PRTN3},{IGHM},
    {CD22},{KIAA0391,PSMA6},{RHOG},{GRN},{DGKA},
    {CD63},{PRDX1},{RBBP4},{TCL1A},{UPP1},{ZYX},
    {ATP2A3},{P4HB},{MEF2C},{CCL3,CCL3L1,CCL3L3},

```
In[88]:= ListPlot[Values@volcanoPlotGolubData,
    PlotTheme → "Scientific",
    PlotLabel →
      "p Value Versus Mean Log Change Plot",
    FrameLabel → {"Mean Log Change", "-Log10 p Value"},
    PlotRange → Full,
    GridLines →
      {{{-1, {Red, Thick, Dashed}}, 0,
        {1, {Red, Thick, Dashed}}},
       {{-Log10[0.05], {Blue, Thick, Dashed}}}}]
```

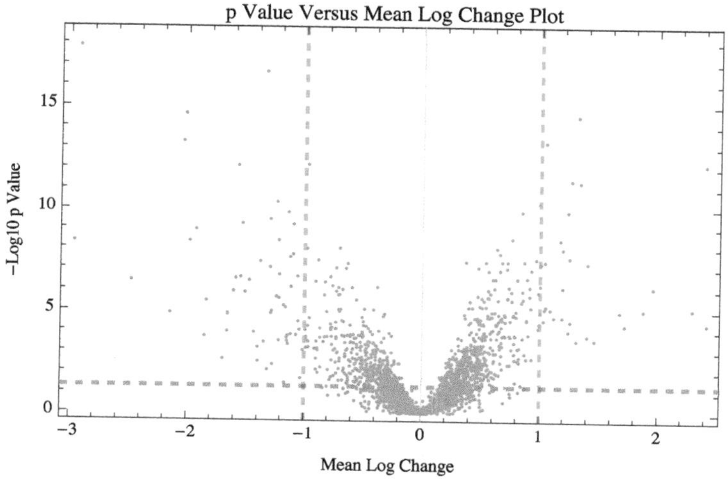

Fig. 6.7 Out[88]= Volcano plot with guide lines

{CXCR4}, {CD24}, {MYB}, {IGH, IGHA1, IGHA2},
{IGHG1, IGHG2, IGHG3, IGHG4, IGHM, IGHV4−31},
{LYZ}, {APLP2}, {LTB}, {ITGB2}, {CXCL8},
{CXCL8}, {ELANE}, {ELANE}, {S100A9}, {CD19},
{CXCL2}, {IGK, IGKC, IGKV1−5, IGKV2−24}, {IFI16},
{CFP}, {CEBPD}, {STOM}, {CORO1B, PTPRCAP},
{LYZ}, {LYZ}, {LYZ}, {TCF3}, {RNASE2}, {ITGB2}}

So we have identified a set of 92 genes that show differential expression based on the criteria we used, with a selected FDR of 0.05 and also displaying a mean Log change greater than 1. The results overlap with what is reported in the original Golub paper - please note that we have analyzed the full set here, and not just the training set.

```
In[89]:= ListPlot[Values@volcanoPlotGolubData,
   PlotTheme → "Scientific",
   PlotLabel →
    "p Value Versus Mean Log Change Plot",
   FrameLabel → {"Mean Log Change", "-Log₁₀ p Value"},
   PlotRange → Full,
   Epilog →
    {Style[ Line[{{-1, -20}, {-1, 20}}],
      Red, Thick, Dashed],
     Style[ Line[{{1, -20}, {1, 20}}],
      Red, Thick, Dashed],
     Style[ Line[{{-10, -Log10[0.05]},
        {10, -Log10[0.05]}}],
      Blue, Thick, Dashed]}]
```

Fig. 6.8 Out[89]= Volcano plot with guide lines using **Epilog**

6.3 Analysis of Variance for Multiple Tests

We often face a situation where we have multiple factors that we must account for in our analysis. Often an analysis of variance (ANOVA) method is required, and additionally accounting for interactions between factors in the analysis is warranted. In this section we use data from Sandberg et al. [15] and follow the analysis by Pavlidis [13, 14].

214   6 Transcriptomics Examples

In[95]:= SetDirectory[NotebookDirectory[]];

In[96]:= dataSandberg=Import["sandberg-sampledata.txt","TSV"];

In[97]:= dataSandberg[[1;;3]]
Out[97]= {{clone,129Amygdala(1a),129Amygdala(1b),
    129Cerebellum(1),129Cerebellum(2),129Cortex(1),
    129Cortex(2),129EntorhinalCortex(1a),
    129EntorhinalCortex(1b),129Hippocampus(1),
    129Hippocampus(2),129Midbrain(1),
    129Midbrain(2),B6Amygdala(1a),B6Amygdala(1b),
    B6Cerebellum(1),B6Cerebellum(2),B6Cortex(1),
    B6Cortex(2),B6EntorhinalCortex(1a),
    B6EntorhinalCortex(1b),B6Hippocampus(1),
    B6Hippocampus(2),B6Midbrain(1),B6Midbrain(2)},
   {aa000148.s.at,298,212,258,205,220,269,
    167,294,218,239,220,157,257,236,256,
    232,240,284,236,264,188,228,241,257},
   {AA000151.at,243,154,228,162,192,173,
    191,125,99,144,186,176,127,54,202,
    166,135,251,185,198,102,145,202,218}}

The data used by Pavlidis et al. consists of 24 microarray samples. The samples are from mouse strains 129 and B6. For each strain there are two duplicates from six different brain regions (amygdala, cerebellum, cortex, entorhinal cortex, hippocampus and midbrain). There are thus 2 factors, the strain (2 levels, 129 and B6) and the brain region (6 levels). We are looking for differentially expressed genes between strains and regions and also to identify interactions between the factors. The data have dimensions 1001x25 (accounting for header column and row):

In[98]:= Dimensions[dataSandberg]
Out[98]= {1001,25}

We can obtain factors for the ANOVA analysis from the header:

In[99]:= dataSandberg[[1]]
Out[99]= {clone,129Amygdala(1a),129Amygdala(1b),
    129Cerebellum(1),129Cerebellum(2),129Cortex(1),
    129Cortex(2),129EntorhinalCortex(1a),
    129EntorhinalCortex(1b),129Hippocampus(1),
    129Hippocampus(2),129Midbrain(1),
    129Midbrain(2),B6Amygdala(1a),B6Amygdala(1b),
    B6Cerebellum(1),B6Cerebellum(2),B6Cortex(1),
    B6Cortex(2),B6EntorhinalCortex(1a),
    B6EntorhinalCortex(1b),B6Hippocampus(1),
    B6Hippocampus(2),B6Midbrain(1),B6Midbrain(2)}

We can use string patterns to extract the information (though we could easily write these down, we are using this as an example). First we can get the mouse strains:

In[100]:= strains=
    StringCases[#,StartOfString~~_~~NumberString]&/@
      dataSandberg[[1]]
Out[100]= {{},{129},{129},{129},{129},{129},{129},{129},

6.3 Analysis of Variance for Multiple Tests

```
            {129},{129},{129},{129},{129},{B6},{B6},{B6},
          {B6},{B6},{B6},{B6},{B6},{B6},{B6},{B6},{B6}}
```

Then we can extract the brain regions of the samples:

```
In[101]:= regions=
          StringCases[#,
            NumberString..~~y:WordCharacter..~~"("~~_->
              y ]&/@dataSandberg[[1]]
Out[101]= {{},{Amygdala},{Amygdala},{Cerebellum},
          {Cerebellum},{Cortex},{Cortex},{EntorhinalCortex},
          {EntorhinalCortex},{Hippocampus},{Hippocampus},
          {Midbrain},{Midbrain},{Amygdala},{Amygdala},
          {Cerebellum},{Cerebellum},{Cortex},{Cortex},
          {EntorhinalCortex},{EntorhinalCortex},
          {Hippocampus},{Hippocampus},{Midbrain},{Midbrain}}
```

And finally we identify the repeat for each sample region:

```
In[102]:= repeats=StringCases[#,__~~"("~~x__~~")"-> x ]&/@
            dataSandberg[[1]]
Out[102]= {{},{1a},{1b},{1},{2},{1},{2},{1a},
          {1b},{1},{2},{1},{2},{1a},{1b},{1},
          {2},{1},{2},{1a},{1b},{1},{2},{1},{2}}
```

We can now set up our data. For example for 1 sample, here are the steps before we put it all together:

```
In[103]:= dataSandberg[[2]]
Out[103]= {aa000148.s.at,298,212,258,205,220,269,
          167,294,218,239,220,157,257,236,256,232,
          240,284,236,264,188,228,241,257}
```

We first transpose the strain and region information (we use **Short** to take a quick look at 3 lines of output):

```
In[104]:= Transpose[{strains,
            regions,dataSandberg[[2]]}]//Short[#,3]&
Out[104]//Short= {{{},{},aa000148.s.at,
                  {{129},{Amygdala},298},<<21>>,
                  {{B6},{Midbrain},241},{{B6},{Midbrain},257}}
```

Then we flatten the inner lists by mapping **Flatten** across the list:

```
In[105]:= Flatten[#]&/@
            Transpose[{strains,
              regions,dataSandberg[[2]]}]//Short[#,3]&
Out[105]//Short= {{aa000148.s.at},{129,Amygdala,298},
                  {129,Amygdala,212},<<20>>,
                  {B6,Midbrain,241},{B6,Midbrain,257}}
```

We can then make this an association:

```
In[106]:= <|#[[1,1]]-> #[[2;;]]|>&@
            (Flatten[#]&/@
              Transpose[{strains,regions,
                dataSandberg[[2]]}])//Short[#,3]&
```

216   6 Transcriptomics Examples

Out[106]//Short= <|aa000148.s.at→
                        {{129,Amygdala,298},{129,Amygdala,212},<<20>>,
                        {B6,Midbrain,241},{B6,Midbrain,257}}|>

And finally, we can put the steps together and do this across all Probes:

In[107]:= associationDataSandberg=
              AssociationThread[#[[**All**,1,1]]->#[[**All**,2;;]]]&@
              ((**Flatten**[#]&/@Transpose[{strains,regions,#}])&/@
              dataSandberg);

Let us look at 2 entries - the first one is the header, and the second one a typical entry, say number 29:

In[108]:= **Query**[{1,29}]@associationDataSandberg
Out[108]= <|clone→{{129,Amygdala,129Amygdala(1a)},
                {129,Amygdala,129Amygdala(1b)},
                {129,Cerebellum,129Cerebellum(1)},
                {129,Cerebellum,129Cerebellum(2)},
                {129,Cortex,129Cortex(1)},
                {129,Cortex,129Cortex(2)},
                {129,EntorhinalCortex,129EntorhinalCortex(1a)},
                {129,EntorhinalCortex,129EntorhinalCortex(1b)},
                {129,Hippocampus,129Hippocampus(1)},
                {129,Hippocampus,129Hippocampus(2)},
                {129,Midbrain,129Midbrain(1)},
                {129,Midbrain,129Midbrain(2)},
                {B6,Amygdala,B6Amygdala(1a)},
                {B6,Amygdala,B6Amygdala(1b)},
                {B6,Cerebellum,B6Cerebellum(1)},
                {B6,Cerebellum,B6Cerebellum(2)},
                {B6,Cortex,B6Cortex(1)},
                {B6,Cortex,B6Cortex(2)},
                {B6,EntorhinalCortex,B6EntorhinalCortex(1a)},
                {B6,EntorhinalCortex,B6EntorhinalCortex(1b)},
                {B6,Hippocampus,B6Hippocampus(1)},
                {B6,Hippocampus,B6Hippocampus(2)},
                {B6,Midbrain,B6Midbrain(1)},
                {B6,Midbrain,B6Midbrain(2)}},
             aa013993.at→{{129,Amygdala,103},
                {129,Amygdala,76},{129,Cerebellum,130},
                {129,Cerebellum,150},{129,Cortex,130},
                {129,Cortex,110},{129,EntorhinalCortex,81},
                {129,EntorhinalCortex,92},{129,Hippocampus,87},
                {129,Hippocampus,111},{129,Midbrain,105},
                {129,Midbrain,117},{B6,Amygdala,81},
                {B6,Amygdala,83},{B6,Cerebellum,74},
                {B6,Cerebellum,63},{B6,Cortex,96},
                {B6,Cortex,132},{B6,EntorhinalCortex,67},
                {B6,EntorhinalCortex,114},
                {B6,Hippocampus,59},{B6,Hippocampus,44},
                {B6,Midbrain,100},{B6,Midbrain,95}}|>

Now we can use these data to build a linear model. First we use the set 29 as an example before we do this across all sets:

6.3 Analysis of Variance for Multiple Tests

In[109]:= lmExample=
 LinearModelFit[associationDataSandberg[[29]],
  {str,rg,str∗rg},{str,rg},
  NominalVariables->{str,rg}]
Out[109]= FittedModel[<<1>>]

We have specified the input for **LinearModelFit** where our data is in the form { strain specification, region specification,expression value}, we are constructing model that is linear in str,rg and an interaction str∗rg, and we specify that {str,rg} are the variables, and that both are nominal variables. The FittedModel has many properties. To get a list of properties we do a "Properties" evaluation, which we list here for reference:

In[110]:= lmExample["Properties"]
Out[110]= {AdjustedRSquared, AIC, AICc, ANOVATable,
 ANOVATableDegreesOfFreedom, ANOVATableEntries,
 ANOVATableFStatistics, ANOVATableMeanSquares,
 ANOVATablePValues, ANOVATableSumsOfSquares,
 BasisFunctions, BetaDifferences, BestFit,
 BestFitParameters, BIC, CatcherMatrix,
 CoefficientOfVariation, CookDistances,
 CorrelationMatrix, CovarianceMatrix,
 CovarianceRatios, Data, DesignMatrix, DurbinWatsonD,
 EigenstructureTable, EigenstructureTableEigenvalues,
 EigenstructureTableEntries, EigenstructureTableIndexes,
 EigenstructureTablePartitions, EstimatedVariance,
 FitDifferences, FitResiduals, **Function**, FVarianceRatios,
 HatDiagonal, MeanPredictionBands,
 MeanPredictionConfidenceIntervals,
 MeanPredictionConfidenceIntervalTable,
 MeanPredictionConfidenceIntervalTableEntries,
 MeanPredictionErrors, ParameterConfidenceIntervals,
 ParameterConfidenceIntervalTable,
 ParameterConfidenceIntervalTableEntries,
 ParameterConfidenceRegion, ParameterErrors,
 ParameterPValues, ParameterTable, ParameterTableEntries,
 ParameterTStatistics, PartialSumOfSquares,
 PredictedResponse, Properties, Response, RSquared,
 SequentialSumOfSquares, SingleDeletionVariances,
 SinglePredictionBands,
 SinglePredictionConfidenceIntervals,
 SinglePredictionConfidenceIntervalTable,
 SinglePredictionConfidenceIntervalTableEntries,
 SinglePredictionErrors, StandardizedResiduals,
 StudentizedResiduals, VarianceInflationFactors}

We can get the "FitResiduals" and "AdjustedRSquared":

In[111]:= lmExample[{"FitResiduals","AdjustedRSquared"}]
Out[111]= {{13.5,−13.5,−10.,10.,10.,−10.,−5.5,5.5,−12.,
 12.,−6.,6.,−1.,1.,5.5,−5.5,−18.,18.,
 −23.5,23.5,7.5,−7.5,2.5,−2.5},0.612559}

And we can get an ANOVA table:

In[112]:= lmExample["ANOVATable"]

Out[112]=

	DF	SS	MS	F-Statistic	P-Value
str	1	3360.67	3360.67	12.905	0.00369748
rg	5	4675.33	935.067	3.59066	0.0323236
rgstr	5	4298.33	859.667	3.30112	0.041804
Error	12	3125.	260.417		
Total	23	15459.3			

Let us look at a few more of the entries:

```
In[113]:= #-> lmExample[#]&/@
    {"AdjustedRSquared","AIC","AICc",
    "ANOVATableDegreesOfFreedom","ANOVATableEntries",
    "ANOVATableFStatistics","ANOVATableMeanSquares"
    ,"ANOVATablePValues","ANOVATableSumsOfSquares",
    "CoefficientOfVariation","Response","RSquared",
    "SequentialSumOfSquares"}
Out[113]= {AdjustedRSquared→0.612559,
    AIC→215.604,AICc→252.004,
    ANOVATableDegreesOfFreedom→{1,5,5,12,23},
    ANOVATableEntries→
     {{1,3360.67,3360.67,12.905,0.00369748},
     {5,4675.33,935.067,3.59066,0.0323236},
     {5,4298.33,859.667,3.30112,0.041804},
     {12,3125.,260.417},{23,15459.3}},
    ANOVATableFStatistics→{12.905,3.59066,3.30112},
    ANOVATableMeanSquares→
     {3360.67,935.067,859.667,260.417},
    ANOVATablePValues→{0.00369748,0.0323236,0.041804},
    ANOVATableSumsOfSquares→
     {3360.67,4675.33,4298.33,3125.,15459.3},
    CoefficientOfVariation→0.168391,
    Response→{103,76,130,150,130,110,81,92,
     87,111,105,117,81,83,74,63,96,132,67,
     114,59,44,100,95},RSquared→0.797857,
    SequentialSumOfSquares→{3360.67,4675.33,4298.33}}
```

We can also get an overall p value for the ANOVA, which is traditionally obtained, from evaluating an **FRatioDistribution** with n numerator and m denominator degrees of freedom, with the degrees of freedom for the factors obtained from the ANOVATable (up to the last two entries that are for the Error and Total respectively), and we can evaluate the CDF at the mean of the F-Statistic for the two factors and interaction. Subtracting the cumulative from 1 gives us the p value:

```
In[114]:= 1-CDF[FRatioDistribution[
    Plus @@
     lmExample["ANOVATableDegreesOfFreedom"][[1;;-3]],
     lmExample["ANOVATableDegreesOfFreedom"][[-2]],
    Mean[lmExample["ANOVATableFStatistics"]]]
Out[114]= 0.00144074
```

We could instead have used the older functionality in the ANOVA package which allows for different treatment of interactions as well as post testing:

```
In[68]:= testGolubConclusions[[1;;3]]
In[115]:= <<ANOVA`
```

6.3 Analysis of Variance for Multiple Tests

We run again ANOVA on the same data, with specifying interactions as **All** and also asking for Post-hoc test for Tukey and Bonferroni corrections:

In[116]:= ?**ANOVA**
        ANOVA[*data*] performs a one−way analysis of variance.
        ANOVA[*data*,*model*,*vars*] performs an analysis of variance
            for model as a function of the categorical
            variables *vars*.?>>

In[117]:= **ANOVA**[associationDataSandberg[[29]],{str,rg,All},
        {str,rg},PostTests −>{Tukey,Bonferroni},
        SignificanceLevel −>.05]

Out[117]= {ANOVA →

	DF	SumOfSq	MeanSq	FRatio	PValue
	1	3360.67	3360.67	12.905	0.00369748
	5	4675.33	935.067	3.59066	0.0323236
	5	4298.33	859.667	3.30112	0.041804
	12	3125.	260.417	23	15459.3

,

CellMeans →

All	95.8333
str(129)	107.667
str(B6)	84.
rg(Amygdala)	85.75
rg(Cerebellum)	104.25
rg(Cortex)	117.
rg(EntorhinalCortex)	88.5
rg(Hippocampus)	75.25
rg(Midbrain)	104.25
str(129)rg(Amygdala)	89.5
rg(Amygdala)str(B6)	82.
str(129)rg(Cerebellum)	140.
str(B6)rg(Cerebellum)	68.5
str(129)rg(Cortex)	120.
str(B6)rg(Cortex)	114.
str(129)rg(EntorhinalCortex)	86.5
str(B6)rg(EntorhinalCortex)	90.5
str(129)rg(Hippocampus)	99.
str(B6)rg(Hippocampus)	51.5
str(129)rg(Midbrain)	111.
str(B6)rg(Midbrain)	97.5

,

PostTests → { str → Bonferroni {129, B6} / Tukey {129, B6} ,
rg → Bonferroni {Cortex, Hippocampus} / Tukey {Cortex, Hippocampus} }}

For our purposes the **LinearModelFit** works well to generalize across the entire measurement set (note that we skip the header line):

In[118]:= linearModelsSandberg=
        Query[2;;,
            LinearModelFit[#,{str,rg,str∗rg},{str,rg},
            NominalVariables −>{str,rg}]&]@
        associationDataSandberg;

Let us again look at the first few entries:

In[119]:= linearModelsSandberg[[1;;4]]

Out[119]= <|aa000148.s.at→FittedModel[<<1>>],
AA000151.at→FittedModel[<<1>>],
aa000380.s.at→
FittedModel[<<16>>+84.
<<1>> DiscreteIndicator[<<1>>]],
aa000467.s.at→FittedModel[<<1>>]|>

We now have an association and a set of fitted models for each. We can then extract all the p values for the model:

In[120]:= pValuesSandberg=
Query[All,Chop[Re[#["ANOVATablePValues"]]]&]
@linearModelsSandberg;

Note that the **Chop** and **Re** are to eliminate any rounding artifacts of small numbers, and they may or not be necessary. We can look at the first three results:

In[121]:= pValuesSandberg[[1;;3]]
Out[121]= <|aa000148.s.at→{0.426513,0.7229,0.782272},
AA000151.at→{0.673324,0.154761,0.242283},
aa000380.s.at→{0.0781434,0.825637,0.434008}|>

Following Pavlidis [13] we can also extract various p values that are lower than a desired cutoff, e.g. p<0.0001, and specific to an effect, such as for example the cases that have a regional effect and do not display an interaction between strain and region:

In[122]:= regionalOnlyGenes=
Query[Select[#[[2]]<0.0001&&#[[3]]>0.1&]]@
pValuesSandberg;

There are 41 genes satisfying this condition:

In[123]:= **Length**[regionalOnlyGenes]
Out[123]= 41

We can extract their accessions:

In[124]:= **Keys**[regionalOnlyGenes]
Out[124]= {AA028280.at,aa028770.i.at,
aa028770.s.at,aa030488.at,aa030488.g.at,
AA030688.s.at,AA033074.s.at,aa033314.s.at,
aa034739.s.at,aa035912.at,aa035912.g.at,
AA038775.s.at,aa105755.at,aa106571.g.at,
aa120563.at,aa122778.s.at,aa123328.s.at,
aa123934.s.at,AA124090.i.at,aa153748.g.at,
aa153942.s.at,aa155529.s.at,AA166452.at,
AA171106.at,aa172673.s.at,aa172864.s.at,
aa174604.s.at,aa175734.s.at,aa175767.f.at,
aa178190.s.at,aa182125.s.at,aa183544.s.at,
aa183623.s.at,AA189677.at,aa189989.at,
AA198947.at,aa204034.s.at,aa204106.s.at,
AA204584.at,aa210359.s.at,aa220788.s.at}

And we can see the first 3 in the results as a verification:

6.3 Analysis of Variance for Multiple Tests

```
In[125]:= regionalOnlyGenes[[1;;3]]
Out[125]= <|AA028280.at→{0.273171,1.63919×10⁻⁷,0.172343},
           aa028770.i.at→{0.404534,0.0000899459,0.602528},
           aa028770.s.at→{0.213727,9.20976×10⁻⁷,0.150637}|>
```

We can now correct for multiple hypothesis testing. Here we are testing 1000 hypotheses. One way to correct would be a Bonferroni correction, setting a cutoff for significance at 0.01/1000. Another way is to use again the Benjamini Hochberg [1] correction available in MathIOmica.

In Pavlidis [13] the FDR calculation is performed for only for the brain region p values. We can also reproduce the results. This corresponds to the second set of p values from the linear model fit:

```
In[126]:= regionFDRResults=
           BenjaminiHochbergFDR[
             Flatten[#[[All,2]]]&@Values@pValuesSandberg,
             SignificanceLevel -> 0.05];
```

The results are a set of original and corrected p-values and a logical True/False on whether the p-value was significant (see also the differential expression analysis section). The p-value cutoff corresponding to the significance level we decided on, and the q-Value are:

```
In[127]:= Query["p-Value Cutoff"]@regionFDRResults
Out[127]= 0.00940035

In[128]:= Query["q-Value Cutoff"]@regionFDRResults
Out[128]= 0.0489602
```

We now make an association of the annotation keys to the values so we can search it:

```
In[129]:= pValuesSandbergRegionFDR=
           Query["Results",
             (AssociationThread[Keys[pValuesSandberg]-> #]&)]@
             regionFDRResults;
```

We can query the association for the results that pass the cutoff, and have an FDR of 0.05 as constructed above:

```
In[130]:= Query[Select[#[[3]]==True&]]@
           pValuesSandbergRegionFDR//Short
Out[130]//Short= <|aa000750.s.at→{<<1>>},
                  <<190>>,aa220788.s.at→{<<1>>}|>
```

We also get 192 genes:

```
In[131]:= Length[%]
Out[131]= 192
```

We can sort these by their corrected p value, by using **SortBy** in operator form to sort by the second value in the list:

```
In[132]:= sortedSandbergRegionFDR=
           Query[Select[#[[3]]==True&]/*SortBy[#[[2]]&]]@
             pValuesSandbergRegionFDR;
```

Let us look at the top 20:

In[133]:= Keys@Query[1;;20]@sortedSandbergRegionFDR
Out[133]= {aa183623.s.at, aa212550.s.at,
aa220788.s.at, AA171106.at, aa123328.s.at,
AA189887.at, AA030688.s.at, aa137436.s.at,
aa153748.g.at, AA028280.at, aa106571.g.at,
aa204106.s.at, aa204034.s.at, aa172864.s.at,
aa189518.s.at, AA038775.s.at, aa028770.s.at,
aa123989.at, aa189989.at, aa172673.s.at}

We can extract the intensity values and cluster the data to generate a matrix heatmap using MathIOmica functions. First we extract the data based on the information for the top 20 hits for region. We use the **Keys** extracted for these 20 sets from a **Query**; for these 20 keys, we extract the 3rd value of the data (i.e. the gene expression) from **All** values in associationDataSandberg with another **Query**:

In[134]:= dataSandbergSorted=
Query[Keys @ Query[1;;20]@sortedSandbergRegionFDR,
All, 3]@associationDataSandberg;

Let us look at the first three entries:

In[135]:= dataSandbergSorted[[1;;3]]
Out[135]= <|aa183623.s.at→{594,708,−15,−2,966,1165,1518,
1481,137,230,81,105,726,535,−8,10,937,950,
1452,1464,205,181,68,240}, aa212550.s.at→
{47,51,252,222,56,43,28,13,50,45,35,29,
59,90,390,377,24,49,35,42,53,53,78,74},
aa220788.s.at→{661,729,8,−66,622,642,
587,567,726,752,447,480,781,743,12,
−25,516,646,541,480,637,720,375,521}|>

We now cluster the data, using MathIOmica's automatic clustering function **MatrixClusters**. We use a standardization of the data (subtraction of the mean and division by the standard deviation). The standardization puts the data on a unit-free scale so we can visualize the genes that may have different ranges of values together, i.e. we look at relative gene expression pattern changes that match across genes.

In[136]:= clustersSandbergSorted=
MatrixClusters[
N[**Standardize**[#]& /@ dataSandbergSorted]];
 Agglomerate::ties: 1 ties have been detected; reordering input may produce a different result.

The **MatrixClusters** returns 1 association:

In[137]:= Keys[clustersSandbergSorted]
Out[137]= {1}

This association in turn has multiple **Keys**:

In[138]:= Keys @ clustersSandbergSorted["1"]
Out[138]= {RowCluster, ColumnCluster,
RowSplitClusters, ColumnSplitClusters,
GroupAssociationsRows, GroupAssociationsColumns}

6.3 Analysis of Variance for Multiple Tests

The "GroupAssociations" gives us the data order of the clusters. For rows we have:

In[139]:= clustersSandbergSorted["1","GroupAssociationsRows"]
Out[139]= <|H:G1→{aa183623.s.at,aa153748.g.at,aa172864.s.at,
           aa028770.s.at,aa189989.at,AA189887.at,
           aa204034.s.at,aa220788.s.at,AA171106.at},
         H:G2→{aa212550.s.at,aa204106.s.at,
           AA038775.s.at,aa189518.s.at,aa106571.g.at,
           aa123989.at,aa123328.s.at,aa172673.s.at,
           aa137436.s.at,AA028280.at,AA030688.s.at}|>

We note here the notation H:G(i) for the Horizontal i^{th} Group. For columns we have:

In[140]:= clustersSandbergSorted["1","GroupAssociationsColumns"]
Out[140]= <|V:G1→{1,2,13,14,5,6,17,18,7,8,19,20},
         V:G2→{21,9,10,22},
         V:G3→{24,23,11,12},V:G4→{15,16,3,4}|>

We see that we have a different notation: V:G(j) for each Vertical j^{th} Group. The numbers in the group values correspond to how the entries are ordered, as they represent the index (list location) of the corresponding input data that we provided to **MatrixClusters**.

Let us also create a set of labels for strains/regions for graphing purposes. We extract the labels:

In[141]:= horizontalLabels=
          Flatten @
            Values @ clustersSandbergSorted["1",
              "GroupAssociationsColumns"]
Out[141]= {1,2,13,14,5,6,17,18,7,8,19,20,
           21,9,10,22,24,23,11,12,15,16,3,4}

We then make shorthand notations for the strains, 129 (1) and B6(2), and for the regions: amygdala (a), cerebellum (c), cortex (cx), entorhinalCortex (e), hippocampus (h), midbrain (m):

In[142]:= regions=
          AssociationThread[**Range**[**Length**[#]]→ #]&@
            Query[1,**All**,{1,2}]@associationDataSandberg /.
          {"129"→ "1","B6"→ "2","Amygdala"→ "a",
           "Cerebellum"→ "c","Cortex"→ "cx",
           "EntorhinalCortex"→ "e","Hippocampus"→ "h",
           "Midbrain"→ "m"}
Out[142]= <|1→{1,a},2→{1,a},3→{1,c},4→{1,c},
           5→{1,cx},6→{1,cx},7→{1,e},8→{1,e},
           9→{1,h},10→{1,h},11→{1,m},12→{1,m},
           13→{2,a},14→{2,a},15→{2,c},16→{2,c},
           17→{2,cx},18→{2,cx},19→{2,e},20→{2,e},
           21→{2,h},22→{2,h},23→{2,m},24→{2,m}|>

Finally, we arrange these region notations in order of the horizontalLabels, by using the horizontalLabels as keys in a **Query**, and we write the paired lists from regions as fractions to conserve space:

In[144]:= `MatrixDendrogramHeatmap[clustersSandbergSorted["1"],`
`FrameName →`
`"Sorted Sandberg Gene Expression Data Clustering",`
`HorizontalAxisName → "Sample Data",`
`HorizontalLabels → heatmapLabels,`
`ImageSize → 350]`

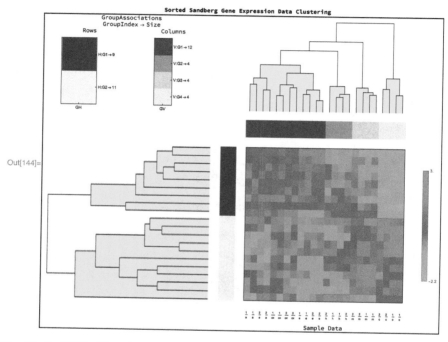

Fig. 6.9 Out[144]= Clustering of top 20 genes showing differences associated with brain region [13, 15]

In[143]:= heatmapLabels=
Values @ **Query**[horizontalLabels, $\frac{\#[[1]]}{\#[[2]]}$&]@regions
Out[143]= $\{\frac{1}{a}, \frac{1}{a}, \frac{2}{a}, \frac{2}{a}, \frac{1}{cx}, \frac{1}{cx}, \frac{2}{cx}, \frac{2}{cx}, \frac{1}{e}, \frac{1}{e}, \frac{2}{e}, \frac{2}{e}, \frac{1}{h}, \frac{1}{h}, \frac{2}{h}, \frac{2}{h}, \frac{1}{m}, \frac{2}{m}, \frac{2}{m}, \frac{1}{c}, \frac{1}{c}, \frac{2}{c}, \frac{2}{c}, \frac{1}{c}, \frac{1}{c}\}$

We generate a heamap with the groupings and dendrograms using the function **MatrixDendrogramHeatmap**, Fig. 6.9. We can see in the figure there is a clear grouping of brain regions, particularly for samples from the hippocampus, the midbrain and the cerebellum. On the other hand for the samples from amygdala, cortex and entorhinal cortex we do not see any difference (at least visually) in expression. Notice also at the top left that there is a color legend, and also an association of grouping to size of the group.

References

1. Benjamini, Y., Hochberg, Y.: Controlling the false discovery rate: a practical and powerful approach to multiple testing. J. R. Stat. Soc. Ser. B (Methodol.), 289–300 (1995)
2. Bertone, P., Stolc, V., Royce, T.E., Rozowsky, J.S., Urban, A.E., Zhu, X., Rinn, J.L., Tongprasit, W., Samanta, M., Weissman, S., Gerstein, M., Snyder, M.: Global identification of human transcribed sequences with genome tiling arrays. Science 306(5705), 2242–6 (2004)
3. Cheng, J., Kapranov, P., Drenkow, J., Dike, S., Brubaker, S., Patel, S., Long, J., Stern, D., Tammana, H., Helt, G., Sementchenko, V., Piccolboni, A., Bekiranov, S., Bailey, D.K., Ganesh, M., Ghosh, S., Bell, I., Gerhard, D.S., Gingeras, T.R.: Transcriptional maps of 10 human chromosomes at 5-nucleotide resolution. Science 308(5725), 1149–54 (2005)
4. Clark, T.A., Sugnet, C.W., Ares Jr., M.: Genomewide analysis of mRNA processing in yeast using splicing-specific microarrays. Science 296(5569), 907–10 (2002)
5. David, L., Huber, W., Granovskaia, M., Toedling, J., Palm, C.J., Bofkin, L., Jones, T., Davis, R.W., Steinmetz, L.M.: A high-resolution map of transcription in the yeast genome. Proc. Natl. Acad. Sci. U.S.A. 103(14), 5320–5 (2006)
6. Dudoit, S., Fridlyand, J., Speed, T.P.: Comparison of discrimination methods for the classification of tumors using gene expression data. J. Am. Stat. Assoc. 97(457), 77–87 (2002)
7. Golub, T.R., Slonim, D.K., Tamayo, P., Huard, C., Gaasenbeek, M., Mesirov, J.P., Coller, H., Loh, M.L., Downing, J.R., Caligiuri, M.A., Bloomfield, C.D., Lander, E.S.: Molecular classification of cancer: class discovery and class prediction by gene expression monitoring. Science 286(5439), 531–537 (1999)
8. Marioni, J.C., Mason, C.E., Mane, S.M., Stephens, M., Gilad, Y.: RNA-seq: an assessment of technical reproducibility and comparison with gene expression arrays. Genome Res. 18(9), 1509–17 (2008)
9. Mias, G., Snyder, M.: Personal genomes, quantitative dynamic omics and personalized medicine. Quant. Biol. 1(1), 71–90 (2013)
10. Mias, G.I., Yusufaly, T., Roushangar, R., Brooks, L.R., Singh, V.V., Christou, C.: Mathiomica: An integrative platform for dynamic Omics. Sci. Rep. 6, 37,237 (2016)
11. Mortazavi, A., Williams, B.A., McCue, K., Schaeffer, L., Wold, B.: Mapping and quantifying mammalian transcriptomes by RNA-seq. Nat. Methods 5(7), 621–8 (2008)
12. Nagalakshmi, U., Wang, Z., Waern, K., Shou, C., Raha, D., Gerstein, M., Snyder, M.: The transcriptional landscape of the yeast genome defined by RNA sequencing. Science 320(5881), 1344–9 (2008)
13. Pavlidis, P.: Using ANOVA for gene selection from microarray studies of the nervous system. Methods 31(4), 282 – 289 (2003). (Candidate Genes from DNA Array Screens: application to neuroscience)
14. Pavlidis, P., Noble, W.S.: Matrix2png: a utility for visualizing matrix data. Bioinformatics 19(2), 295–296 (2003)
15. Sandberg, R., Yasuda, R., Pankratz, D.G., Carter, T.A., Del Rio, J.A., Wodicka, L., Mayford, M., Lockhart, D.J., Barlow, C.: Regional and strain-specific gene expression mapping in the adult mouse brain. Proc. Natl. Acad. Sci. 97(20), 11038–11043 (2000)
16. Wang, Z., Gerstein, M., Snyder, M.: RNA-seq: a revolutionary tool for transcriptomics. Nat. Rev. Genet. 10(1), 57–63 (2009)
17. Wilhelm, B.T., Marguerat, S., Watt, S., Schubert, F., Wood, V., Goodhead, I., Penkett, C.J., Rogers, J., Bahler, J.: Dynamic repertoire of a eukaryotic transcriptome surveyed at single-nucleotide resolution. Nature 453(7199), 1239–43 (2008)
18. Wolfram, S.: An Elementary Introduction to the Wolfram Language, Wolfram Media (2015)
19. Wolfram Alpha LLC: Wolfram|Alpha (2017). Accessed Nov 2017

20. Wolfram Research, Inc.: Mathematica, Version 11.2. Champaign, IL (2017)
21. Yamada, K., Lim, J., Dale, J.M., Chen, H., Shinn, P., Palm, C.J., Southwick, A.M., Wu, H.C., Kim, C., Nguyen, M., Pham, P., Cheuk, R., Karlin-Newmann, G., Liu, S.X., Lam, B., Sakano, H., Wu, T., Yu, G., Miranda, M., Quach, H.L., Tripp, M., Chang, C.H., Lee, J.M., Toriumi, M., Chan, M.M., Tang, C.C., Onodera, C.S., Deng, J.M., Akiyama, K., Ansari, Y., Arakawa, T., Banh, J., Banno, F., Bowser, L., Brooks, S., Carninci, P., Chao, Q., Choy, N., Enju, A., Goldsmith, A.D., Gurjal, M., Hansen, N.F., Hayashizaki, Y., Johnson-Hopson, C., Hsuan, V.W., Iida, K., Karnes, M., Khan, S., Koesema, E., Ishida, J., Jiang, P.X., Jones, T., Kawai, J., Kamiya, A., Meyers, C., Nakajima, M., Narusaka, M., Seki, M., Sakurai, T., Satou, M., Tamse, R., Vaysberg, M., Wallender, E.K., Wong, C., Yamamura, Y., Yuan, S., Shinozaki, K., Davis, R.W., Theologis, A., Ecker, J.R.: Empirical analysis of transcriptional activity in the arabidopsis genome. Science **302**(5646), 842–6 (2003)

Chapter 7
Proteomic Data

7.1 Amino Acids

Based on the central dogma of molecular biology, our DNA is transcribed to RNA, which is in turn translated to proteins. The translation takes place in the ribosome, where RNA is processed in 3-base steps, in terms of codons that correspond to particular amino acids. There is some code redundancy in terms of the information as we have multiple codons that correspond to one amino acid, and while there are 4^3 codon possibilities (corresponding to a different base pair choice at each point) there are only 20 amino acids.

Following translation, the resulting macromolecules of multiple amino acids are the proteins. Short sequences of amino acids are called peptides. The Wolfram Language [1] has **ChemicalData** associated with the 20 amino acid molecules. Additionally, there is a "Chemical" **EntityClass**, with "AminoAcids" included:

In[1]:= aminoAcids=**EntityClass**["Chemical", "AminoAcids"];

The amino acids have many "Properties":

In[2]:= **ChemicalData**["AminoAcids","Properties"]//**Short**
Out[2]//**Short**= {AcidityConstants, AdjacencyMatrix,
    AlternateNames, AtomPositions,<<92>>,
    VertexCoordinates, VertexTypes, Viscosity}

Here are the first few included:

In[3]:= **ChemicalData**["AminoAcids","Properties"][[1;;20]]
Out[3]= {AcidityConstants, AdjacencyMatrix, AlternateNames,
    AtomPositions, AutoignitionPoint, BeilsteinNumber,
    BlackStructureDiagram, BoilingPoint,
    BondTally, BoundaryMeshRegion, CASNumber,
    CHBlackStructureDiagram, CHColorStructureDiagram,

Electronic supplementary material The online version of this chapter (https://doi.org/10.1007/978-3-319-72377-8_7) contains supplementary material, which is available to authorized users.

228   7 Proteomic Data

CIDNumber, Codons, ColorStructureDiagram,
CombustionHeat, CompoundFormulaDisplay,
CompoundFormulaString, CriticalPressure}

Let us get a list of the names of the `"AminoAcids"`:

In[4]:= **ChemicalData**[`"AminoAcids"`,`"Name"`]
Out[4]= {glycine, L–alanine, L–serine, L–proline,
L–valine, L–threonine, L–cysteine, L–isoleucine,
L–leucine, L–asparagine, L–aspartic acid,
L–glutamine, L–lysine, L–glutamic acid,
L–methionine, L–histidine, L–phenylalanine,
L–arginine, L–tyrosine, L–tryptophan}

We will use this information to create a list of the codons for each amino acid. The codons are the codings sequence in RNA of length 3 each, that are used for translation of RNA to protein (i.e. each codon represents one amino acid). We can generate an association of the codons using **ChemicalData**:

In[5]:= aminoAcid2Codon=
   AssociationThread[
      ChemicalData[`"AminoAcids"`,`"Name"`]–>
      ChemicalData[`"AminoAcids"`,`"Codons"`]]
Out[5]= <|glycine→{GGU,GGC,GGA,GGG},
L–alanine→{GCU,GCC,GCA,GCG},
L–serine→{UCU,UCC,UCA,UCG,AGU,AGC},
L–proline→{CCU,CCC,CCA,CCG},
L–valine→{GUU,GUC,GUA,GUG},
L–threonine→{ACU,ACC,ACA,ACG},
L–cysteine→{UGU,UGC},
L–isoleucine→{AUU,AUC,AUA},
L–leucine→{UUA,UUG,CUU,CUC,CUA,CUG},
L–asparagine→{AAU,AAC},
L–aspartic acid→{GAU,GAC},
L–glutamine→{CAA,CAG}, L–lysine→{AAA,AAG},
L–glutamic acid→{GAA,GAG},
L–methionine→{AUG}, L–histidine→{CAU,CAC},
L–phenylalanine→{UUU,UUC},
L–arginine→{CGU,CGC,CGA,CGG,AGA,AGG},
L–tyrosine→{UAU,UAC}, L–tryptophan→{UGG}|>

Additionally, we often want to know what a particular sequence translates to. So we create a codon to amino acid association:

In[6]:= codon2AminoAcid=
   Join[
      Join @@
      (**AssociationThread**[#[[2]],
         ConstantArray[#[[1]],**Length**[#[[2]]]]]&/@
      Transpose[
         {**ChemicalData**[`"AminoAcids"`,`"Name"`],
          ChemicalData[`"AminoAcids"`,`"Codons"`]}]) , <|
      `"UAA"`–> `"Stop Codon"`, `"UAG"`–> `"Stop Codon"`,
      `"UGA"`–> `"Stop Codon"`|>]
Out[6]= <|GGU→glycine, GGC→glycine, GGA→glycine,

7.1 Amino Acids

```
GGG→glycine ,GCU→L-alanine ,GCC→L-alanine ,
GCA→L-alanine ,GCG→L-alanine ,UCU→L-serine ,
UCC→L-serine ,UCA→L-serine ,UCG→L-serine ,
AGU→L-serine ,AGC→L-serine ,CCU→L-proline ,
CCC→L-proline ,CCA→L-proline ,CCG→L-proline ,
GUU→L-valine ,GUC→L-valine ,GUA→L-valine ,
GUG→L-valine ,ACU→L-threonine ,ACC→L-threonine ,
ACA→L-threonine ,ACG→L-threonine ,
UGU→L-cysteine ,UGC→L-cysteine ,
AUU→L-isoleucine ,AUC→L-isoleucine ,
AUA→L-isoleucine ,UUA→L-leucine ,
UUG→L-leucine ,CUU→L-leucine ,CUC→L-leucine ,
CUA→L-leucine ,CUG→L-leucine ,AAU→L-asparagine ,
AAC→L-asparagine ,GAU→L-aspartic acid ,
GAC→L-aspartic acid ,CAA→L-glutamine ,
CAG→L-glutamine ,AAA→L-lysine ,AAG→L-lysine ,
GAA→L-glutamic acid ,GAG→L-glutamic acid ,
AUG→L-methionine ,CAU→L-histidine ,
CAC→L-histidine ,UUU→L-phenylalanine ,
UUC→L-phenylalanine ,CGU→L-arginine ,
CGC→L-arginine ,CGA→L-arginine ,CGG→L-arginine ,
AGA→L-arginine ,AGG→L-arginine ,UAU→L-tyrosine ,
UAC→L-tyrosine ,UGG→L-tryptophan ,
UAA→Stop Codon,UAG→Stop Codon,UGA→Stop Codon|>
```

Code Notes: Here we took them together in pairs: {amino acid name,{list of codons}}. Next we created labels for the amino acid (first list element), equal to the number of codons, which is the length of the second list. using (**ConstantArray**[#[[1]], **Length**[#[[2]]]] &/@). Using the **AssociationThread** we created a codon to name association for each amino acid, and joined these. Finally, we used the external **Join** to add in the association the `"Stop Codons"`.

There are 64 possible entries, corresponding to 64 coding possibilities.

In[7]:= **Length**[%]
Out[7]= 64

The amino acid sequence in a protein or peptide can be represented in the simplest form as a linear word sequence. It is often more convenient to work with abbreviations for the amino acid. There are actually both 1-letter and 3-letter abbreviations. We create an association here for reference purposes from a list containing {amino acid, 1 letter abbreviation, 3 letter abbreviation}:

```
In[8]:= aminoAcidAbbreviations=<|
         "1letter"->AssociationThread[
           #[[All,1]]-> #[[All,2]]] ,
         "3letter"-> AssociationThread[
           #[[All,1]]-> #[[All,3]]]|>&@
       {{"glycine","G","Gly"},
        {"L-alanine","A","Ala"},
        {"L-serine","S","Ser"},
        {"L-proline","P","Pro"},
        {"L-valine","V","Val"},
        {"L-threonine","T","Thr"},
```

```
{"L-cysteine","C","Cys"},
{"L-leucine","L","Leu"},
{"L-asparagine","N","Asn"},
{"L-aspartic acid","D","Asp"},
{"L-glutamine","Q","Gln"},
{"L-lysine","K","Lys"},
{"L-glutamic acid","E","Glu"},
{"L-methionine","M","Met"},
{"L-histidine","H","His"},
{"L-phenylalanine","F","Phe"},
{"L-arginine","R","Arg"},
{"L-tyrosine","Y","Tyr"},
{"L-tryptophan","W","Trp"},
{"L-isoleucine","I","Ile"},
{"Stop Codon","*","*"}};
```

We can use this to create our own codon translation associations to the alphabet, by using **Query** and evaluate the association values for each amino acid in the codon2AminoAcid association using them as keys for the aminoAcidAbbreviations:

In[9]:= codon2AA=
   Query[All, aminoAcidAbbreviations["1letter"][#]&]@
   codon2AminoAcid

Out[9]= <|GGU→G,GGC→G,GGA→G,GGG→G,GCU→A,
   GCC→A,GCA→A,GCG→A,UCU→S,UCC→S,UCA→S,
   UCG→S,AGU→S,AGC→S,CCU→P,CCC→P,CCA→P,
   CCG→P,GUU→V,GUC→V,GUA→V,GUG→V,ACU→T,
   ACC→T,ACA→T,ACG→T,UGU→C,UGC→C,AUU→I,
   AUC→I,AUA→I,UUA→L,UUG→L,CUU→L,CUC→L,
   CUA→L,CUG→L,AAU→N,AAC→N,GAU→D,GAC→D,
   CAA→Q,CAG→Q,AAA→K,AAG→K,GAA→E,GAG→E,
   AUG→M,CAU→H,CAC→H,UUU→F,UUC→F,CGU→R,
   CGC→R,CGA→R,CGG→R,AGA→R,AGG→R,UAU→Y,
   UAC→Y,UGG→W,UAA→*,UAG→*,UGA→*|>

We can extract a lot of properties of the amino acids, as with any chemical.

In[10]:= **CanonicalName**[#]&/@
   EntityValue["Chemical","Properties"][[1;;10]]
Out[10]= {AcidityConstants, AdjacencyMatrix,
   AlternateNames, AtomPositions, AutoignitionPoint,
   BeilsteinNumber, BlackStructureDiagram,
   BoilingPoint, BondCounts, BondEnergies}

We can draw the structure diagram for each amino acid, as in Fig. 7.1. We can also get a list of the amino acids included using **EntityList**, Fig. 7.2.

Additionally we can look up amino acids using Wolfram|Alpha [2], Fig. 7.3. In fact, if we enter a coding (DNA) sequence, Wolfram|Alpha will generate an amino acid sequence as well, Fig. 7.4.

As you can see in Fig. 7.4 there is a "(5' to 3' frame 1)" label in the "Amino acid sequence" box. There are actually 3 coding frames corresponding to which letter is used first in the translation (i.e. start from the first letter in the sequence for frame 1, start from the second letter for frame 2, and the 3rd letter for

7.1 Amino Acids

```
In[11]:= EntityValue[#, "BlackStructureDiagram", "EntityAssociation"] & @
   aminoAcids
```

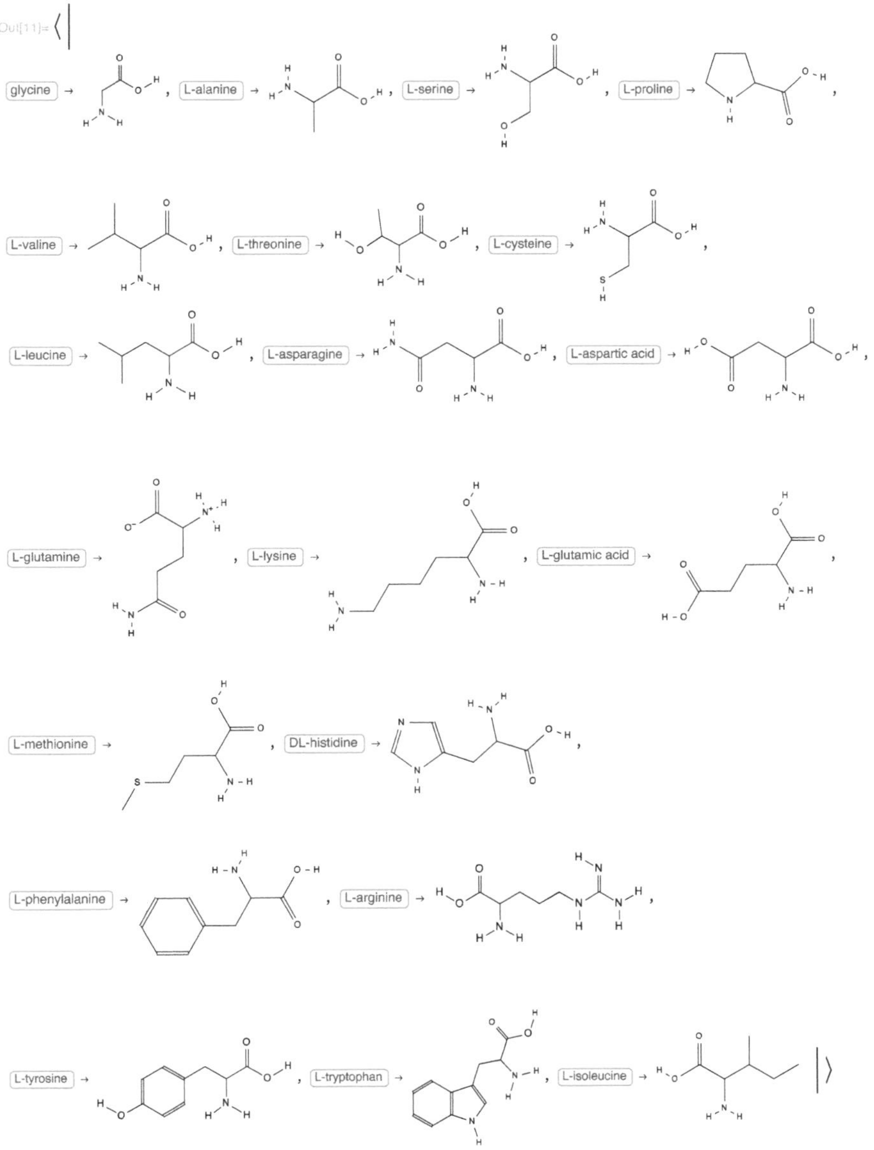

Fig. 7.1 Out[11]= Amino acid structures

In[12]:= **EntityList**[▦ **amino acids** CHEMICALS]

Out[12]= { glycine , L-alanine , L-serine , L-proline , L-valine , L-threonine , L-cysteine , L-leucine , L-asparagine , L-aspartic acid , L-glutamine , L-lysine , L-glutamic acid , L-methionine , DL-histidine , L-phenylalanine , L-arginine , L-tyrosine , L-tryptophan , L-isoleucine }

Fig. 7.2 Out[12]= Amino acid **EntityList**

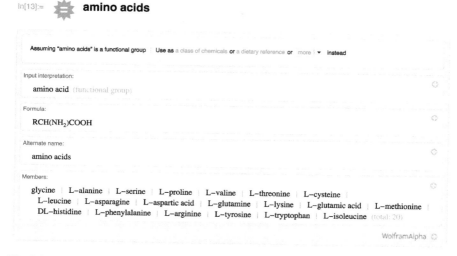

Fig. 7.3 Out[13]= Wolfram|Alpha amino acids search

frame 3). You can visualize all the frames by clicking the "All reading frames" button in the Wolfram|Alpha output. Additionally, you can press the + button on the right to generate code for retrieving this information by selecting from a drop down menu: Selecting "Formatted data" will give you:

In[15]:= **WolframAlpha**["AATCATTTGGCAGCTGAAAAAGGATGTGGATA",{{"AminoAcidSequence",1},"FormattedData"},PodStates->{"AminoAcidSequence__All reading frames"}]

7.1 Amino Acids

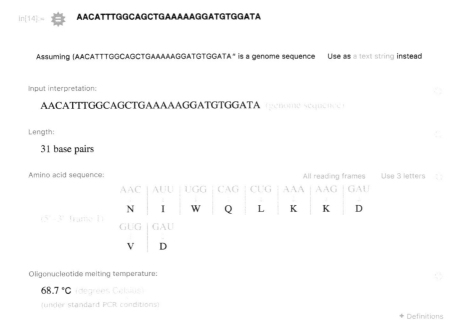

Fig. 7.4 Out[14]= Wolfram|Alpha DNA sequence interpretation generated in mathematica

```
                    UCA UUU GGC AGC UGA AAA AGG AUG
                     ↓   ↓   ↓   ↓   ↓   ↓   ↓   ↓
                     S   F   G   S   –   K   R   M

        5'-3' frame 3  UGG AUA
                        ↓   ↓
                        W   I

                    UAU CCA CAU CCU UUU UCA GCU GCC
                     ↓   ↓   ↓   ↓   ↓   ↓   ↓   ↓
                     Y   P   H   P   F   S   A   A

        3'-5' frame 1  AAA UGA
                        ↓   ↓
                        K   –

                    AUC CAC AUC CUU UUU CAG CUG CCA
                     ↓   ↓   ↓   ↓   ↓   ↓   ↓   ↓
                     I   H   I   L   F   Q   L   P

        3'-5' frame 2  AAU GAU
                        ↓   ↓
                        N   D

                    UCC ACA UCC UUU UUC AGC UGC CAA
                     ↓   ↓   ↓   ↓   ↓   ↓   ↓   ↓
                     S   T   S   F   F   S   C   Q

        3'-5' frame 3  AUG AUU
                        ↓   ↓
                        M   I
```

The computable data version is

In[16]:= **WolframAlpha**["AATCATTTGGCAGCTGAAAAAGGATGTGGATA",
    {{"AminoAcidSequence",1},"ComputableData"},
    PodStates→{"AminoAcidSequence__All reading frames"}]

Out[16]= {{5'−3' frame 1,
    {{{AAU,CAU,UUG,GCA,GCU,GAA,AAA,GGA},
        {↓,↓,↓,↓,↓,↓,↓,↓},{N,H,L,A,A,E,K,G}},
      {{UGU,GGA},{↓,↓},{C,G}}}},{5'−3' frame 2,
    {{{AUC,AUU,UGG,CAG,CUG,AAA,AAG,GAU},
        {↓,↓,↓,↓,↓,↓,↓,↓},{I,I,W,Q,L,K,K,D}},
      {{GUG,GAU},{↓,↓},{V,D}}}},{5'−3' frame 3,
    {{{UCA,UUU,GGC,AGC,UGA,AAA,AGG,AUG},
        {↓,↓,↓,↓,↓,↓,↓,↓},{S,F,G,S,−,K,R,M}},
      {{UGG,AUA},{↓,↓},{W,I}}}},{3'−5' frame 1,
    {{{UAU,CCA,CAU,CCU,UUU,UCA,GCU,GCC},
        {↓,↓,↓,↓,↓,↓,↓,↓},{Y,P,H,P,F,S,A,A}},

7.1 Amino Acids   235

{{AAA,UGA},{↓,↓},{K,−}}}},{3'−5' frame 2,
{{{AUC,CAC,AUC,CUU,UUU,CAG,CUG,CCA},
{↓,↓,↓,↓,↓,↓,↓,↓},{I,H,I,L,F,Q,L,P}},
{{AAU,GAU},{↓,↓},{N,D}}}},{3'−5' frame 3,
{{{UCC,ACA,UCC,UUU,UUC,AGC,UGC,CAA},
{↓,↓,↓,↓,↓,↓,↓,↓},{S,T,S,F,F,S,C,Q}},
{{AUG,AUU},{↓,↓},{M,I}}}}}}

We can also calculate this using our dictionary above. We first partition the string in 3-mer strings, then we replace Ts with Us, and then we use our codon2AA dictionary:

In[17]:= codon2AA/@
           StringReplace[
           (StringPartition["AATCATTTGGCAGCTGAAAAAGGATGTGGATA",
           3]),"T"−>"U"]
Out[17]= {N,H,L,A,A,E,K,G,C,G}

In[19]:= **ProteinData["JUN", "MoleculePlot"]**

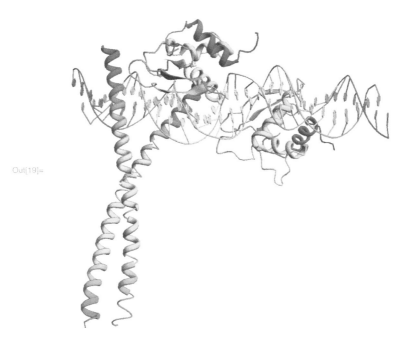

Fig. 7.5 Out[19]= JUN "MoleculePlot"

Fig. 7.6 Out[20]=
Wolfram|Alpha
`"JUN protein"`
interpretation

7.2 Protein Information

In the Wolfram Language we can extract information regarding various proteins directly using **ProteinData**. For example, if we are interested in the protein JUN, we can get all the following information:

```
In[18]:= ProteinData["JUN","Properties"]
Out[18]= {AdditionalAtomPositions, AdditionalAtomTypes,
        AtomPositions, AtomRoles, AtomTypes,
        BiologicalProcesses, CellularComponents,
        ChainLabels, ChainSequences, DihedralAngles,
        DNACodingSequence, DNACodingSequenceLength,
        DomainIDs, DomainPositions, Domains, Gene, GeneID
        , GyrationRadius, Memberships, MolecularFunctions,
        MolecularWeight, MoleculePlot, Name, NCBIAccessions,
        PDBIDList, PrimaryPDBID, SecondaryStructureRules,
        Sequence, SequenceLength, StandardName}
```

For example, we can use "MoculePlot" to see what the protein looks like, Fig. 7.5. Similarly we can access information regarding proteins on Wolfram|Alpha, Fig. 7.6.

7.3 UniProt

For the most up-to-date information we can get online data from protein databases, such as UniProt or from NCBI. We have discussed NCBI E-utilities extensively in Chap. 4. Here instead we focus on how we can retrieve information regarding proteins directly from UniProt [3, 4] using their RESTful API
(http://www.uniprot.org/help/api 11/2017).

7.3.1 Individual Entry Retrieval

If we know the accession for our protein we can get the information directly. We can use **Import**, or **URLExecute** to get the sequence in FASTA format for the accession P19838 in this example:

```
In[21]:= uniprotP19838FASTA=
         URLExecute[
         "http://www.uniprot.org/uniprot/P19838.fasta"];

In[22]:= uniprotP19838FASTA//Short[#,6]&
Out[22]//Short= >sp|P19838|NFKB1_HUMAN Nuclear factor NF-kappa-B
                p105 subunit OS = Homo sapiens GN = NFKB1 PE=1 SV=2

                MAEDDPYLGRPEQMFHLDPSL...
                PAPSKTLMDNYEVSGGTVRELVEALRQMGYTEAIEVIQAASSPVKTTSQ
```

```
AHSLPLSPASTRQQIDELRDSDSVCDSGVETSFRKLSFTESLTSGASLLTLN
KMPHDYGQ
EGPLEGKI
```

Import recognizes the file to be a FASTA file, so will only import the sequence, unless we specify the format. We can also specify the format and what to import in **URLExecute**:

```
In[23]:= URLExecute[
           "http://www.uniprot.org/uniprot/P19838.fasta",
           {"FASTA","LabeledData"}]
Out[23]= {sp|P19838|NFKB1_HUMAN Nuclear
           factor NF-kappa-B p105 subunit
           OS = Homo sapiens GN=NFKB1 PE=1 SV=2→
         MAEDDPYLGRPEQMFHLDPSLTHTIFNPEVFQPQMALPTDGPYLQILEQP
         KQRGFRFRYVCEGPSHGGLPGASSEKNKKSYPQVKICNYVGPAKV
         IVQLVTNGKNIHLHAHSLVGKHCEDGICTVTAGPKDMVVGFANLG
         ILHVTKKKVFETLEARMTEACIRGYNPGLLVHPDLAYLQAEGGGD
         RQLGDREKELIRQAALQQTKEMDLSVVRLMFTAFLPDSTGSFTRR
         LEPVVSDAIYDSKAPNASNLKIVRMDRTAGCVTGGEEIYLLCDKV
         QKDDIQIRFYEEEENGGVWEGFGDFSPTDVHRQFAIVFKTPKYKD
         INITKPASVFVQLRRKSDLETSEPKPFLYYPEIKDKEEVQRKRQK
         LMPNFSDSFGGGSGAGAGGGGMFGSGGGGGGTGSTGPGYSFPHYG
         FPTYGGITFHPGTTKSNAGMKHGTMDTESKKDPEGCDKSDDKNTV
         NLFGKVIETTEQDQEPSEATVGNGEVTLTYATGTKEESAGVQDNL
         FLEKAMQLAKRHANALFDYAVTGDVKMLLAVQRHLTAVQDENGDS
         VLHLAIIHLHSQLVRDLLEVTSGLISDDIINMRNDLYQTPLHLAV
         ITKQEDVVEDLLRAGADLSLLDRLGNSVLHLAAKEGHDKVLSILL
         KHKKAALLLDHPNGDGLNAIHLAMMSNSLPCLLLLVAAGADVNAQ
         EQKSGRTALHLAVEHDNISLAGCLLLEGDAHVDSTTYDGTTPLHI
         AAGRGSTRLAALLKAAGADPLVENFEPLYDLDDSWENAGEDEGVV
         PGTTPLDMATSWQVFDILNGKPYEPEFTSDDLLAQGDMKQLAEDV
         KLQLYKLLEIPDPDKNWATLAQKLGLGILNNAFRLSPAPSKTLMD
         NYEVSGGTVRELVEALRQMGYTEAIEVIQAASSPVKTTSQAHSLP
         LSPASTRQQIDELRDSDSVCDSGVETSFRKLSFTESLTSGASLLT
         LNKMPHDYGQEGPLEGKI}
```

We can also import gff files:

```
In[24]:= nfkbGFF=URLExecute["http://www.uniprot.org/uniprot/P19838.gff"
         ,"TSV"];
```

We can see that the data is gff version 3 format:

```
In[25]:= nfkbGFF[[1;;2]]
Out[25]= {{##gff-version 3},{##sequence-region P19838 1 968}}
```

See also: ("https://github.com/The-Sequence-Ontology/Specifications/blob/master/gff3.md").

The format has 9 columns. The columns correspond to:

```
{"seqid","source","type","start","end",
 "score","strand","phase","attributes"}.
```

For the protein we just extracted we get the information:

```
In[26]:= nfkbGFF[[3;;5]]
Out[26]= {{P19838,UniProtKB,Chain,1,968,
```

7.3 UniProt

```
. , . , . , ID=PRO_0000030310 ; Note=Nuclear
    factor NF-kappa-B p105 subunit , } ,
{ P19838 , UniProtKB , Chain , 1 , 433 , . , . , . ,
    ID=PRO_0000030311 ; Note=Nuclear
    factor NF-kappa-B p50 subunit , } ,
{ P19838 , UniProtKB , Domain , 42 , 367 , . , . , . ,
    Note=RHD ; Ontology_term=ECO:0000255 ; evidence = ECO:
    0000255|PROSITE-ProRule : PRU00265 , } }
```

7.3.2 Query Entry Retrieval

We can use **URLExecute** as for other databases to perform queries and retrieve data. The database can take several parameters: `"query"`, `"format"`, `"columns"`, `"include"`, `"compress"`, `"limit"`, `"offset"`.

7.3.2.1 Parameters in the UniProt API

The following parameters and assigned values are used routinely in the RESTful UniProt API:

- `"query"` is a string search that can also use Boolean values (AND, NOT, OR), and special matching characters, *. Entries in quotations " " have to be exact matches. Other characters are also available as well as fields, specified with *field*: query, for example gene:JUN to search all proteins where the gene matches JUN.
- `"format"` is the required returned format and can take as value any of the following values:

  `{"html", "tab", "xls", "fasta", "gff", "txt",`
  `"xml", "rdf", "list", "rss"}.`

- `"columns"` can be specified if the format as `"xml"` or `"tab"`, to include any of:

  `{"citation", "clusters", "comments", "domains", "domain",`
  `"ec", "id", "entry name", "existence", "families",`
  `"features", "genes", "go", "go-id", "interactor",`
  `"keywords", "last-modified", "length", "organism",`
  `"organism-id", "pathway", "protein names", "reviewed",`
  `"sequence", "3d", "version", "virus hosts"}`

  if available.
- `"include"` is a Boolean with values {`"yes"`,`"no"`} for `"fasta"` or `"rdf"` formats only. It specifies whether or to include isoform sequences in the output for fasta formats. For `"rdf"` format the choice is whether or not to include a description of referenced data.
- `"compress"` sets whether or not to compress results and return them in gzip format.
- `"limit"` sets the maximum result number.
- `"offset"` can be used to shift the first result retrieved by the specified integer.

7.3.3 Executing Queries

Let us look at some examples. We can decide to return certain columns, for example for P19838 (NFKB1), in a tab format, limiting the entries to 5:

```
In[27]:= URLExecute["http://www.uniprot.org/uniprot/?",
           {"query"-> "P19838",
            "format"-> "tab",
            "limit"-> "5",
            "columns"-> "id,entry name,families,length"},"TSV"]
Out[27]= {{Entry,Entry name,Protein families,Length},
          {P35606,COPB2_HUMAN,WD repeat COPB2 family,906},
          {P35222,CTNB1_HUMAN,Beta-catenin family,781},
          {P03372,ESR1_HUMAN,
           Nuclear hormone receptor family, NR3 subfamily,
           595},{P08238,HS90B_HUMAN,
           Heat shock protein 90 family,724},
          {Q13547,HDAC1_HUMAN,
           Histone deacetylase family, HD type 1 subfamily,
           482}}
```

We can also search for NFKB1 directly, here we return the "id" and "entry name" columns, and sort the results by "score", which gives a better match:

```
In[28]:= URLExecute["http://www.uniprot.org/uniprot/?",
           {"query"-> "NFKB1","sort"-> "score",
            "columns"-> "id,entry name","format"-> "tab",
            "limit"->"10"},"TSV"]
Out[28]= {{Entry,Entry name},
          {P19838,NFKB1_HUMAN},{P25799,NFKB1_MOUSE},
          {Q63369,NFKB1_RAT},{Q969V5,MUL1_HUMAN},
          {Q8VCM5,MUL1_MOUSE},{P03372,ESR1_HUMAN},
          {Q6F3J0,NFKB1_CANLF},{P35222,CTNB1_HUMAN},
          {Q04206,TF65_HUMAN},{P17612,KAPCA_HUMAN}}
```

The "limit" parameter sets the maximum returned results. If a "limit" is set, we can use "offset" so we do not start from the first entry. In our example below, say we start on the 8th entry by setting "offset"-> "8" and get 10 entries back:

```
In[29]:= URLExecute["http://www.uniprot.org/uniprot/?",
           {"query"-> "NFKB",
            "sort"-> "score",
            "columns"-> "id,entry name",
            "format"-> "tab",
            "limit"->"10",
            "offset"-> "8"},"TSV"]
Out[29]= {{Entry,Entry name},
          {P05067,A4_HUMAN},{Q9Y6Q6,TNR11_HUMAN},
          {Q9WTK5,NFKB2_MOUSE},{Q8VCM5,MUL1_MOUSE},
          {O35305,TNR11_MOUSE},{P0CG48,UBC_HUMAN},
          {Q12933,TRAF2_HUMAN},{Q04861,NFKB1_CHICK},
          {Q9BT67,NFIP1_HUMAN},{O00255,MEN1_HUMAN}}
```

7.3 UniProt

We should note that the database is constantly updated, so the order and/or results you may get when limiting the output can be different.

We can obtain a list of identifiers as a list by specifying the `"format"` parameter:

```
In[30]:= URLExecute["http://www.uniprot.org/uniprot/?",
         {"query"-> "NFKB",
          "format"-> "list",
          "limit"->"10"},
         "TSV"]
Out[30]= {{P05067},{E1BKA3},{O43823},{P12023},{P08592},
         {B5TTX7},{Q3V096},{Q5U2Z2},{Q5VK71},{Q9DBR0}}
```

We can also specify `"format"->"fasta"` to obtain the FASTA sequence:

```
In[31]:= uniprotQueryP19838Fasta=URLExecute["http://www.uniprot.org/
         uniprot/?",{"query"-> "P19838","format"-> "fasta"}];
```

This is the same as we recovered above in Sect. 7.3.1 with the direct method. We can check a few lines:

```
In[32]:= uniprotQueryP19838Fasta//Short[#,6]&
Out[32]//Short= >sp|P19838|NFKB1_HUMAN Nuclear factor NF-kappa-B
                p105 subunit OS=Homo sapiens GN=NFKB1 PE=1 SV=2

                MAEDDPYLGRPEQMFHLDPSL...
                PAPSKTLMDNYEVSGGTVRELVEALRQMGYTEAIEVIQAASSPVKTTSQ

AHSLPLSPASTRQQIDELRDSDSVCDSGVETSFRKLSFTESLTSGASLLTLN
                KMPHDYGQ
                EGPLEGKI
```

Since we did not specify it was a `"FASTA"` file, it was parsed as a string. We can instead specify explicitly to the **URLExecute** that it is a `"FASTA"` file, and get the `"Header"`:

```
In[33]:= URLExecute["http://www.uniprot.org/uniprot/?",
         {"query"-> "P19838",
          "format"-> "fasta"},
         {"FASTA","Header"}]
Out[33]= {sp|P19838|NFKB1_HUMAN Nuclear factor NF-kappa-B
         p105 subunit OS=Homo sapiens GN=NFKB1 PE=1 SV=2}
```

For this case we can choose if we want to include isoform information (in the cases `"fasta"` is the selected output). We import these as `"LabeledData"` to get header to sequence associations:

```
In[34]:= isoformsNFKB1=
         URLExecute["http://www.uniprot.org/uniprot/?",
         {"query"-> "P19838",
          "include"-> "yes",
          "format"-> "fasta"},{"FASTA","LabeledData"}];
```

We can see that we have 3 entries:

```
In[35]:= Keys@ isoformsNFKB1
Out[35]= {sp|P19838|NFKB1_HUMAN Nuclear factor NF-kappa-B
```

p105 subunit OS=Homo sapiens GN=NFKB1 PE=1 SV=2,
sp|P19838−2|NFKB1_HUMAN Isoform 2 of
Nuclear factor NF-kappa-B p105
subunit OS=Homo sapiens GN=NFKB1,
sp|P19838−3|NFKB1_HUMAN Isoform 3 of
Nuclear factor NF-kappa-B p105
subunit OS=Homo sapiens GN=NFKB1}

7.3.4 Random Entry Generation

UniProt even lets us retrieve a random entry in the query through the `"random"` parameter:

In[36]:= **URLExecute**["http://www.uniprot.org/uniprot/?",
        {"format"−> "fasta","random"−> "yes"}]
Out[36]= >tr|A0A0D9X2A9|A0A0D9X2A9_9ORYZ Uncharacterized
        protein OS=Leersia perrieri PE=4 SV=1
        MQVQSDSTPNYLSSAKSTIQIPCGFQEDYGCRVNEIILGTEHGGVAAEEDDGG
        PGEGQRT
        EGVSRTLHQSHHAYETIQPLLRRGGYLLPSPATALRLSRFATVFSAGDVRH

Note that more than likely you will get a different entry than the above when you execute the same code.

7.3.5 Identifier Mapping

Identifiers can be mapped from one kind of accession to another - which is useful, particularly when integrating information from multiple databases. The available databases for identifier mapping can be found online:

In[37]:= **SystemOpen**["http://www.uniprot.org/help/api_idmapping"]

Let us convert two IDs from UniProt accessions, `"ACC"` to RefSeq [5] protein accessions: `"P_REFSEQ_CA"`. We need to define two parameters: the `"from"` and the `"to"` parameters for the database the query IDs are from to the database we are converting to:

In[38]:= **URLExecute**["http://www.uniprot.org/uploadlists/",
        {"from"−> "ACC",
         "to"−> "P_REFSEQ_AC",
         "format" −> "tab",
         "query"−> "P97793 P19838"},
        "TSV"]
Out[38]= {{From,To},{P97793,NP_031465.2},
        {P19838,NP_001158884.1},
        {P19838,NP_001306155.1},{P19838,NP_003989.2}}

7.4 NCBI Entrez Utils

As an example, we can use the Entrez Utils as discussed in Chap. 4 to retrieve information for proteins. Here let us retrieve an identifier of interest, we look for protein P19838 (which is NFKB1), using ESearch:

In[39]:= **URLExecute**[
    "https://eutils.ncbi.nlm.nih.gov/entrez/eutils/
        esearch.fcgi",
    {"RequestMethod"−>"POST",
    "retmode"−>"json",
    "db"−> "protein",
    "term"−> "P19838"},"RawJSON"]
Out[39]= <|header→ <|type→esearch, version→0.3|>,
    esearchresult→ <|count→1,retmax→1,
        retstart→0,idlist→{21542418},
        translationset −>{},querytranslation→|>|>

7.4.1 Sequence Retrieval

We can retrieve the sequence of a protein in FASTA format using EFetch. We use the id obtained above, "21542418":

In[40]:= efetch21542418=
    URLExecute[
        "https://eutils.ncbi.nlm.nih.gov/entrez/eutils/
            efetch.fcgi?",
        {"RequestMethod"−>"POST",
        "db"−> "protein",
        "id"−>"21542418",
        "retmode"−>"text",
        "rettype"−> "fasta"},{"FASTA","LabeledData"}];

In[41]:= efetch21542418//**Short**[#,10]&
Out[41]//**Short**= {sp|P19838.2|NFKB1_HUMAN RecName: Full=Nuclear
    factor NF-kappa-B p105 subunit; AltName:
    Full=DNA-binding factor KBF1; AltName:
    Full=EBP-1; AltName: Full=Nuclear
    factor of kappa light polypeptide gene
    enhancer in B-cells 1; Contains: RecName:
    Full=Nuclear factor NF-kappa-B p50 subunit→
MAEDDPYLGRPEQMFHLDPSLTHTIFNPEVFQPQMALPTDGPYL
    QILEQ
    PKQRGF...
QIDELRDSDSVCDSGVETSFRKLSFTESLTSGASLLTLNKMPHDYG
    QEGPLEGKI}

7.4.2 From Sequence to Protein

Let us look at another example. We look for the protein for JUN:

In[42]:= **URLExecute**[
  "https://eutils.ncbi.nlm.nih.gov/entrez/eutils/
    esearch.fcgi",
  {"RequestMethod"->"POST",
   "retmode"->"json",
   "db"-> "gene",
   "term"->
     "Jun[gene] AND proto-oncogene AND
       Human[organism]"},"RawJSON"]

Out[42]= <|header-> <|type->esearch, version->0.3|>,
  esearchresult->
    <|count->1,retmax->1,retstart->0,idlist->{3725},
     translationset->{<|from->Human[organism],
       to->"Homo sapiens"[Organism]|>},
     translationstack->{<|term->Jun[gene],
       field->gene,count->230,explode->N|>,
     <|term->proto-oncogene[All Fields],
       field->All Fields,count->15871,explode->N|>,
     AND, <|term->"Homo sapiens"[Organism],
       field->Organism,count->221908,explode->Y|>,
     AND},querytranslation->
       Jun[gene] AND proto-oncogene[All Fields]
         AND "Homo sapiens"[Organism]|>|>

We will use the ID to retrieve a gene table and retrieve the exon locations and accessions:

In[43]:= junGeneTable=
  URLExecute[
    "https://eutils.ncbi.nlm.nih.gov/entrez/eutils/
     efetch.fcgi?",
    {"RequestMethod"->"POST",
     "db"-> "gene",
     "id"->"3725",
     "rettype"->"gene_table",
     "retmode"-> "text"},"TSV"]

Out[43]= {{JUN Jun proto-oncogene, AP-1
     transcription factor subunit[Homo sapiens]},
  {Gene ID: 3725, updated on 3-Dec-2017},
  {},{},
  {Reference GRCh38.p7 Primary Assembly NC_000001.11
    (minus strand) from: 58784113 to: 58780791},
  {mRNA NM_002228.3, 1 exon, total annotated
    spliced exon length: 3323},
  {protein NP_002219.1 (CCDS610.1), 1 coding
    exon, annotated AA length: 331},
  {},{Exon table for mRNA NM_002228.3
    and protein NP_002219.1},
  {Genomic **Interval** Exon,,Genomic **Interval** Coding,
    ,Gene **Interval** Exon,,

7.4 NCBI Entrez Utils   245

                 Gene **Interval** Coding,,Exon **Length**,
                 Coding **Length**,Intron **Length**},

             {————————————————————————————
              ———————————————————————
              ———————————————————
              ——————————————},
             {58784113−58780791,,58783070−58782075,
              ,1−3323,,1044−2039,,3323,,996},{}}

The coordinates for the exons are at the second to last line:

In[44]:= junGeneTable[[−2]]
Out[44]= {58784113−58780791,,58783070−58782075,,
              1−3323,,1044−2039,,3323,,996}

We use the information above to extract the mRNA, using the "NM_002228.3" accession as ID:

In[45]:= junCodingFasta=
            URLExecute[
             "https://eutils.ncbi.nlm.nih.gov/entrez/eutils/
               efetch.fcgi?",
             {"RequestMethod"−>"POST",
             "db"−> "nucleotide",
             "id"−>"NM_002228.3",
             "retmode"−>"text",
             "rettype"−> "fasta"},{"FASTA","LabeledData"}];

We extract the sequence from locations 1044–2039 and convert to RNA by changing "T"s to "U"s. Then we use our codon to amino acid association from earlier in the chapter, codon2AA, to convert to a protein sequence:

In[46]:= junConverted=
            StringJoin[
              codon2AA/@StringReplace[
                (StringPartition[**StringTake**[junCodingFasta[[1,2]],
                  {1044,2039}],3]),"T"−>"U"]]
Out[46]= MTAKMETTFYDDALNASFLPSESGPYGYSNPKILKQSMTLNLADPVGSLKPH
            LRAKNSDLLTSPDVGLLKLASPELERLIIQSSNGHITTTPTPTQFLC
            PKNVTDEQEGFAEGFVRALAELHSQNTLPSVTSAAQPVNGAGMVAPA
            VASVAGGSGSGGFSASLHSEPPVYANLSNFNPGALSSGGGAPSYGAA
            GLAFPAQPQQQQQPPHHLPQQMPVQHPRLQALKEEPQTVPEMPGETP
            PLSPIDMESQERIKAERKRMRNRIAASKCRKRKLERIARLEEKVKTL
            KAQNSELASTANMLREQVAQLKQKVMNHVNSGCQLMLTQQLQTF*

Notice the stop codon noted by * at the end. Let us drop this:

In[47]:= junProteinSequence=**StringDrop**[junConverted,−1];

Now let us retrieve the protein sequence directly from NCBI:

In[48]:= junNCBI=
            URLExecute[
             "https://eutils.ncbi.nlm.nih.gov/entrez/eutils/
               efetch.fcgi?",

```
{"RequestMethod"->"POST","db"-> "protein",
 "id"->"NP_002219.1","retmode"->"text",
 "rettype"-> "fasta"},"FASTA"];
```

Let us see how well we matched:

```
In[49]:= junProteinSequence==junNCBI[[1]]
Out[49]= True
```

7.4.3 Search by Molecular Weight

We may want to search for a list of proteins corresponding to a particular molecular weight. This is often the case in mass spectrometry. We can search for a range of molecular weights, by specifying a term of molecular weight ranges, and a field "MOLWT". For example, searching between 7000 and 7001:

```
In[50]:= URLExecute[
           "https://eutils.ncbi.nlm.nih.gov/entrez/eutils/
              esearch.fcgi",
           {"RequestMethod"->"POST",
            "retmode"->"json",
            "db"-> "protein",
            "term"-> "7000:7001",
            "field"->"MOLWT"},"RawJSON"]
Out[50]= <|header-> <|type->esearch,version->0.3|>,
           esearchresult->
             <|count->14927,retmax->20,retstart->0,
               idlist->{1280048552,1280040868,1280022610,
                 1279982930,915612992,1279703157,1279701127,
                 1279685109,1279670416,446296271,
                 259646910,46395925,1279662217,1279661964,
                 1279635796,1279522783,757800860,
                 658435822,1279482056,1279475578},
               translationset->{},
               translationstack->{<|term->000007000[MOLWT],
                 field->MOLWT,count->0,explode->N|>,
                 <|term->000007001[MOLWT],field->MOLWT,count->
                 0,explode->N|>,RANGE},querytranslation->
                 000007000[MOLWT]  :  000007001[MOLWT]|>|>
```

7.5 Proteins Sequence Alignment

Similar to DNA sequences, we may want to align Protein Sequences. We can also use BLASTp [6–8] to do so, which allows us to align protein sequences - see also the BLAST discussion in Chap. 5.

7.5 Proteins Sequence Alignment

7.5.1 SequenceAlignment

The alignment through **SequenceAlignment** uses substitution matrix rules, which are scores that calculate the likelihood that characters mutate over time to each other. Each matrix is square, and the elements represent a score for substituting letter i and letter j. Various substitution matrices are available in the Wolfram Language, including PAM (Point Accepted Mutation) and BLOSUM (BLOcks SUbstitution Matrix) matrices [9, 10]. Let us extract the sequences for human JUN and mouse Jun using the UniProt API. First we retrieve the entries for each query:

```
In[51]:= URLExecute["http://www.uniprot.org/uniprot/?",
         {"query"-> "JUN Human",
          "sort"-> "score",
          "columns"->
          "id,entry name",
          "format"-> "tab",
          "limit"->"10"},"TSV"]
Out[51]= {{Entry, Entry name},
          {P04637,P53_HUMAN},{P05412,JUN_HUMAN},
          {P63244,RACK1_HUMAN},{P05067,A4_HUMAN},
          {Q14524,SCN5A_HUMAN},{P00451,FA8_HUMAN},
          {P09450,JUNB_MOUSE},{P38398,BRCA1_HUMAN},
          {P68871,HBB_HUMAN},{Q96EB6,SIR1_HUMAN}}
```

```
In[52]:= URLExecute["http://www.uniprot.org/uniprot/?",
         {"query"-> "Jun Mouse",
          "sort"-> "score",
          "columns"-> "id,entry name",
          "format"-> "tab",
          "limit"->"10"},"TSV"]
Out[52]= {{Entry, Entry name},
          {P05627,JUN_MOUSE},{P22725,WNT5A_MOUSE},
          {P09450,JUNB_MOUSE},{P12023,A4_MOUSE},
          {Q923E4,SIR1_MOUSE},{P31750,AKT1_MOUSE},
          {Q91Y86,MK08_MOUSE},{P06804,TNFA_MOUSE},
          {P15066,JUND_MOUSE},{Q9ESN9,JIP3_MOUSE}}
```

From the entries above, we can pick the accessions of interest and directly obtain the FASTA sequence:

```
In[53]:= junHumanProt=
         URLExecute[
           "http://www.uniprot.org/uniprot/P05412.fasta",
           "FASTA"][[1]]
Out[53]= MTAKMETTFYDDALNASFLPSESGPYGYSNPKILKQSMTLNLADPVGSLKPH
         LRAKNSDLLTSPDVGLLKLASPELERLIIQSSNGHITTTPTPTQFLCPK
         NVTDEQEGFAEGFVRALAELHSQNTLPSVTSAAQPVNGAGMVAPAVASV
         AGGSGSGGFSASLHSEPPVYANLSNFNPGALSSGGGAPSYGAAGLAFPA
         QPQQQQQPPHHLPQQMPVQHPRLQALKEEPQTVPEMPGETPPLSPIDME
         SQERIKAERKRMRNRIAASKCRKRKLERIARLEEKVKTLKAQNSELAST
         ANMLREQVAQLKQKVMNHVNSGCQLMLTQQLQTF
```

In[54]:= junMouseProt=
          URLExecute[
          "http://www.uniprot.org/uniprot/P05627.fasta",
          "FASTA"][[1]]

Out[54]= MTAKMETTFYDDALNASFLQSESGAYGYSNPKILKQSMTLNLADPVGSLKPH
         LRAKNSDLLTSPDVGLLKLASPELERLIIQSSNGHITTTPTPTQFLCPK
         NVTDEQEGFAEGFVRALAELHSQNTLPSVTSAAQPVSGAGMVAPAVASV
         AGAGGGGYSASLHSEPPVYANLSNFNPGALSSGGGAPSYGAAGLAFPS
         QPQQQQQPPQPPHHLPQQIPVQHPRLQALKEEPQTVPEMPGETPPLSPI
         DMESQERIKAERKRMRNRIAASKCRKRKLERIARLEEKVKTLKAQNSEL
         ASTANMLREQVAQLKQKVMNHVNSGCQLMLTQQLQTF

Now we use SequenceAlignent to align the protein sequences:

In[55]:= **SequenceAlignment**[junHumanProt,junMouseProt]
Out[55]= {MTAKMETTFYDDALNASFL,{P,Q},SESG,{P,A},
         YGYSNPKILKQSMTLNLADPVGSLKPHLRAKNSDLLTSPDVGLLKLASPEL
         ERLIIQSSNGHITTTPTPTQFLCPKNVTDEQEGFAEGFVRALAELHS
         QNTLPSVTSAAQPV,{N,S},GAGMVAPAVASVAG,
         {,A},G,{S,},G,{S,G},GG,{F,Y},
         SASLHSEPPVYANLSNFNPGALSSGGGAPSYGAAGLAFP,
         {A,S},QPQQQQ,{,QPP},QPPHHLPQQ,{M,I},
         PVQHPRLQALKEEPQTVPEMPGETPPLSPIDMESQERIKAERKRMRNRIAA
         SKCRKRKLERIARLEEKVKTLKAQNSELASTANMLREQVAQLKQKVM
         NHVNSGCQLMLTQQLQTF}

We can see that the sequences are very well aligned. Let us use the BLOSUM62 matrix for the alignment:

In[56]:= **SequenceAlignment**[junHumanProt,junMouseProt,
          SimilarityRules->"BLOSUM62"]
Out[56]= {MTAKMETTFYDDALNASFL,{P,Q},SESG,{P,A},
         YGYSNPKILKQSMTLNLADPVGSLKPHLRAKNSDLLTSPDVGLLKLASPEL
         ERLIIQSSNGHITTTPTPTQFLCPKNVTDEQEGFAEGFVRALAELHS
         QNTLPSVTSAAQPV,{N,S},
         GAGMVAPAVASVAG,{GS,AG},G,{S,G},GG,{F,Y},
         SASLHSEPPVYANLSNFNPGALSSGGGAPSYGAAGLAFP,
         {A,S},QPQQQQ,{,QPP},QPPHHLPQQ,{M,I},
         PVQHPRLQALKEEPQTVPEMPGETPPLSPIDMESQERIKAERKRMRNRIAA
         SKCRKRKLERIARLEEKVKTLKAQNSELASTANMLREQVAQLKQKVM
         NHVNSGCQLMLTQQLQTF}

The alignment is mostly similar to before, but notice the subtle difference where a few gaps were eliminated when using the BLOSUM62.

7.5.2 BLASTp

We can also use the BLAST Web interface to align any protein [6–8]. Let us do the simplest example. We will upload a Jun mouse protein sequence and see how it aligns:

7.5 Proteins Sequence Alignment

In[57]:= junMouseProtFASTARecord=
    URLExecute[
      "http://www.uniprot.org/uniprot/P05627.fasta"];

We will use the "blastp" as "PROGRAM" and the non-redundant protein database as "nr":

In[58]:= blastpJunMouse=
    URLExecute[
      "https://blast.ncbi.nlm.nih.gov/Blast.cgi?",
      {"QUERY"-> junMouseProtFASTARecord,
      "PROGRAM"-> "blastp",
      "DATABASE"-> "nr",
      "CMD"-> "Put"},"Text"];

We can extract our request ID:

In[59]:= blastpInfo=**StringCases**[blastpJunMouse,
    "QBlastInfoBegin"~~__~~"QBlastInfoEnd"]
Out[59]= {QBlastInfoBegin
    RID = 2HBY6YPU015
    RTOE = 28
  QBlastInfoEnd}

In[60]:= ridBlastp=**StringSplit**[blastpInfo][[1,4]]
Out[60]= 2HBY6YPU015

Now we can check on the status (N.B. if you are running the commands yourself your RID will be different):

In[61]:= blastpReturnStatus=
    URLExecute[
      "https://blast.ncbi.nlm.nih.gov/Blast.cgi?",
      {"CMD"-> "Get","RID"-> ridBlastp,
      "FORMAT_TYPE"->"Tabular",
      "FORMAT_OBJECT"-> "status"},"Text"];

When we receive the Status as "READY" we can retrieve our data:

In[62]:= **StringCases**[#,"QBlastInfoBegin"~~__~~
    "QBlastInfoEnd"]&@blastpReturnStatus
Out[62]= {QBlastInfoBegin
    Status=READY
  QBlastInfoEnd}

If your status is "WAITING" please wait a while and run the above command **In[61]**–**In[62]** again. If the results are "READY" you can go ahead and extract the information:

In[63]:= blastpReturnCSV=
    URLExecute["https://blast.ncbi.nlm.nih.gov/Blast.cgi?",
      {"CMD"-> "Get",
      "RID"-> ridBlastp,
      "FORMAT_TYPE"->"CSV",
      "FORMAT_OBJECT"-> "Alignment",

```
                "ALIGNMENT_VIEW"-> "Tabular",
                "DESCRIPTIONS"-> "3"}]
Out[63]= {{sp|P05627|JUN_MOUSE,
                gi|6754402|ref|NP_034721.1|;gi|589922342|ref|XP_
                006974083.1|;gi|135299|sp|P05627.3|JUN_MOUSE;
                gi|52763|emb|CAA31236.1|;gi|309169|gb|AAA37419
                .1|;gi|12805239|gb|AAH02081.1|;gi|21284397|gb|
                AAH21888.1|;gi|62825871|gb|AAH94032.1|;gi|
                74149179|dbj|BAE22389.1|;gi|74192749|dbj|
                BAE34891.1|;gi|226132|prf||1411300A,
                100.,100.,334,0,0,1,334,1,334,0.,681},
            {sp|P05627|JUN_MOUSE,gi|74204894|dbj|BAE20944.1|,
                99.701,100.,334,1,0,1,334,1,334,0.,679},
            {sp|P05627|JUN_MOUSE,
                gi|1195550672|ref|XP_021056061.1|;gi|1195714545|ref
                |XP_021015381.1|,99.701,100.,334,
                1,0,1,334,1,334,0.,679},{},{},{},{}}
```

As expected the top hits are the actual mouse Jun protein. You can extract additional information as needed, and also alignments as we discussed in Chap. 5 on BLAST.

References

1. Wolfram Research, Inc.: Mathematica, Version 11.2. Champaign, IL (2017)
2. Wolfram Alpha LLC: Wolfram|Alpha (2017). Accessed Nov 2017
3. Nucleic Acids Res. Uniprot: the universal protein knowledgebase. **45**(D1), D158–D169 (2017)
4. UniProt, C.: Uniprot: a hub for protein information. Nucleic Acids Res **43**(Database issue), D204–12 (2015)
5. Pruitt, K.D., Tatusova, T., Maglott, D.R.: Ncbi reference sequences (refseq): a curated non-redundant sequence database of genomes, transcripts and proteins. Nucleic Acids Research **35**(Database issue), D61–5 (2007)
6. Altschul, S.F., Gish, W., Miller, W., Myers, E.W., Lipman, D.J.: Basic local alignment search tool. J. Mol. Biol. **215**(3), 403–410 (1990)
7. Altschul, S.F., Madden, T.L., Schaffer, A.A., Zhang, J., Zhang, Z., Miller, W., Lipman, D.J.: Gapped blast and psi-blast: a new generation of protein database search programs. Nucleic Acids Res. **25**(17), 3389–402 (1997)
8. Schäffer, A.A., Aravind, L., Madden, T.L., Shavirin, S., Spouge, J.L., Wolf, Y.I., Koonin, E.V., Altschul, S.F.: Improving the accuracy of psi-blast protein database searches with composition-based statistics and other refinements. Nucleic Acids Res. **29**(14), 2994–3005 (2001)
9. Dayhoff, M., Schwartz, R., Orcutt, B.: A model of evolutionary change in proteins. Atlas Protein Seq. Struct. **5**, 345–352 (1978)
10. Henikoff, S., Henikoff, J.G.: Amino acid substitution matrices from protein blocks. Proc. Natl. Acad. Sci. **89**(22), 10915–10919 (1992)

Chapter 8
Metabolomics Example

8.1 Metabolomics Data

Mass spectrometry has greatly aided high throughput investigations of the study of small molecules [24]. Small molecules are involved in multiple cellular processes, biological pathways, and can be used as biomarkers for diseases [22]. The collective set of small molecules in cells is termed the metabolome and its study metabolomics. Metabolites are identified by mass, and their biological involvement is ascertained through interactions both computationally and experimentally, and establishing their membership in pathways and networks [3, 5, 8, 9, 13, 16, 18, 20, 23, 26]. Applications also now include looking at their role in precision individualized medicine [4, 14, 15, 21], pharmacogenomics [25], cancer profiling [6, 7, 19] and many more [10, 11, 17].

In this chapter we will show an example of analysis of metabolomics data from small molecules intensity data using the Wolfram Language [27, 28]. The approach will involve normalizing the data, carrying out differential expression analysis and obtaining putative identities for mass features of interest.

8.2 Germ Free and Inoculated Mice Data

The study by Marcobal et al. [12], introduced in Chap. 3 used two groups of Swiss-Webster germ free (GF) mice (in gnotobiotic isolators). A group of 3 mice was kept germ free, and the second group's gut was colonized with two bacterial species, *Bacteroides thetaiotaomicron* and *Bifidobacterium longum*. The data we will use come from an analysis of small molecules from urine collected for the GF mice and

---

Electronic supplementary material The online version of this chapter (https://doi.org/10.1007/978-3-319-72377-8_8) contains supplementary material, which is available to authorized users.

at days 5, 10, 15, 20 and 25 from the inoculated mice. The data contain a mass feature (in Daltons) and the intensities for each sample. We first import the data, and analyze further in the following sections:

In[1]:= **SetDirectory**[**NotebookDirectory**[]];

In[2]:= importGFMice=**Import**["GFMice.csv"];

8.2.1 Processing Imported Data

The imported data contains both negative and positive mode acquired results from mass spectrometry, separated by lines indicating "Negative Mode" and "Positive Mode". Let us look at the first few lines:

In[3]:= importGFMice[[1;;4]]
Out[3]= {{**Negative Mode**,,,,,,,
,,,,,,,,,,,,,},{mzmed,GF1,GF2,GF3,
Day5_BTBL_1,Day5_BTBL_2,Day5_BTBL_3,
Day5_BTBL_4,Day10_BTBL_1,Day10_BTBL_2,
Day10_BTBL_3,Day10_BTBL_4,Day15_BTBL_1,
Day15_BTBL_2,Day15_BTBL_3,Day15_BTBL_4,
Day20_BTBL_1,Day20_BTBL_2,Day20_BTBL_3,
Day25_BTBL_1,Day25_BTBL_2,Day25_BTBL_3},
{,0(GF),0(GF),0(GF),5,5,5,5,10,10,10,
10,15,15,15,15,20,20,20,25,25,25},
{269.125,4.77187×10^6,1.83739×10^6,2.69081×10^6,
1.78109×10^6,1.95988×10^6,1.61111×10^6,
1.84043×10^6,3.46027×10^6,7.17564×10^6,
4.24283×10^6,5.18863×10^6,3.44216×10^6,
5.37601×10^6,5.75372×10^6,5.62317×10^6,
5.60079×10^6,4.82033×10^6,2.78184×10^6,
2.68937×10^6,3.73472×10^6,5.98949×10^6}}

We can find the location for the positive mode in the file:

In[4]:= positiveModeLocation=
Position[importGFMice,"Positive Mode"]
Out[4]= {{1559,1}}

We can use **Extract** directly on the above **Position** output:

In[5]:= **Extract**[importGFMice,positiveModeLocation]
Out[5]= {Positive Mode}

8.2 Germ Free and Inoculated Mice Data

Now we can see if this is correct by looking at data around this position:

In[6]:= importGFMice[[1558;;1559+3]]
Out[6]= {{233.097,486634.,236327.,17058.,
    72046.,66187.,48386.7,100225.,86517.2,
    174761.,152207.,182583.,194218.,
    91068.7,185375.,195549.,182746.,173075.,
    125166.,149478.,156302.,110914.},
    {Positive Mode,,,,,,,,,,,,,,,
    ,,,,,,},{mzmed,GF1,GF2,GF3,
    Day5_BTBL_1,Day5_BTBL_2,Day5_BTBL_3,
    Day5_BTBL_4,Day10_BTBL_1,Day10_BTBL_2,
    Day10_BTBL_3,Day10_BTBL_4,Day15_BTBL_1,
    Day15_BTBL_2,Day15_BTBL_3,Day15_BTBL_4,
    Day20_BTBL_1,Day20_BTBL_2,Day20_BTBL_3,
    Day25_BTBL_1,Day25_BTBL_2,Day25_BTBL_3},
    {,GF,GF,GF,5,5,5,5,10,10,10,10,
    15,15,15,15,20,20,20,25,25,25},
    {400.148,151323.,111751.,62803.1,37103.3,
    38007.8,40273.7,62485.1,45894.4,
    100245.,91827.7,103192.,110570.,
    64440.1,88330.9,79779.5,90142.1,77907.6,
    62669.8,51059.6,47668.2,61009.8}}

We can drop all lines we do not want to get the list of all metabolites. The **Drop** removes lines 1559 to 1561 and the subsequent take ignores the first 3 lines and the first column:

In[7]:= metabolitesGFRawData=
    Drop[importGFMice,{1559,1561}][[4;;,2;;]];

We also extract the annotations:

In[8]:= metabolitesGFAnnotations=
    importGFMice[[1561,2;;]]
Out[8]= {GF,GF,GF,5,5,5,5,10,10,10,
    10,15,15,15,15,20,20,20,25,25,25}

And finally we can obtain the corresponding mass features:

In[9]:= metabolitesGFFeatures=
    Drop[importGFMice,{1559,1561}][[4;;,1]];
In[10]:= metabolitesData=**Drop**[importGFMice,{1559,1561}][[
    4;;,2;;]];

We will first look at data profiles. We can use **SmoothHistogram** to plot a smooth kernel histogram of the intensities for each sample, Fig. 8.1. The **Transpose** gives us the samples as lists. To account for the various ranges we took the **Log2** of the values x_i. You can pick any base for the logarithm: common bases are 2, 10, and occasionally natural log is also used. Though natural log is mathematically more intuitive, log base 2 is more convenient when discussing changes in terms of fold changes.

```
In[11]:= SmoothHistogram[
    Log2[#+1] &@Transpose[metabolitesData],
    PlotRange → All, PlotTheme → "Scientific",
    FrameLabel → {"Log2 Intensity", "Frequency"},
    PlotLabel →
        "Smooth Histogram of Raw Intensity Data"]
```

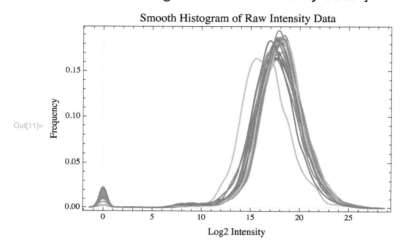

Out[11]=

Fig. 8.1 Out[11]= Smooth histogram of intensity data for metabolites [12]

We have also added 1 to each value, to avoid zero values, which accounts for the secondary peak seen in Fig. 8.1. This is one method often used to avoid zeros, and in this case the intensities are large so the addition has minimal effect (The data for this accounts for the bump around zero in the histogram, Fig. 8.1). As an alternative, we can tag 0 values as **Missing**[] and delete them, Fig. 8.2.

The data is not normal, and also we notice that not all means are aligned. We can transform the data in different ways, for example standardize the **Log2** values of the data. We use a **MathIOmica** function that can handle **Missing** values, **StandardizedExtended**, subtracting the mean and dividing by the **MeanDeviation**.

In[13]:= <<MathIOmica`

```
In[14]:= standardizationLogs=
    SeriesApplier[
        StandardizeExtended[Log2[#],Mean,
            MeanDeviation]&,
        Transpose@metabolitesData /. 0|0. -> Missing[]];
```

The smooth histogram now is centered around zero, Fig. 8.3.

8.2 Germ Free and Inoculated Mice Data

In[12]:= `SmoothHistogram[`
  `Log2[DeleteMissing[#] + 1] & /@`
    `(Transpose[metabolitesData] /.`
      `0 | 0. -> Missing[]), PlotRange -> All,`
  `PlotTheme -> "Scientific",`
  `FrameLabel -> {"Log2 Intensity", "Frequency"},`
  `PlotLabel ->`
    `"Smooth Histogram of Raw Intensity Data"]`

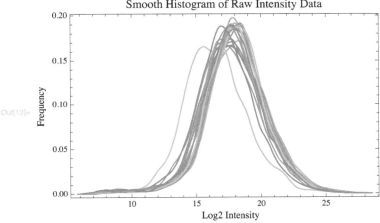

Out[12]=

Fig. 8.2 Out[12]= Smooth histogram of intensity data for metabolites, zero values removed

Here we perform for the purpose of illustration a power transformation to get the data distributions to be close to normal. The Box-Cox [2] power transformation is implemented in MathIOmica.

In[16]:= **?ApplyBoxCoxTransform**
  ApplyBoxCoxTransform[data] for a given
    data set, computes the Box–Cox transformation
    at the maximum likelihood λ parameter.

We apply this Box-Cox transformation to the metabolite data for each sample. As all the values need to be positive for the transformation, we tag 0 data as Missing by a substitution. The optimized $\hat{\lambda}$ parameter for each sample is printed out for reference:

In[17]:= boxCoxTransformedMetaboliteData=
  ApplyBoxCoxTransform[#]&/@
    Transpose[metabolitesData/. 0|0.-> **Missing**[]];

  (*The lines below will be printed during evaluation*)
  Calculated Box–Cox parameter $\hat{\lambda} = -0.00130934$
  Calculated Box–Cox parameter $\hat{\lambda} = 0.0126927$

```
In[15]:= SmoothHistogram[
    DeleteMissing[#] & /@ standardizationLogs,
    PlotRange → All, PlotTheme → "Scientific",
    FrameLabel → {"Standardized Log2 Intensity",
      "Frequency"},
    PlotLabel →
    "Smooth Histogram of Standardized Log2 Intensity Data"]
```

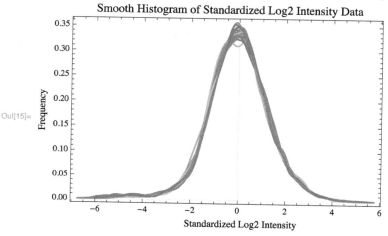

Fig. 8.3 Out[15]= Smooth histogram of standardized intensity data for metabolites, zero values removed

Calculated Box–Cox parameter $\hat{\lambda} = -0.0109138$
...
Calculated Box–Cox parameter $\hat{\lambda} = 0.0403079$
Calculated Box–Cox parameter $\hat{\lambda} = 0.0406148$

Now we can plot the transformed data, in Fig. 8.4.

Next, we standardize the data, using again **StandardizeExtended** that ignores **Missing[]** or text values:

```
In[19]:= standardizedMetaboliteData=
    StandardizeExtended[#,Mean, StandardDeviation]&/@
    boxCoxTransformedMetaboliteData;
```

As we can see, compared to standardizing the Log values, the distributions spread is closer to a normal distribution, Fig. 8.5.

```
In[18]:= SmoothHistogram[
    DeleteMissing[#] & /@
      boxCoxTransformedMetaboliteData, PlotRange → All,
    PlotTheme → "Scientific",
    FrameLabel → {"Box-Cox Transformed Intensity",
      "Frequency"},
    PlotLabel →
  "Smooth Histogram of Box-Cox Transformed Intensity Data"]
```

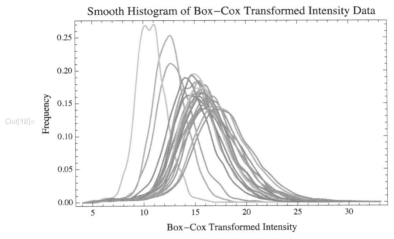

Fig. 8.4 Out[18]= Smooth histogram of Box-Cox transformed intensity data for metabolites

8.3 Principal Component Analysis

A typical method used in viewing metabolomics mass spectrometry data is to look at Principal Component Analysis (PCA). This can be carried out in the Wolfram Language. The function we will use is **PrincipalComponents** which acts on a matrix to transform into unscaled principal components, and with **Method**–> "Correlation" to standardized/scaled principal components. The data needs to be numeric, so we set **Missing** values to 0:

```
In[21]:= principalComponentsMetabolites=
    PrincipalComponents[
      standardizedMetaboliteData/._Missing-> 0,
      Method-> "Correlation"];
```

The dimensions are the same as before:

```
In[22]:= Dimensions[principalComponentsMetabolites]
Out[22]= {21,3247}
```

```
In[20]:= SmoothHistogram[
    DeleteMissing[#] & /@ standardizedMetaboliteData,
    PlotRange → All, PlotTheme → "Scientific",
    FrameLabel →
     {"Box-Cox Transformed Standardized Intensity",
      "Frequency"},
    PlotLabel →
     "Smooth Histogram of Box-Cox Transformed
      Standardized Intensity Data"]
```

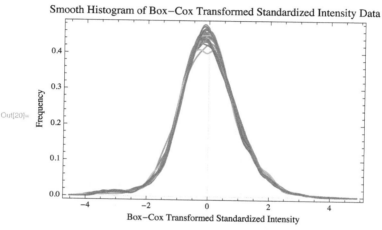

Fig. 8.5 Out[18]= Smooth histogram of Standardized Box-Cox tranformed intensity data for metabolites

We can calculate the variance across the components as a percent of the total variance (the **Plus** @@ adds up the list components of the principalComponentsMetabolites):

```
In[23]:= varianceRatios=
    Variance[principalComponentsMetabolites]/
    Plus @@ Variance[principalComponentsMetabolites];
```

We look at the first 5:

```
In[24]:= varianceRatios[[1;;5]]
Out[24]= {0.404371,0.207679,0.061707,0.0462776,0.0373017}
```

We see that the first couple of components account for most of the variance.

Next we want to build a labeled plot of the information. We need to create annotations for the plot, so we first extract the indexed position of all the sample information:

```
In[25]:= positionsGFAnnotations=
    Thread[{Range[Length[metabolitesGFAnnotations]],
     metabolitesGFAnnotations}]
Out[25]= {{1,GF},{2,GF},{3,GF},{4,5},{5,5},{6,5},
```

8.3 Principal Component Analysis

```
In[27]:=  ListPointPlot3D[
            principalComponentsMetabolites[[All,
                {1, 2, 3}]][[#]] & /@
              (Values@groupsGFAnnotations),
            PlotStyle → ({#, PointSize[0.025]} & /@ #),
            PlotLegends →
              SwatchLegend[#,
                Style[#, FontFamily → "Arial"] & /@
                  {"GF", "5 Days", "10 Days", "15 Days",
                   "20 Days", "25 Days"},
                LegendLabel → Style["Mouse Groups", Italic,
                  FontFamily → "Arial"],
                LegendMarkers → "Bubble"],
            PlotLabel →
              Style[
                "Principal Component Analysis of Normalized
                  Feature Data",
                Bold, FontFamily → "Arial"],
            AxesLabel → {"P1", "P2", "P3"}] &@
          {Green, Blue, Cyan, Red, Orange, Purple}
```

Out[27]=

Fig. 8.6 Out[27]= Principal Component Analysis (PCA) of standardized Box-Cox transformed intensity data for metabolites

In[28]:= `ListPlot[`
    `principalComponentsMetabolites[[All,`
        `{1, 2}]][[#]] & /@`
     `(Values@groupsGFAnnotations),`
    `PlotStyle → ({#, PointSize[0.025]} & /@ #),`
    `PlotLegends →`
     `SwatchLegend[#,`
      `Style[#, FontFamily → "Arial"] & /@`
       `{"GF", "5 Days", "10 Days", "15 Days",`
        `"20 Days", "25 Days"},`
      `LegendLabel → Style["Mouse Groups", Italic,`
       `FontFamily → "Arial"],`
      `LegendMarkers → "Bubble"],`
    `PlotLabel → Style["Principal Components P1-P2",`
     `Bold, FontFamily → "Arial"],`
    `FrameLabel → {"P1", "P2"},`
    `PlotRange → All,`
    `PlotTheme → "Scientific"] &@`
   `{Green, Blue, Cyan, Red, Orange, Purple}`

Out[28]=

Fig. 8.7 Out[28]= Principal Component Analysis (PCA) of standardized Box-Cox transformed intensity data for metabolites, P1–P2

8.3 Principal Component Analysis

```
In[29]:= ListPlot[
    principalComponentsMetabolites[[All,
        {1, 3}]][[#]] & /@
    (Values@groupsGFAnnotations),
    PlotStyle → ({#, PointSize[0.025]} & /@ #),
    PlotLegends →
      SwatchLegend[#,
        Style[#, FontFamily → "Arial"] & /@
          {"GF", "5 Days", "10 Days", "15 Days",
          "20 Days", "25 Days"},
        LegendLabel → Style["Mouse Groups", Italic,
          FontFamily → "Arial"],
        LegendMarkers → "Bubble"],
      PlotLabel → Style["Principal Components P1-P3",
        Bold, FontFamily → "Arial"],
      FrameLabel → {"P1", "P3"},
      PlotRange → All,
      PlotTheme → "Scientific"] &@
    {Green, Blue, Cyan, Red, Orange, Purple}
```

Fig. 8.8 Out[29]= Principal Component Analysis (PCA) of standardized Box-Cox transformed intensity data for metabolites, P1–P3

{7,5},{8,10},{9,10},{10,10},{11,10},
{12,15},{13,15},{14,15},{15,15},{16,20},
{17,20},{18,20},{19,25},{20,25},{21,25}}

We can extract the positions of the groups

In[26]:= groupsGFAnnotations=
  GroupBy[positionsGFAnnotations, **Last**-> **First**]
Out[26]= <|GF→{1,2,3},5→{4,5,6,7},
  10→{8,9,10,11},15→{12,13,14,15},
  20→{16,17,18},25→{19,20,21}|>

We can now use the above indexing to plot the groups and see how they look in Fig. 8.6 - we separated out the colors, so you can try different ones). Notice how the GF mice and the 5 Day inoculated groups are very well separated from the rest.

Some notes on our code in Fig. 8.6: We group the points to be plotted in sets based on their group annotation so we can apply different colors to the points. The colors are applied to each group using the **PlotStyle** option, to pass a list of styles, which is a list of {Color, PointSize} for each group. Each point is given a size 0.025. Next, the legend is generated using a SwatchLegend, that associates swatches of colors with the specified labels for our groups { "GF",...., "25 Days"}.

We can plot individual components separately, of example the P1–P2 plane projection, in Fig. 8.7 and the P1–P3 plane projection in Fig. 8.8.

8.4 Differential Analysis GF Versus 5 Days After Inoculation

We have seen from the PCA analysis that GF mice have a different overall urine metabolome profile. We will investigate this further in this section, to show how we can carry out differential metabolite analysis for GF versus Day5 mice. First we get the relevant data, the first 7 columns, with columns 1–3 corresponding to the GF mice, and columns 4–7 corresponding to D5 mice. We transpose the sets so that the features are in rows, and create a feature to data association using AssociationThread:

In[30]:= gfD5Micedata=**AssociationThread**[
  metabolitesGFFeatures,
  {#[[1;;3]],#[[4;;7]]}&/
  Transpose @standardizedMetaboliteData];
In[31]:= gfD5Micedata[[1;;3]]
Out[31]= <|269.125→{{1.59827,1.3155,1.9353},
  {1.20028,1.23387,1.18128,1.10904}},
  225.083→{{0.513168,0.734814,1.04897},
  {0.626587,0.702056,0.671343,0.689867}},
  323.134→{{0.320948,0.200707,0.51853},
  {0.275585,0.254827,0.257726,0.419021}}|>

We have 3247 features.

In[32]:= **Length**[gfD5Micedata]

8.4 Differential Analysis GF Versus 5 Days After Inoculation

Out[32]= 3247

Some data can have **Missing** values. In our analysis here we want to remove data that have less than 3 numeric values. In each value in our association we have a list for GF mice, and one for Day 5. For GF mice we have 3 data points. Additionally, we could be missing data in the Day 5 features. For the second list in each value (the Day 5 intensities for a given feature) we have 4 points. We can remove data that have less than or equal to 2 points in either list after we delete **Missing** data. First we can delete all missing values:

In[33]:= **Query[All, All, DeleteMissing]**@gfD5Micedata;

Then we can use a second **Query** to select the entries with 3 values in the first list and at least three values in the second list:

In[34]:= filteredGFd5Data=
  **Query[
    Select[
      (Length[#[[1]]]==3)&&Length[#[[2]]] >= 3&]]@
    (Query[All, All, DeleteMissing]@gfD5Micedata);**

In[35]:= **Length[filteredGFd5Data]**
Out[35]= 3175

This corresponds to dropping 72 data points compared to the original. We will apply a location test to each small metabolite using **LocationTest** (see also Chaps. 3 and 6 for gene expression data). For example:

In[36]:= **LocationTest[filteredGFd5Data[[1]], Automatic,
  {"TestDataTable", All}]**

Out[36]=
	Statistic	P-Value
Mann-Whitney	12.	0.0215563
T	2.40355	0.132893
Z	2.40355	0.0162366

The t-test is probably more robust, and we will use that, which also the Wolfram Language returns. We can get a "TestConclusion":

In[37]:= **LocationTest[filteredGFd5Data[[1]], Automatic,
  "TestConclusion"]**
Out[37]= The null hypothesis that
  the mean difference is 0 is not rejected at the
  5 percent level based on the T test.

We will calculate the test p value and conclusion for all our sets (similarly to Chap. 6 for gene expression). We will obtain the "PValue" and the shorter test conclusion for each test:

In[38]:= testGFd5Conclusions=
  **Query[All,
    LocationTest[#, Automatic,
      {"PValue","ShortTestConclusion"}]&]@
    filteredGFd5Data;**

We get an association of mass feature to the results. Let us look at the first few:

```
In[39]:=  testGFd5Conclusions[[1;;3]]
Out[39]=  <|269.125 → {0.132893,Do not reject},
          225.083 → {0.610388,Do not reject},
          323.134 → {0.640274,Do not reject}|>
```

We get a lot of possible hits:

```
In[40]:=  Length[#]&/@ GroupBy[testGFd5Conclusions, Last]
Out[40]=  <|Do not reject → 1724, Reject → 1451|>
```

Here we have grouped the entries in the association by the last value, which is the test conclusion, and then did a count to get the totals.

This means, in 1451 entries the **LocationTest** would have us reject the null hypothesis. However we are testing many hypotheses, so we have to correct for that. Using a Bonferroni correction we can set the p value cutoff at 0.05/3175 (i.e. divide 0.05 significance cutoff by the number of tests cf. Chap. 6):

```
In[41]:=  testGFd5ConclusionsBFCorrected=
          Query[All,
            LocationTest[#,Automatic,
              {"PValue","ShortTestConclusion"},
              SignificanceLevel ->(0.05/3175)]&]@
          filteredGFd5Data;
```

In this approach only 87 results are listed as rejecting the null hypothesis:

```
In[42]:=  Length[#]&/@GroupBy[testGFd5ConclusionsBFCorrected, Last]
Out[42]=  <|Do not reject → 3088, Reject → 87|>
```

We can plot the p values to check the distribution shape, Fig. 8.9. We may think that the Bonferroni correction is a bit too strict. We now use the Benjamini–Hochberg [1] False Discovery Rate (FDR) correction instead:

```
In[44]:=  testGFd5FDRResults=
          BenjaminiHochbergFDR[
            Values @testGFd5ConclusionsBFCorrected[[All,1]],
            SignificanceLevel → 0.05];
```

The SignificanceLevel is set to 0.05 as a statistical significance cutoff. The output has **Keys**:

```
In[45]:=  Keys @ testGFd5FDRResults
Out[45]=  {Results, p-Value Cutoff, q-Value Cutoff}
```

We take a look at the first few results:

```
In[46]:=  Query["Results",1;;5]@testGFd5FDRResults
Out[46]=  {{0.0361271,0.0723224,False},
          {0.510254,0.59083,False},
          {0.640274,0.70732,False},
          {0.00260507,0.0106799,True},
          {0.137148,0.204915,False}}
```

We also obtain the "p-Value Cutoff" value for the p value corresponding to the significance level we decided on, and the "q-Value Cutoff" for the corresponding corrected cutoff value:

8.4 Differential Analysis GF Versus 5 Days After Inoculation

```
In[43]:= Histogram[testGFd5ConclusionsBFCorrected[[All, 1]],
    PlotTheme → "Scientific",
    PlotLabel →
     "P Value Distribution from 3175 Location Tests",
    FrameLabel → {"P Value", "Count"}]
```

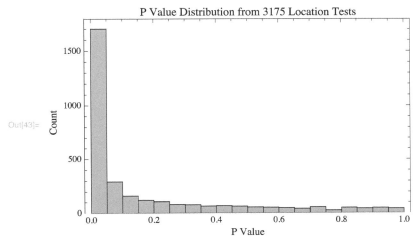

Out[43]=

Fig. 8.9 Out[43]= Distribution of p values from hypotheses testing for metabolites

```
In[47]:= testGFd5FDRResults["p-Value Cutoff"]
Out[47]= 0.0217436

In[48]:= testGFd5FDRResults["q-Value Cutoff"]
Out[48]= 0.0497377
```

We can next make an association for the mass features to the p value results:

```
In[49]:= pValuesGFd5FDR=
          Query["Results",
           (AssociationThread[
             Keys[testGFd5ConclusionsBFCorrected]->
               #]&)]@testGFd5FDRResults;
         pValuesGFd5FDR // Short[#,4]&
Out[50]//Short= <|269.125→{0.0361271,0.0723224,False},
                 <<3173>>,235.178→{0.418964,0.505976,False}|>
```

We select the results that have "True" for significance based on the FDR:

```
In[51]:= significantQuery=
          Query[Select[#[[3]]==True&]]@pValuesGFd5FDR;

In[52]:= significantQuery[[1;;3]]
Out[52]= <|86.0247→{0.00260507,0.0106799,True},
          301.056→{0.00411965,0.0145214,True},
```

```
In[55]:= Histogram[deltaGFd5Data,
    PlotTheme → "Scientific",
    PlotLabel →
     "Transformed Intensity Difference between
     Germ Free and Day 5 Feature Intensities",
    FrameLabel → {"Absolute Mean Transformed Change",
     "Count"}]
```

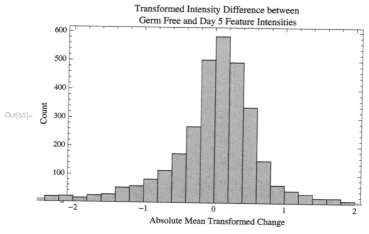

Fig. 8.10 Out[55]= Distribution of p values from hypotheses testing for metabolites

130.039 → {0.000253831, 0.00220311, **True**}|>

There are in total 1388 features passing the FDR at 5% (corresponding to 70 false positive possible calls, 69.4):

In[53]:= **Length**[significantQuery]
Out[53]= 1388

We can extract the values and also calculate an effect size for all these values. As we have already transformed the data to normal distributions, we look at the shift of the data means as an effective fold size:

In[54]:= deltaGFd5Data=**Query**[All ,**Mean**[#[[2]]] −**Mean**[#[[1]]] &]@
    filteredGFd5Data;

We can see the distribution of the changes, which is centered about zero in Fig. 8.10.

We also calculate the relevant corrected p values negative Log base 10 values:

In[56]:= pValueNegLogGFd5Data=
    −**Log10** @ **Query**[All ,2]@pValuesGFd5FDR;

We use **Merge** to combine together the effect data and the p values:

8.4 Differential Analysis GF Versus 5 Days After Inoculation

```
In[58]:= ListPlot[Values@volcanoPlotGFd5Data,
    PlotTheme → "Scientific",
    PlotRange → Full,
    PlotLabel →
      "p Value Versus Mean Change Volcano Plot",
    FrameLabel → {"Mean Transformed Change",
      "-Log10 p Value"},
    GridLines →
      {{{-1, {Red, Thick, Dashed}}, 0,
        {1, {Red, Thick, Dashed}}},
       {{-Log10[0.05], {Blue, Thick, Dashed}}}}]
```

Out[58]=

Fig. 8.11 Out[58]= Volcano plot of results for metabolites

In[57]:= volcanoPlotGFd5Data=
    Merge[{ deltaGFd5Data , pValueGFd5Data } , **Identity**];

Now we can construct a volcano plot, Fig. 8.11 with reference lines corresponding to cutoffs, at ±1 for transformed mean change, and also at $-\mathrm{Log10}[0.05]$ for significance (see also Chap. 6).

We can select features that have transformed mean changes greater than 1 in absolute terms, and corrected p values smaller than 0.05:

In[59]:= selectedFeaturesGFd5=
    Query[
      Select[
        Abs[#[[1]]]>1 && #[[2]] > (−**Log10**[0.05])&]]@
      volcanoPlotGFd5Data ;

The result is 366 mass features:

In[62]:= **chemSpider = ServiceConnect["ChemSpider"]**

Out[62]= **ServiceObject**[⇌ ChemSpider Connected]

Fig. 8.12 Out[62]= Connecting to ChemSpider

In[60]:= **Length[selectedFeaturesGFd5]**
Out[60]= 366

In[61]:= selectedFeaturesGFd5[[1;;10]]
Out[61]= <|238.045 → {−1.65905, 3.09779},
338.146 → {2.71869, 4.37504},
388.136 → {−2.3985, 2.57451},
279.038 → {−1.88966, 1.9473},
128.028 → {−1.35472, 2.17012},
178.983 → {−1.43019, 2.45974},
420.109 → {−2.2604, 1.73732},
128.034 → {−1.3964, 2.23483},
353.085 → {−2.00294, 1.85289},
178.034 → {−1.21014, 2.1053}|>

We can use the mass feature to see if we can figure out which putative IDs we have for compounds in the next sections.

8.5 Identifying Compounds Using ChemSpider

Here we show how can use the Wolfram Language to connect to the ChemSpider API. There is an inbuilt service to access ChemSpider database information, and we will use it to match our mass features.

First we use **ServiceConnect** to connect to ChemSpider, Fig. 8.12. (N.B. You will be asked to sign in with your Wolfram User ID, as well as to sign in to ChemSpider to obtain a security token by a popup window.)

ChemSpider uses many databases. We can check these by calling the service to get a list:

In[63]:= chemSpiderDatabases=chemSpider["Databases"];
In[64]:= chemSpiderDatabases[[1;;10]]
Out[64]= {4C Pharma Scientific, A&A Life Science,
A1 BioChem Labs, A2Z Chemical, Abblis Chemicals, Abcam,
abcr, ACB Blocks, Accel Pharmtech, Accela ChemBio}

In[65]:= chemSpiderDatabases//**Short**
Out[65]//**Short**= {4C Pharma Scientific, <<574>>, Zylexa Pharma}

We can use **ServiceExecute**["ChemSpider", "request", parameters] to send a request with specified parameters. Since we have named the service above we can go ahead

8.5 Identifying Compounds Using ChemSpider

In[67]:= `idExample = chemSpider["Search",
        "Mass" → 238.0452933, "Range" → 0.001, MaxItems → 5]`

Out[67]=
ID
63109
71899
81385
90104
91088

Fig. 8.13 Out[67]= Searching ChemSpider by mass

In[69]:= `id63109Info = chemSpider["CompoundInformation",
        "ID" → "63109"]`

Out[69]=
CSID	63109
InChI	InChI=1S/C15H10OS/c16-13-10-15(11-6-2-1-3-7-11)17-14-9-5-4-8-12(13)14/h1-10H
InChIKey	GIQPSSZMIZARDW-UHFFFAOYSA-N
SMILES	c1ccc(cc1)c2cc(=O)c3ccccc3s2

Fig. 8.14 Out[69]= Searching ChemSpider by ID

directly. The search looks like chemSpider["Search","Mass" –> m, "Range" –> r, MaxItems –> n]. This corresponds to a "Search" request, to search by "Mass" equal to m, with m being within range of r, and to return n maximum number of identifiers. Please see the documentation for additional parameters and values:

In[66]:= **SystemOpen**["paclet:ref/service/ChemSpider"]

The search returns a dataset, Fig. 8.13.
The returned **Dataset** is a list of IDs:

In[68]:= Normal@idExample
Out[68]= { <|ID→63109|>,<|ID→71899|>,
         <|ID→81385|>,<|ID→90104|>,<|ID→91088|>}

Once we have an ID we can go ahead and look the information up, Fig. 8.14.
If you also prefer to work with associations, you can suppress the dynamic output with ;. The information is still available if we take the variable directly, and also we can look at the **Normal** representation of the dataset:

In[70]:= **Head**[id63109Info]
Out[70]= **Dataset**

In[71]:= **Normal** @ id63109Info

In[72]:= `extended63109Info =`
`chemSpider["ExtendedCompoundInformation",`
`"ID" → "63109"]`

Out[72]=

CSID	63109
MF	C_{15}H_{10}OS
SMILES	c1ccc(cc1)c2cc(=O)c3ccccc3s2
InChI	InChI=1/C15H10OS/c16-13-10-15(11-6-2-1-3-7-11)17-14-9-5-4-8-12(13)14/h1-10H
InChIKey	GIQPSSZMIZARDW-UHFFFAOYAA
AverageMass	238.3043
MolecularWeight	238.3043
MonoisotopicMass	238.045242
NominalMass	238
ALogP	0
XLogP	3.9
CommonName	2-Phenyl-4H-thiochromen-4-one

Fig. 8.15 Out[72]= Extended information from ChemSpider by ID

In[73]:= `chemSpider["CompoundThumbnail", "ID" → "63109"]`

Out[73]=

Fig. 8.16 Out[73]= Compound thumbnail from ChemSpider by ID

Out[71]= <|CSID→63109, InChI→
InChI=1S/C15H10OS/c16-13-10-15(11-6-2-1-3-7-11)17-14
-9-5-4-8-12(13)14/h1-10H,
InChIKey→GIQPSSZMIZARDW-UHFFFAOYSA-N,
SMILES→c1ccc(cc1)c2cc(=O)c3ccccc3s2|>

We can also get extended information, Fig. 8.15, and a `"CompoundThumbnail"`, Fig. 8.16.

As an aside, we could do the lookup directly if we have a name of a compound:

In[74]:= **Normal** @ chemSpider["Search",
"Query"->"2-Phenyl-4H-thiochromen-4-one"]
Out[74]= <|ID →63109|>

Now let us look at our various entries. We construct a ServiceRequest as a pure function to search the **Keys** for our selected features, within a range of 5 parts per million (ppm). The search again will return up to 5 items (please note this can take a while depending on your connection and list of compounds):

In[75]:= resultsIDs=
chemSpider["Search","Mass"-> #,

8.5 Identifying Compounds Using ChemSpider

```
        "Range"->(5*10^(-6)*#),MaxItems->5]&/@
      (Keys @ selectedFeaturesGFd5);
```

We can check how many unique IDs we got, using a **Tally**:

```
In[76]:= Tally[Length[#]&/@resultsIDs]
Out[76]= {{5,308},{1,37},{0,21}}
```

We notice that 308 IDs are not unique (5 items each, which is the maximum), 37 are unique and 21 have no matches. We are interested in the unique ones, so let us extract them:

```
In[77]:= positionsChemSpiderUnique=
            Position[resultsIDs,x_/;Length[x]==1]
Out[77]= {{5},{14},{15},{17},{38},{42},{44},{46},
          {49},{56},{58},{66},{67},{83},{85},{111},
          {113},{131},{174},{175},{192},{202},{211},
          {216},{217},{221},{231},{234},{262},{263},
          {284},{286},{292},{296},{300},{359},{361}}
```

We use **Query** to extract the IDs by selecting first the elements that have length equal to 1, and then selecting the first part of these which is the identifier.

```
In[78]:= chemSpiderUniqueIDs=
            Query[(Select[Length[#]==1&]),1]@resultsIDs
Out[78]= {10695492,57522026,8117740,16127232,4484268,
          10816284,183101,120057,9963870,2062433,2849023,
          183101,10531457,8139662,455565,57532061,
          9052908,4890256,57269218,2050613,364974,
          1544479,9193986,376658,5496216,10816284,
          1544479,2796974,9228003,49071248,23136140,
          71333,109782,559241,9732048,9140563,1554167}
```

We can make an **Association** of our mass feature to ID:

```
In[79]:= featuresGFd5ChemSpiderID=
            AssociationThread[
              Extract[Keys @selectedFeaturesGFd5,
              positionsChemSpiderUnique],chemSpiderUniqueIDs]
Out[79]= <|128.028->10695492,147.011->57522026,
          99.0075->8117740,150.989->16127232,
          87.0074->4484268,92.9827->10816284,94.9783->183101,
          106.998->120057,86.0234->9963870,176.001->2062433,
          96.0442->2849023,94.9784->183101,84.0442->10531457,
          92.0573->8139662,91.054->455565,74.0234->57532061,
          173.1->9052908,132.048->4890256,198.91->57269218,
          195.913->2050613,89.1077->364974,134.027->1544479,
          82.0155->9193986,193.162->376658,192.159->5496216,
          92.9831->10816284,134.027->1544479,130.032->2796974,
          207.115->9228003,107.049->49071248,170.98->23136140,
          167.926->71333,71.0861->109782,99.0807->559241,
          76.0398->9732048,191.12->9140563,176.094->1554167|>
```

Now we can use **Query** to extract per ID the extended information from ChemSpider as the new value for each mass feature:

In[80]:= featuresGFd5ChemSpiderIDExtended=
    Query[All,
      Normal@chemSpider["ExtendedCompoundInformation",
        "ID"->#]&]@featuresGFd5ChemSpiderID;

Next, for every feature we can extract any of the following information:

In[81]:= Query[1,Keys]@featuresGFd5ChemSpiderIDExtended
Out[81]= {CSID,MF,SMILES,InChI,InChIKey,AverageMass,
    MolecularWeight,MonoisotopicMass,
    NominalMass,ALogP,XLogP,CommonName}

Here we have the "InChI" (IUPAC [International Union of Pure and Applied Chemistry] International Chemical Identifier) or "InChIKey" (a hash version of InChI) - see "https://iupac.org/who-we-are/divisions/division-details/inchi/" and "http://www.inchi-trust.org".

For each feature we get the names of the potential compound:

In[82]:= Query[All,"CommonName"]@featuresGFd5ChemSpiderIDExtended
Out[82]= <|128.028→2−Furyl dihydrogen borate,
    147.011→2,4,6,8,10−Undecapentaynenitrile,
    99.0075→Beryllium diformate,150.989→
    3−Mercapto−2−mercaptomethylpropanoate,87.0074→
    (2R)−2−Hydroxy−4−oxo−1−oxoniabicyclo[1.1.0]butane,
    92.9827→SODIUM NITRILOACETATE,
    94.9783→Phosphoramidate,
    106.998→Acrylonitrile, trifluoro−,
    86.0234→Hydroxy(2−oxopropylidyne)ammonium,176.001→
    3−(Chloromethyl)−1,1,2,2−tetrafluorocyclobutane,
    96.0442→3−Pyridinyloxonium,94.9784→Phosphoramidate,
    84.0442→3−Methyl−1,3−oxazol−3−ium,
    92.0573→DL−Alanine−13C3,91.054→Phenylmethylium,
    74.0234→4,5−Dihydro−1,3,2−dioxazol−1−ium,
    173.1→Cyclohexyl(dimethoxy)silyl,
    132.048→1−Oxo−2−sulfanylpiperidinium,198.91→
    1−Butene, zinc salt, hydrobromide (1:1:1),195.913→
    2,2,2−Trichloroethanimidamide hydrochloride (1:1),
    89.1077→1−Ethyl−1,1−dimethylhydrazinium,
    134.027→2,5−Dihydro−3−thiophenaminium 1,1−dioxide,
    82.0155→2−Chloro(1,1−~2~H_2_)ethanol,
    193.162→1,2−Ethanediamine −
    2−(butylsulfanyl)ethanamine (1:1),192.159→
    (2S)−2−Hydroxy−N,N−bis[(2R)−2−hydroxypropyl]−1−
    propanaminium,
    92.9831→SODIUM NITRILOACETATE,
    134.027→2,5−Dihydro−3−thiophenaminium 1,1−dioxide,
    130.032→3−(2−Hydroxyethyl)−1,3−thiazol−3−ium,
    207.115→Butyl(diethoxymethyl)oxophosphonium,
    107.049→L−(1,2−~13−C_2_)Serine,
    170.98→3−Amino−1−propaneseleninic acid,
    167.926→TRIFLUOROMETHANESULFONYLCHLORIDE,
    71.0861→Pentyl,99.0807→1−Oxohex−1−ylium,
    76.0398→Ethyl (~14~C)formate,
    191.12→Methoxy(octyl)oxophosphonium,176.094→

8.5 Identifying Compounds Using ChemSpider

```
            7-Chloro-1,3,5-triazoniatricyclo[3.3.1.1~3,7~]decane
          |>
```

For each feature we get the potential CSID:

```
In[83]:= Query[All,"CSID"]@featuresGFd5ChemSpiderIDExtended
Out[83]= <|128.028→10695492,147.011→57522026,
          99.0075→8117740,150.989→16127232,
          87.0074→4484268,92.9827→10816284,94.9783→183101,
          106.998→120057,86.0234→9963870,176.001→2062433,
          96.0442→2849023,94.9784→183101,84.0442→10531457,
          92.0573→8139662,91.054→455565,74.0234→57532061,
          173.1→9052908,132.048→4890256,198.91→57269218,
          195.913→2050613,89.1077→364974,134.027→1544479,
          82.0155→9193986,193.162→376658,192.159→5496216,
          92.9831→10816284,134.027→1544479,130.032→2796974,
          207.115→9228003,107.049→49071248,170.98→23136140,
          167.926→71333,71.0861→109782,99.0807→559241,
          76.0398→9732048,191.12→9140563,176.094→1554167|>
```

Or we can extract the SMILES (Simplified Molecular Input Line Entry Specification).

```
In[84]:= Query[{7},"SMILES"]@featuresGFd5ChemSpiderIDExtended
```

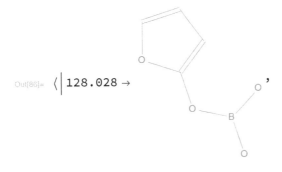

Fig. 8.17 Out[86]= Importing MOL for each identifier

Out[84]= <|94.9783→NP(=O)([O-])[O-]|>

We can convert all our identifiers to `"MOL"` (molecular model MDL files - Elsevier Molecular Design Limited). Here we only do it for the first 3, in Fig. 8.17.

Finally, we can disconnect from chemSpider using **ServiceDisconnect**:

In[87]:= **ServiceDisconnect**[chemSpider]

8.6 Identifying Compounds and Pathway Analysis: KEGG

We may be interested to find the membership of particular identifiers to pathway information. The KEGG pathway database contains many pathways. The compound identifiers for KEGG [9] are restricted to what is in their database and have the form cpd:*number*. We can use MathIOmica's **MassMatcher** to find matching compounds to our features list. The **MassMatcher** will assign putative mass identification to input data based on monoisotopic mass (using MathIOmica's **MassDictionary** to KEGG identities), using the accuracy in parts per million (ppm). For example, using a 3 ppm cutoff:

In[88]:= putativeIDs=**MassMatcher**[#,3]&/@
        Keys@selectedFeaturesGFd5;

Some features have no identity returned, and some can have multiple. Our cutoff is fairly strict so we do not get many hits:

In[89]:= putativeIDs//**Short**
Out[89]//Short= {{},{},<<362>>,{},{}}

We can see that 346 mass features did not match:

In[90]:= **Tally**[putativeIDs]
Out[90]= {{{},346},{{cpd:C01852,cpd:C09781,cpd:C09802},1},
        {{cpd:C18409},1},
        {{cpd:C16698,cpd:C19770,cpd:C21027},1},
        {{cpd:C18922},1},
        {{cpd:C08503,cpd:C14538},1},
        {{cpd:C05983},1},{{cpd:C09123},1},
        {{cpd:C18421},1},{{cpd:C14648},1},
        {{cpd:C18553},1},{{cpd:C06862},1},
        {{cpd:C17132,cpd:C17133,cpd:C17136},1},
        {{cpd:C05366,cpd:C08831,cpd:C09077,cpd:C09515,
          cpd:C09516,cpd:C10317,cpd:C10558,cpd:C10682,
          cpd:C17414,cpd:C17508,cpd:C17808,cpd:C20455},1},
        {{cpd:C00062,cpd:C00792,cpd:C02385},1},
        {{cpd:C10545,cpd:C10563,cpd:C10640},1},
        {{cpd:C09098},1},{{cpd:C17704},1},
        {{cpd:C15193,cpd:C18927},1},
        {{cpd:C17364},1},{{cpd:C15660},1}}

8.6 Identifying Compounds and Pathway Analysis: KEGG

Depending on your cutoff you can get more identities. The more relaxed the cutoff the more the identities that can be found, but also a lot of them will not be unique. For example, we set the ppm to 10:

```
In[91]:= ppm10putativeIDs=
          MassMatcher[#,10]&/@Keys@selectedFeaturesGFd5;
```

We can now see that we get more hits:

```
In[92]:= ppm10putativeIDs[[1;;45]]
Out[92]= {{},{cpd:C19333},
         {cpd:C01852,cpd:C04541,cpd:C09781,cpd:C09802},
         {},{cpd:C16472,cpd:C16473},{},{},{cpd:C08734},
         {cpd:C18592},{},{},{},{},{},{},{},{},{},
         {},{},{cpd:C08477,cpd:C10583},{},{},{},{},
         {cpd:C20313},{},{cpd:C08549},{},{},{},{},{},
         {},{cpd:C18409},{},{},{},{},{},{},{},{},{}}
```

We will filter here the unique ID positions:

```
In[93]:= positionsPutativeIDs=
          Position[putativeIDs,x_/;Length[x]==1]
Out[93]= {{36},{75},{161},{181},{186},{187},
         {206},{230},{314},{346},{360},{364}}
```

Fore example, at position 360 we have a mass feature 315.144.

```
In[94]:= selectedFeaturesGFd5[[{360}]]
Out[94]= <|315.144→{1.91576,2.275}|>
```

And we also get the corresponding KEGG IDs:

```
In[95]:= accessionPutativeIDs=
          Flatten@Extract[putativeIDs,positionsPutativeIDs]
Out[95]= {cpd:C18409,cpd:C18922,cpd:C05983,cpd:C09123,
          cpd:C18421,cpd:C14648,cpd:C18553,cpd:C06862,
          cpd:C09098,cpd:C17704,cpd:C17364,cpd:C15660}
```

We can actually obtain an annotation using **KEGGDictionary**. **N.B.** KEGG Pathways and databases have restrictive rules regarding use for non-academic purposes. Please consult the KEGG license before using such information in your research to make sure you have the appropriate permissions, or request these as necessary.

We can use MathIOmica's inbuilt dictionary:

```
In[96]:= compoundDictionary=
          KEGGDictionary[KEGGQuery1→ "cpd",KEGGQuery2→ ""];
In[97]:= compoundDictionary[[1;;10]]
Out[97]= <|cpd:C00001→H2O; Water,cpd:C00002→
          ATP; Adenosine 5'−triphosphate,cpd:C00003→
          NAD+; NAD; Nicotinamide adenine dinucleotide; DPN;
              Diphosphopyridine nucleotide; Nadide; beta−NAD+,
          cpd:C00004→NADH; DPNH; Reduced nicotinamide
              adenine dinucleotide,
          cpd:C00005→NADPH; TPNH; Reduced
              nicotinamide adenine dinucleotide phosphate,
          cpd:C00006→NADP+; NADP; Nicotinamide adenine
```

dinucleotide phosphate; beta−Nicotinamide
adenine dinucleotide phosphate; TPN;
Triphosphopyridine nucleotide; beta−NADP+,
cpd:C00007→Oxygen; O2, cpd:C00008→
ADP; Adenosine 5'−diphosphate,
cpd:C00009→Orthophosphate; Phosphate;
Phosphoric acid; Orthophosphoric acid,
cpd:C00010→CoA; Coenzyme A; CoA−SH|>

We query our identifiers against the compound dictionary we just created:

In[98]:= **Query**[accessionPutativeIDs]@compoundDictionary
Out[98]= <|cpd:C18409→Cycloprothrin, cpd:C18922→Dinocton 6,
cpd:C05983→Propionyladenylate; Propionyl−adenosine
monophosphate, cpd:C09123→Athamantin,
cpd:C18421→Propamocarb hydrochloride,
cpd:C14648→6alpha−Chloro−17−acetoxyprogesterone;
d6alpha−Chloro−17−hydroxypregn−4−ene−3,20−dione
acetate, cpd:C18553→Spirodiclofen,
cpd:C06862→Busulfan, cpd:C09098→Canthin−6−one;
6H−Indolo(3,2,1−de)(1,5)naphthyridin−6−one,
cpd:C17704→Antibiotic JI−20A; JI−20A,
cpd:C17364→Clavamycin **D**,
cpd:C15660→A 77003|>

We can obtain any information for the molecules above as we have discussed in other sections. We can for example look them up, for example using "ctrl"+"=" "Coenzyme A" or through direct entry using **Entity**, that will can the entity representation in Fig. 8.18.

We can look at different properties, as for any chemical, listed here for your reference:

In[101]:= CanonicalName[#]&/@Entity["Chemical"]["Properties"]
Out[101]= {AcidityConstants, **AdjacencyMatrix**,
AlternateNames, AtomPositions, AutoignitionPoint,
BeilsteinNumber, BlackStructureDiagram,
BoilingPoint, BondCounts, BondEnergies,
BondLengths, CASNumber, CHBlackStructureDiagram,
CHColorStructureDiagram, CIDNumber, Codons,
ColorStructureDiagram, CombustionHeat,
CriticalPressure, CriticalTemperature, Density,

Fig. 8.18 Out[99]=
Entering "ctrl"+"=""", or
Out[100]= **Entity**[] to get an
entity in the Wolfram
Language

In[99]:= [**busulfan** CHEMICAL]

Out[99]= [busulfan]

In[100]:= **Entity["Chemical", "Busulfan"]**

Out[100]= [busulfan]

8.6 Identifying Compounds and Pathway Analysis: KEGG   277

DielectricConstant, DipoleMoment, DOTHazardClass,
DOTNumbers, EdgeRules, EdgeTypes, EGECNumber,
ElectronAffinity, ElementCounts, ElementMassFraction,
ElementTypes, EUNumber, FlashPoint, FormalCharges,
FormattedName, Formula, FormulaString, FusionHeat,
GmelinNumber, HBondAcceptorCount, HBondDonorCount,
HenryLawConstant, HildebrandSolubility, HillFormula,
HillFormulaString, InChI, IonCounts, IonEquivalents,
Ions, IsoelectricPoint, IsomericSMILES, Isomers,
IUPACName, LewisDotStructureDiagram, LightSpeed,
LogAcidityConstants, LowerExplosiveLimit,
MDLNumber, MeanFreePath, MeltingPoint, Memberships,
MolarMass, MolarVolume, MolecularMass, MoleculePlot,
MultiBondMoleculePlot, MultiBondStickMoleculePlot,
Name, NetCharge, NFPAFireRating, NFPAHazards,
NFPAHealthRating, NFPALabel, NFPAReactivityRating,
NonHydrogenCount, NonStandardIsotopeCount,
NonStandardIsotopeCounts, NonStandardIsotopeNumbers,
NSCNumber, OdorThreshold, OdorType,
PartitionCoefficient, pH, Phase, ProtonAffinity,
RefractiveIndex, RelativeMolecularMass,
Resistivity, RotatableBondCount, RTECSClasses,
RTECSNumber, SideChainLogAcidityConstant, SMILES,
Solubility, SpaceFillingMoleculePlot, StandardName,
StickMoleculePlot, SurfaceTension, TautomerCount,
ThermalConductivity, TopologicalPolarSurfaceArea,
UpperExplosiveLimit, VanDerWaalsConstants,
VaporDensity, VaporizationHeat, VaporPressure,
VertexCoordinates, VertexTypes, Viscosity}

We can get chemical specific multi-lists of properties, as well as a molecule plot, Fig. 8.19.

In[102]:= **Entity**["Chemical","Busulfan"][
          {"CASNumber","CIDNumber","Formula","IUPACName",
           "MolarMass","MolecularMass"}]
Out[102]= {CAS55−98−1,CID2478,$H_3SO_2O(CH_2)_4OSO_2CH_3$,
          methanesulfonic acid 4−methylsulfonyloxybutyl ester,
          246.302g/mol,246.302u}

For a given ID, we can also use Wolfram|Alpha [27], as shown for example for NADH in Fig. 8.20

We can also see if we get any over representation in known KEGG pathways for these features, using **KEGGAnalysis** from MathIOmica by specifying the AnalysisType to "Molecular":

In[105]:= **KEGGAnalysis**[accessionPutativeIDs,
          FilterSignificant−> **False**, AnalysisType−> "Molecular"]
Out[105]= <|path:map07218→
          {{0.00256629,0.0102652,**True**},{3,5,5841,1},
           {HIV protease inhibitors,{{cpd:C15660}}}},
          path:map00640→{{0.0244554,0.0489109,**True**},
           {3,48,5841,1},{Propanoate metabolism,
            {{cpd:C05983}}}}, path:map00524→

In[103]:= `Entity["Chemical", "Busulfan"]["StickMoleculePlot"]`

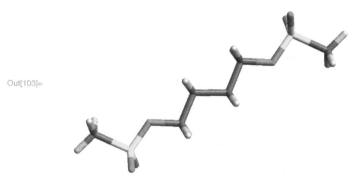

Out[103]=

Fig. 8.19 Out[103]= We obtain a stick molecule plot for a chemical entity

{{0.0410351,0.0547135,**False**},{3,81,5841,1},
{Neomycin, kanamycin and gentamicin biosynthesis,
{{cpd:C17704}}}},path:map01130→
{{0.369022,0.369022,**False**},{3,831,5841,1},
{Biosynthesis of antibiotics,{{cpd:C17704}}}}|>

The analysis returns an association, with KEGG pathways as keys, and each pathway n has a multi-list value in the form:

KEGG: pathway$_n$ → {{p–value$_n$,
  multiple hypothesis adjusted p–value$_n$,
  True/False for statistical significance},
  {{number of members in group being tested,
  number of successes for term$_n$ in population,
  total number of members in population,
  number of members (or more) in current group
    being tested associated to pathway$_n$}},
  {KEGG pathway$_n$} description,
  {input IDs associated to pathway$_n$}}}}

For the low number of IDs we just get a couple of over representation analysis (ORA) results. We will relax the ppm matcher cutoff for the purposes of illustrating that this can give us different positive results:

In[106]:= ppm10positionsPutativeIDs=
  Position[ppm10putativeIDs,x_/;**Length**[x]==1];

In[107]:= ppm10accessionPutativeIDs=
  Flatten@Extract[ppm10putativeIDs,
    ppm10positionsPutativeIDs];

In[108]:= ppm10accessionPutativeIDs **// Short**
Out[108]//**Short**= {cpd:C19333,cpd:C08734,<<47>>,
  cpd:C08264,cpd:C17364}

8.6 Identifying Compounds and Pathway Analysis: KEGG

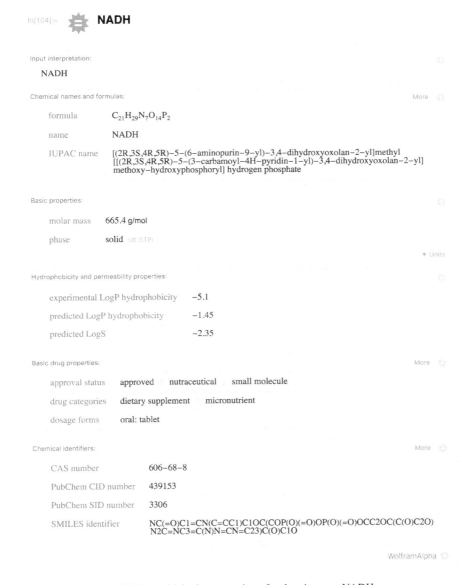

Fig. 8.20 Out[104]= Wolfram|Alpha interpretation of a chemica, e.g. NADH

In[109]:= keggPathways10ppm =
    KEGGAnalysis[ppm10accessionPutativeIDs,
    FilterSignificant -> **False**,
    AnalysisType -> "Molecular"];

We now get 15 pathways:

In[110]:= **Length**[%]
Out[110]= 15

Here are the first few:

In[111]:= keggPathways10ppm[[1 ;; 3]]
Out[111]= <|path:map00791 -> {{0.000795184, 0.0119278, **True**},
    {12, 21, 5841, 2}, {Atrazine degradation,
    {{cpd:C08734}, {cpd:C06553}}}},
    path:map00331 -> {{0.0203711, 0.152783, **False**},
    {12, 10, 5841, 1}, {Clavulanic acid biosynthesis,
    {{cpd:C06656}}}},
    path:map00450 -> {{0.0541309, 0.174115, **False**},
    {12, 27, 5841, 1}, {Selenocompound metabolism,
    {{cpd:C05690}}}}|>

If we are interested in any pathway, we can look it up using the **KEGGPathwayVisual** function:

In[112]:= **KEGGPathwayVisual**["path:map00791"]
Out[112]= <|Pathway ->
    path:map00791, Results ->
{http://www.kegg.jp/kegg-bin/show_pathway?map=map00791}|>

The link can take us to the pathway on KEGG's website.

References

1. Benjamini, Y., Hochberg, Y.: Controlling the false discovery rate: a practical and powerful approach to multiple testing. J. R. Stat. Soc. Ser. B (Methodological) **57**, 289–300 (1995)
2. Box, G., Cox, D.: An analysis of transformations. J. R. Stat. Soc. Ser. B (Methodological) **26**(2), 211–252 (1964)
3. Caspi, R., Altman, T., Dreher, K., Fulcher, C.A., Subhraveti, P., Keseler, I.M., Kothari, A., Krummenacker, M., Latendresse, M., Mueller, L.A., Ong, Q., Paley, S., Pujar, A., Shearer, A.G., Travers, M., Weerasinghe, D., Zhang, P., Karp, P.D.: The metacyc database of metabolic pathways and enzymes and the biocyc collection of pathway/genome databases. Nucleic Acids Res. **40**((Database issue)), D742–D753 (2012)
4. Chen, R., Mias, G.I., Li-Pook-Than, J., Jiang, L., Lam, H.Y., Chen, R., Miriami, E., Karczewski, K.J., Hariharan, M., Dewey, F.E., Cheng, Y., Clark, M.J., Im, H., Habegger, L., Balasubramanian, S., O'Huallachain, M., Dudley, J.T., Hillenmeyer, S., Haraksingh, R., Sharon, D., Euskirchen, G., Lacroute, P., Bettinger, K., Boyle, A.P., Kasowski, M., Grubert, F., Seki, S., Garcia, M., Whirl-Carrillo, M., Gallardo, M., Blasco, M.A., Greenberg, P.L., Snyder, P., Klein, T.E., Altman, R.B., Butte, A.J., Ashley, E.A., Gerstein, M., Nadeau, K.C., Tang, H., Snyder, M.: Personal omics profiling reveals dynamic molecular and medical phenotypes. Cell **148**(6), 1293–307 (2012)

5. Croft, D., O'Kelly, G., Wu, G., Haw, R., Gillespie, M., Matthews, L., Caudy, M., Garapati, P., Gopinath, G., Jassal, B., Jupe, S., Kalatskaya, I., Mahajan, S., May, B., Ndegwa, N., Schmidt, E., Shamovsky, V., Yung, C., Birney, E., Hermjakob, H., D'Eustachio, P., Stein, L.: Reactome: a database of reactions, pathways and biological processes. Nucleic Acids Res. **39**(Database issue), D691–D697 (2011)
6. Griffin, J.L., Shockcor, J.P.: Metabolic profiles of cancer cells. Nat. Rev. Cancer **4**(7), 551–561 (2004)
7. Jain, M., Nilsson, R., Sharma, S., Madhusudhan, N., Kitami, T., Souza, A.L., Kafri, R., Kirschner, M.W., Clish, C.B., Mootha, V.K.: Metabolite profiling identifies a key role for glycine in rapid cancer cell proliferation. Science **336**(6084), 1040–4 (2012)
8. Joshi-Tope, G., Gillespie, M., Vastrik, I., D'Eustachio, P., Schmidt, E., de Bono, B., Jassal, B., Gopinath, G.R., Wu, G.R., Matthews, L., Lewis, S., Birney, E., Stein, L.: Reactome: a knowledgebase of biological pathways. Nucleic Acids Res. **33**(Database issue), D428–D432 (2005)
9. Kanehisa, M., Goto, S.: Kegg: kyoto encyclopedia of genes and genomes. Nucleic Acids Res. **28**(1), 27–30 (2000)
10. Li, X., Gianoulis, T.A., Yip, K.Y., Gerstein, M., Snyder, M.: Extensive in vivo metabolite-protein interactions revealed by large-scale systematic analyses. Cell **143**(4), 639–50 (2010)
11. Luxon, B.A.: Metabolomics in asthma. Adv. Exp. Med. Biol. **795**, 207–20 (2014)
12. Marcobal, A., Yusufaly, T., Higginbottom, S., Snyder, M., Sonnenburg, J.L., Mias, G.I.: Metabolome progression during early gut microbial colonization of gnotobiotic mice. Sci. Rep. **5**, 11589 (2015)
13. Matthews, L., Gopinath, G., Gillespie, M., Caudy, M., Croft, D., de Bono, B., Garapati, P., Hemish, J., Hermjakob, H., Jassal, B., Kanapin, A., Lewis, S., Mahajan, S., May, B., Schmidt, E., Vastrik, I., Wu, G., Birney, E., Stein, L., D'Eustachio, P.: Reactome knowledgebase of human biological pathways and processes. Nucleic Acids Res. **37**(Database issue), D619–D622 (2009)
14. Mias, G., Snyder, M.: Personal genomes, quantitative dynamic omics and personalized medicine. Quant. Biol. **1**(1), 71–90 (2013)
15. Mias, G.I., Snyder, M.: Multimodal dynamic profiling of healthy and diseased states for future personalized health care. Clin. Pharmacol. Ther. **93**(1), 29–32 (2013)
16. NCBI Resource Coordinators: Nucleic Acids Res. Database resources of the national center for biotechnology information. **45**(D1), D12–D17 (2017)
17. Newgard, C.B.: Interplay between lipids and branched-chain amino acids in development of insulin resistance. Cell Metab. **15**(5), 606–14 (2012)
18. Psychogios, N., Hau, D.D., Peng, J., Guo, A.C., Mandal, R., Bouatra, S., Sinelnikov, I., Krishnamurthy, R., Eisner, R., Gautam, B., Young, N., Xia, J., Knox, C., Dong, E., Huang, P., Hollander, Z., Pedersen, T.L., Smith, S.R., Bamforth, F., Greiner, R., McManus, B., Newman, J.W., Goodfriend, T., Wishart, D.S.: The human serum metabolome. PLoS One **6**(2), e16957 (2011)
19. Serkova, N.J., Glunde, K.: Metabolomics of cancer. Methods Mol. Biol. **520**, 273–95 (2009)
20. Smith, C.A., O'Maille, G., Want, E.J., Qin, C., Trauger, S.A., Brandon, T.R., Custodio, D.E., Abagyan, R., Siuzdak, G.: Metlin: a metabolite mass spectral database. Ther. Drug Monit. **27**(6), 747–51 (2005)
21. Stanberry, L., Mias, G.I., Haynes, W., Higdon, R., Snyder, M., Kolker, E.: Integrative analysis of longitudinal metabolomics data from a personal multi-omics profile. Metabolites **3**(3), 741–60 (2013)
22. Suhre, K., Gieger, C.: Genetic variation in metabolic phenotypes: study designs and applications. Nat. Rev. Genet. **13**(11), 759–69 (2012)
23. Tautenhahn, R., Cho, K., Uritboonthai, W., Zhu, Z., Patti, G.J., Siuzdak, G.: An accelerated workflow for untargeted metabolomics using the metlin database. Nat. Biotechnol. **30**(9), 826–8 (2012)
24. Theodoridis, A.A., Eich, C., Figdor, C.G., Steinkasserer, A.: Infection of dendritic cells with herpes simplex virus type 1 induces rapid degradation of cytip, thereby modulating adhesion and migration. Blood **118**(1), 107–15 (2011)

25. Thorn, C.F., Klein, T.E., Altman, R.B.: Pharmacogenomics and bioinformatics: Pharmgkb. Pharmacogenomics **11**(4), 501–5 (2010)
26. Wang, Y., Xiao, J., Suzek, T.O., Zhang, J., Wang, J., Zhou, Z., Han, L., Karapetyan, K., Dracheva, S., Shoemaker, B.A., Bolton, E., Gindulyte, A., Bryant, S.H.: Pubchem's bioassay database. Nucleic Acids Res. **40**(Database issue), D400–D412 (2012)
27. Wolfram Alpha LLC: Wolfram|Alpha (2017). Accessed November 2017
28. Wolfram Research, Inc.: Mathematica, Version 11.2. Champaign, IL (2017)

Chapter 9
Machine Learning

9.1 A Taste of Clustering

Let us go back to our MYL8B example from Chaps. 3 and 6. We want to see if we can get a natural clustering in the leukemia dataset [2, 4], AML or ALL based on the gene expression intensities of this gene. We get the golubAssociation and annotation as before:

In[1]:= **SetDirectory**[**NotebookDirectory**[]];

In[2]:= golubAssociation=<<"golubAssociation";

In[3]:= hu6800IDtoAnnotation=<<"hu6800IDtoAnnotation";

In[4]:= myosinExample=**Query**[**All**,"M31211_s_at"]@
golubAssociation
Out[4]= <|AML→ {−0.929698,−1.21439,
−1.36559,−1.03054,−1.29915,−0.401122,
−0.263245,−1.27241,−0.651176,−0.323125,
−1.43473,−0.203518,−1.04336,−1.22702,
−1.28071,0.260207,−1.05801,−0.720331,
−0.809905,−0.588208,−0.513632,
−0.247296,−1.42036,−1.48163,−1.3111},
ALL→ {0.21673,−0.0512012,0.0512196,
0.0658298,0.355842,−0.0764156,−0.298932,
−0.46567,0.490372,0.651373,0.715212,
0.148149,0.354363,−0.466193,0.576175,
0.417461,−0.407586,0.388138,0.266817,
0.634913,0.707898,−0.775871,0.0765331,
0.108658,0.224704,0.466898,−0.87136,
0.193097,−0.37865,−0.0896119,0.284866,
0.51995,−0.204948,0.560927,−0.0758743,

Electronic supplementary material The online version of this chapter (https://doi.org/10.1007/978-3-319-72377-8_9) contains supplementary material, which is available to authorized users.

© Springer International Publishing AG 2018
G. Mias, *Mathematica for Bioinformatics*,
https://doi.org/10.1007/978-3-319-72377-8_9

284   9 Machine Learning

$$0.726773, 0.337038, 0.374954, 0.146782,$$
$$-0.474574, 0.170777, 0.136991, 0.107438,$$
$$-0.55069, 0.279121, -0.665758, 0.545882\}|>$$

We generate a rule from the values of the subjects to their location in the list to their class:

In[5]:= labelsMYL6BAML=**Thread**[#-> "AML"&@Range[1,25]]
Out[5]= {1→AML, 2→AML, 3→AML, 4→AML, 5→AML,
6→AML, 7→AML, 8→AML, 9→AML, 10→AML,
11→AML, 12→AML, 13→AML, 14→AML, 15→AML,
16→AML, 17→AML, 18→AML, 19→AML, 20→AML,
21→AML, 22→AML, 23→AML, 24→AML, 25→AML}

In[6]:= labelsMYL6BALL=**Thread**[#-> "ALL"&@Range[1,47]]
Out[6]= {1→ALL, 2→ALL, 3→ALL, 4→ALL, 5→ALL, 6→ALL,
7→ALL, 8→ALL, 9→ALL, 10→ALL, 11→ALL, 12→ALL,
13→ALL, 14→ALL, 15→ALL, 16→ALL, 17→ALL,
18→ALL, 19→ALL, 20→ALL, 21→ALL, 22→ALL,
23→ALL, 24→ALL, 25→ALL, 26→ALL, 27→ALL,
28→ALL, 29→ALL, 30→ALL, 31→ALL, 32→ALL,
33→ALL, 34→ALL, 35→ALL, 36→ALL, 37→ALL,
38→ALL, 39→ALL, 40→ALL, 41→ALL, 42→ALL,
43→ALL, 44→ALL, 45→ALL, 46→ALL, 47→ALL}

We then use the **FindClusters** function:

In[7]:= ?**FindClusters**
    FindClusters[{$e_1, e_2, \ldots$}] partitions
        the e_i into clusters of similar elements.
    FindClusters[{$e_1 \to v_1, e_2 \to v_2, \ldots$}] returns the v_i
        corresponding to the e_i in each cluster.
    FindClusters[{$e_1, e_2, \ldots$} → {$v_1, v_2, \ldots$}] gives the same result.
    FindClusters[<|label$_1$ → e_1, label$_2$ → $e_2, \ldots$|>] returns
        the label$_i$ corresponding to the e_i in each cluster.
    FindClusters[data,n] partitions data into
        at most n clusters. >>

Notice how we use a rule to assign elements to the labels we just generated:

In[8]:= **Flatten** @ **Values** @myosinExample->
        Values @ **Join**[labelsMYL6BAML, labelsMYL6BALL]//
        Short
Out[8]//**Short**= {-0.929698,<<70>>,0.545882}-><<1>>

Now we use the **FindClusters** function:

In[9]:= **FindClusters**[Flatten@Values@myosinExample->
        Join[labelsMYL6BAML, labelsMYL6BALL]]
Out[9]= {{1→AML, 2→AML, 3→AML, 4→AML, 5→AML, 8→AML,
9→AML, 11→AML, 13→AML, 14→AML, 15→AML, 17→AML,
18→AML, 19→AML, 20→AML, 23→AML,
24→AML, 25→AML, 22→ALL, 27→ALL, 46→ALL},
{6→AML, 7→AML, 10→AML, 12→AML, 21→AML,
22→AML, 2→ALL, 3→ALL, 6→ALL, 7→ALL,
8→ALL, 14→ALL, 17→ALL, 29→ALL, 30→ALL,

9.1 A Taste of Clustering   285

$$33 \to \text{ALL}, 35 \to \text{ALL}, 40 \to \text{ALL}, 44 \to \text{ALL}\},$$
$$\{16 \to \text{AML}, 1 \to \text{ALL}, 4 \to \text{ALL}, 5 \to \text{ALL}, 9 \to \text{ALL},$$
$$10 \to \text{ALL}, 11 \to \text{ALL}, 12 \to \text{ALL}, 13 \to \text{ALL},$$
$$15 \to \text{ALL}, 16 \to \text{ALL}, 18 \to \text{ALL}, 19 \to \text{ALL},$$
$$20 \to \text{ALL}, 21 \to \text{ALL}, 23 \to \text{ALL}, 24 \to \text{ALL}, 25 \to \text{ALL},$$
$$26 \to \text{ALL}, 28 \to \text{ALL}, 31 \to \text{ALL}, 32 \to \text{ALL}, 34 \to \text{ALL},$$
$$36 \to \text{ALL}, 37 \to \text{ALL}, 38 \to \text{ALL}, 39 \to \text{ALL}, 41 \to \text{ALL},$$
$$42 \to \text{ALL}, 43 \to \text{ALL}, 45 \to \text{ALL}, 47 \to \text{ALL}\}\}$$

We can see how many clusters we get by counting up the classes in each separated list using **Counts**:

In[10]:= Counts[Values @ #]&/@ %
Out[10]= {<|AML→18,ALL→31|>,<|AML→6,ALL→13|>,<|AML→1,ALL→31|>}

We see a very clear separation of the points in the first and third cluster, but a bit of mixture in the second cluster.

9.2 Dimensional Reduction

We would like to use the Golub AML/ALL dataset and see how data are associated with phenotype further. We will need the phenotype annotation. Let us first import the Golub **OmicsObject**

In[11]:= golubOmicsObject=<<golubOmicsObject;

We next need to annotate the information using the phenotype data:

In[12]:= golubPhenotypeData=<<golubPhenotypeData;
In[13]:= golubPhenotypeData // Short
Out[13]// Short= <|Samples→{ALL/AML,<<8>>,Source},
        <<71>>,72→{<<1>>}|>

We will use the ALL/AML data for disease annotation. We query from the second entry, skipping the header, and select the first element in each value list which is the ALL or AML indication:

In[14]:= sampleToCondition=
        Query[2;;,1]@golubPhenotypeData;
        sampleToCondition // Short
Out[15]// Short= <|1→ALL,2→ALL,<<68>>,71→ALL,72→ALL|>

We next create a list of rules for all gene expression for each sample:

In[16]:= genesAllGolub=
        #[[2]]-> sampleToCondition[#[[1]]]&/@
        Normal@Query[All, Values/*Flatten]@golubOmicsObject;

Each member is a rule from a list of gene expression to condition:

In[17]:= genesAllGolub[[1]] // Short[#,3]&
Out[17]// Short= {-0.78835,-0.756913,
        <<3567>>,-0.331179,-0.825661}→ALL

We may want to reduce the dimensionality of a dataset. We have seen how to use Principal Component Analysis (PCA) directly in the context of looking at how samples are similar when we discussed metabolites in Chap. 8.

```
In[18]:= golub2Dimensions=
         DimensionReduce[Keys@genesAllGolub,2,
         Method->"PrincipalComponentsAnalysis"];
```

We see that now we have a projection into 2-dimensional coordinates (pairs of values):

```
In[19]:= golub2Dimensions//Short[#,5]&
Out[19]//Short= {{4.25183,23.2184},{-7.23657,-8.71078},<<68>>,
         {35.1694,14.8217},{11.9888,-15.5859}}
```

We can visualize the data by grouping the information by disease so we can color code it and plot it in in Fig. 9.1. In the figure code, first create a rule from the 2D projections to the disease using Thread. Then, by using a rule **Last**->**First** with **GroupBy**, we group by the **Last** element, the disease annotations, and return the **First**, the coordinates, in each list. Finally, we plot the coordinates using **ListPlot**.

We can also try the "TSNE" method as another example. The "TSNE" method uses a t-distributed stochastic neighbor embedding algorithm:

```
In[22]:= golub2DimensionsTSNE=
         DimensionReduce[Keys@genesAllGolub,2,
         Method->"TSNE"];
```

We can again plot the results by grouping by disease as we did above for the PCA method. The results for "TSNE" are shown in Fig. 9.2.

In both dimensional reduction methods we can see a clear separation of samples, but the Wolfram Language selects the PCA to report.

9.3 Classification

Next, let us create a random annotation for the leukemia dataset for training a model with various classifiers.

```
In[25]:= SeedRandom[3333];
         randomGolub=RandomSample[genesAllGolub];
```

We take the first 51 (70%) elements of the data for a training set, and use 21 (30%) for testing:

```
In[27]:= trainGolub=randomGolub[[1;;51]];
         testGolub=randomGolub[[52;;]];
```

We can use the function **Classify** to get classifications for multiple types of methods. We will use the "LogisticRegression" with standard options just for this example:

```
In[29]:= golubClassifier=
         Classify[trainGolub,Method->"LogisticRegression"];
```

9.3 Classification

```
In[20]:= golubByDisease = GroupBy[Thread[golub2Dimensions →
            Values@genesAllGolub], Last → First];
        ListPlot[Values[golubByDisease],
         PlotLegends → Keys[golubByDisease],
         PlotTheme → "Scientific",
         PlotLabel →
          "Dimensional Reduction of ALL/AML Data
        Using Principal Component Analysis"]
```

Out[21]= *(scatter plot: Dimensional Reduction of ALL/AML Data Using Principal Component Analysis, legend: ALL, AML)*

Fig. 9.1 Out[21]= Dimensional reduction of leukemia dataset using **DimensionReduce** with the method "PrincipaComponentsAnalysis"

You will notice a graphical dynamic output indicating the progress of the result as the command is running. Please wait until the training is complete. The result is the creation of a ClassifierFunction function:

```
In[30]:= Head @ golubClassifier
Out[30]= ClassifierFunction
```

The generated ClassifierFunction can be used to classify any other data. For example, let us pick from the testing set the first entry:

```
In[31]:= testGolub[[1]]// Short[#,5]&
Out[31]//Short= {-1.31426,-1.31426,-1.31426,-0.772743,<<3563>>,
                 -1.01648,0.603772,0.238744,-0.221022}→AML
```

We can use this as the input argument for our ClassifierFunction, and we actually get the same disease prediction as the real value, AML:

```
In[32]:= golubClassifier[testGolub[[1,1]]]
Out[32]= AML
```

```
In[23]:= golubByDiseaseTSNE =
    GroupBy[Thread[golub2DimensionsTSNE →
                Values@genesAllGolub], Last → First];
  ListPlot[Values[golubByDiseaseTSNE],
    PlotLegends → Keys[golubByDiseaseTSNE],
    PlotTheme → "Scientific",
    PlotLabel →
    "Dimensional Reduction of ALL/AML Data
Using t-distributed Stochastic Neighbor Embedding Algorithm"]
```

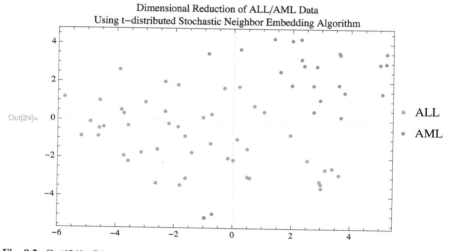

Fig. 9.2 Out[24]= Dimensional reduction of leukemia dataset using **DimensionReduce** with the method "TSNE"

The ClassifierFunction has many "Properties":

```
In[33]:= golubClassifier[testGolub[[1,1]], "Properties"]
Out[33]= {Decision, ExpectedUtilities, LogProbabilities,
         Probabilities, Properties, TopProbabilities}
```

We can get the classification probabilities for our specific example:

```
In[34]:= golubClassifier[testGolub[[1,1]], "Probabilities"]
Out[34]= <|ALL→0.0538275, AML→0.946173|>
```

We can carry out a **ClassifierMeasurements** analysis on the test set for obtaining the "Accuracy":

```
In[35]:= ClassifierMeasurements[golubClassifier, testGolub,
     "Accuracy"]
Out[35]= 0.952381
```

9.3 Classification

In[37]:= `ClassifierMeasurements[golubClassifier, testGolub, "ConfusionMatrixPlot"]`

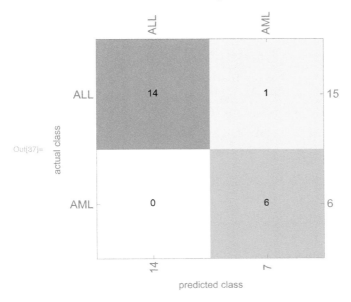

Out[37]=

Fig. 9.3 Out[37]= Confusion matrix for leukemia classification

The classifier is very accurate at 0.95. We can also get the associated confusion matrix information:

In[36]:= **ClassifierMeasurements**[golubClassifier, testGolub, "ConfusionMatrix"]
Out[36]= {{14,1,0},{0,6,0}}

We can look at the "ConfusionMatrixPlot" for a graphical representation, Fig. 9.3.

As we can see in this limited example, the classifier performs well, and misses only one call, misclassifying an ALL case as AML.

9.4 The Iris Data Classified Across Methods

For an expanded classification example, we will now use the Fisher/Anderson [1, 3] *Iris* dataset, which is a prototypical set used for Machine Learning introductions, as well as algorithmic testing. The set is available through **ExampleData**, under the "MachineLearning" dataset listing with the Wolfram Language/Mathematica. There are actually other standard machine learning data that you may be interested in:

```
In[38]:= ExampleData["MachineLearning"]
Out[38]= {{MachineLearning, BostonHomes},
          {MachineLearning, FisherIris},
          {MachineLearning, MNIST},
          {MachineLearning, MovieReview},
          {MachineLearning, Mushroom},
          {MachineLearning, Satellite},
          {MachineLearning, Titanic},
          {MachineLearning, UCILetter},
          {MachineLearning, WineQuality}}
```

We will continue with the Iris set. Let us see what kind of properties we can get:

```
In[39]:= ExampleData[{"MachineLearning","FisherIris"},
          "Properties"]
Out[39]= {Data, Description, Data, Dimensions, LearningTask,
          LongDescription, MissingData, Name, Source, TestData,
          TrainingData, VariableDescriptions, VariableTypes}
```

The data originates from the work of Fisher, back in 1936, which is listed under the "Source" property in the **ExampleData**, as well as Anderson [1]:

```
In[40]:= ExampleData[{"MachineLearning","FisherIris"},
          "Source"]
Out[40]= Fisher, R.A. "The use of multiple
          measurements in taxonomic problems" Annual
          Eugenics, 7, Part II, 179-188 (1936);
          also in "Contributions to Mathematical Statistics"
          (John Wiley, NY, 1950).
```

We can also read the "LongDescription" that gives details on the measurements:

```
In[41]:= ExampleData[{"MachineLearning","FisherIris"},
          "LongDescription"]
Out[41]=    The data set consists of 50 samples
          from each of three species of iris flowers
          (setosa, versicolor and virginica). Four
          features were measured from each flower,
            the length and the width of the sepal and petal.

            The test and training sets were constructed by
          stratified random sampling, using 30
          for the test set and the rest for the training set.
```

There are 150 samples, with 4 measurements each and a class association:

```
In[42]:= ExampleData[{"MachineLearning","FisherIris"},
          "Dimensions"]
Out[42]= <|NumberClasses→3,
          NumberFeatures→4,
          NumberExamples→150|>
```

The measurements correspond to physical characteristics of sepals and petals, for three different species, Iris-Setosa, Iris-versicolor and Iris-virginica. We can look up "Iris" using Wolfram|Alpha, Fig. 9.4.

The columns of the data have headings corresponding to these measures:

9.4 The Iris Data Classified Across Methods

Fig. 9.4 Out[43]= Wolfram|Alpha "Iris" interpretation

```
In[44]:=  irisData=
          ExampleData[{"MachineLearning","FisherIris"},"Data"];
In[45]:=  irisData[[1;;3]]
Out[45]=  {{5.1,3.5,1.4,0.2}→setosa,
           {4.9,3.,1.4,0.2}→setosa,
           {4.7,3.2,1.3,0.2}→setosa}
```

We can also get the column descriptions for these:

```
In[46]:=  ExampleData[{"MachineLearning","FisherIris"},
          "VariableDescriptions"]
Out[46]=  {Sepal length in cm.,Sepal width in cm.,
           Petal length in cm.,Petal width in cm.,
           Species of iris}→species of iris
```

Even though the example data has training and testing datasets, we will create our own. We start by first performing a pseudorandom permutation of the data using **RandomSample** (we use the **SeedRandom** for reproducibility for this manuscript, but you should be getting - most likely - different results in your own code):

```
In[47]:=  SeedRandom[7777];
          randomIris=RandomSample[irisData];

In[49]:=  randomIris//Short[#,5]&
Out[49]//Short=  {{6.4,2.8,5.6,2.1}->virginica,
                  {5.,3.6,1.4,0.2}->setosa,
                  {7.2,3.6,6.1,2.5}->virginica,
                  <<145>>,
                  {5.4,3.,4.5,1.5}->versicolor,
                  {7.1,3.,5.9,2.1}->virginica}
```

We then take the first 105 (70%) of the data for a training set, and use 45 (30%) for testing

```
In[50]:=  trainIris=randomIris[[1;;105]];
          testIris=randomIris[[106;;]];
```

We will use the function **Classify** to get classifications for multiple types of methods. We define the methods we will use in a variable:

```
In[52]:=  classifyMethods=
          {"DecisionTree","GradientBoostedTrees",
           "LogisticRegression","Markov","NaiveBayes",
           "NearestNeighbors","NeuralNetwork",
           "RandomForest","SupportVectorMachine"};
```

We will use all the classifiers with standard options just for this example, Fig. 9.5.

```
In[53]:=  trainIrisClassifiers=
          Association @@
          (#-> Classify[trainIris,Method->{#}]&/@
          classifyMethods)
Out[53]=  The output is shown in Fig. 9.5.
```

The generated ClassifierFunctions can be used to classify any other data. For example, let us pick from the test set the 4th entry:

9.4 The Iris Data Classified Across Methods

```
In[53]:= trainIrisClassifiers =
    Association @@ (# → Classify[trainIris,
              Method → {#}] & /@ classifyMethods)
```

Out[53]= ⟨| DecisionTree → ClassifierFunction[... Input type: NumericalVector (length: 4) Classes: setosa, versicolor, virginica],

GradientBoostedTrees → ClassifierFunction[... Input type: NumericalVector (length: 4) Classes: setosa, versicolor, virginica],

LogisticRegression → ClassifierFunction[... Input type: NumericalVector (length: 4) Classes: setosa, versicolor, virginica],

Markov → ClassifierFunction[... Input type: NumericalVector (length: 4) Classes: setosa, versicolor, virginica],

NaiveBayes → ClassifierFunction[... Input type: NumericalVector (length: 4) Classes: setosa, versicolor, virginica],

NearestNeighbors → ClassifierFunction[... Input type: NumericalVector (length: 4) Classes: setosa, versicolor, virginica],

NeuralNetwork → ClassifierFunction[... Input type: NumericalVector (length: 4) Classes: setosa, versicolor, virginica],

RandomForest → ClassifierFunction[... Input type: NumericalVector (length: 4) Classes: setosa, versicolor, virginica],

SupportVectorMachine → ClassifierFunction[... Input type: NumericalVector (length: 4) Classes: setosa, versicolor, virginica] |⟩

Fig. 9.5 Out[53] Iris set classification across methods

```
In[54]:= testIris[[4]]
Out[54]= {6.1, 2.9, 4.7, 1.4} → versicolor
```

We can get the results across classifiers:

```
In[55]:= Query[All, #[{6.1, 2.9, 4.7, 1.4}] &]@
             trainIrisClassifiers
Out[55]= <|DecisionTree → versicolor,
          GradientBoostedTrees → versicolor,
          LogisticRegression → versicolor,
          Markov → versicolor,
          NaiveBayes → versicolor,
          NearestNeighbors → versicolor,
          NeuralNetwork → virginica,
          RandomForest → versicolor,
          SupportVectorMachine → versicolor|>
```

We can see that various classifiers give different results, some correctly or incorrectly classifying our test example. We can use **ClassifierMeasurements** to get the accuracy of the set of classifiers:

In[56]:= **Query[All,**
 ClassifierMeasurements[#,testIris,"Accuracy"]&]@
 trainIrisClassifiers

Out[56]= <|DecisionTree→0.933333,
 GradientBoostedTrees→0.977778,
 LogisticRegression→0.977778,
 Markov→0.888889,
 NaiveBayes→0.955556,
 NearestNeighbors→0.955556,
 NeuralNetwork→0.733333,
 RandomForest→0.955556,
 SupportVectorMachine→0.933333|>

We can also get confusion matrix information for all the classifiers:

In[57]:= **Query[All,**
 ClassifierMeasurements[#,testIris,
 "ConfusionMatrix"]&]@trainIrisClassifiers

Out[57]= <|DecisionTree→
 {{14,1,0,0},{0,13,2,0},{0,0,15,0}},
 GradientBoostedTrees→{{15,0,0,0},{0,14,1,0},
 {0,0,15,0}},
 LogisticRegression→{{15,0,0,0},{0,15,0,0},
 {0,1,14,0}},
 Markov→{{14,1,0,0},{0,12,3,0},{0,1,14,0}},
 NaiveBayes→{{15,0,0,0},{0,13,2,0},{0,0,15,0}},
 NearestNeighbors→{{15,0,0,0},{0,14,1,0},
 {0,1,14,0}},
 NeuralNetwork→{{15,0,0,0},{2,3,10,0},
 {0,0,15,0}},
 RandomForest→{{15,0,0,0},{0,13,2,0},
 {0,0,15,0}},
 SupportVectorMachine→{{15,0,0,0},{0,13,2,0},
 {0,1,14,0}}|>

We can instead look at the "ConfusionMatrixPlot" for each classifier. We plot as an example the first 2 in Fig. 9.6. We can see very clear classifications between the various classes.

These classification approaches briefly considered in this chapter can be generalized to different problems in bioinformatics where we would like to build a classifier into relevant biological categories based on experimental or predicted measurements [5]. As machine learning algorithms can now efficiently be implemented with increased computational power, we expect broader usage in data sciences, and particularly genomics, which is an area of research producing large datasets that are amenable to this kind of analysis. Selection of an appropriate classifier should take into account the data that is the input, and also the limitations of each classifier algorithm should be carefully investigated for their predictions, as well as stated in

9.4 The Iris Data Classified Across Methods

```
In[58]:= Query[1 ;; 2,
         ClassifierMeasurements[#, testIris,
           "ConfusionMatrixPlot"] & /@ trainIrisClassifiers
```

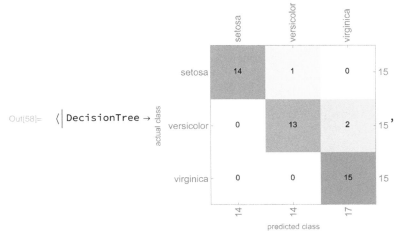

Fig. 9.6 Out[58] Iris set confusion matrices

results. Realistically, the predictive power of a classifier is only as good as the input data used to train the model.

References

1. Anderson, E.: The irises of the Gaspe Peninsula. Bull. Am. Iris Soc. **59**, 2–5 (1935)
2. Dudoit, S., Fridlyand, J., Speed, T.P.: Comparison of discrimination methods for the classification of tumors using gene expression data. J. Am. Stat. Assoc. **97**(457), 77–87 (2002)
3. Fisher, R.A.: The use of multiple measurements in taxonomic problems. Ann. Eugen. **7**(2), 179–188 (1936)
4. Golub, T.R., Slonim, D.K., Tamayo, P., Huard, C., Gaasenbeek, M., Mesirov, J.P., Coller, H., Loh, M.L., Downing, J.R., Caligiuri, M.A., Bloomfield, C.D., Lander, E.S.: Molecular classification of cancer: class discovery and class prediction by gene expression monitoring. Science **286**(5439), 531–537 (1999)
5. Libbrecht, M.W., Noble, W.S.: Machine learning applications in genetics and genomics. Nat. Rev. Genet. **16**, 321 EP (2015)

Chapter 10
Graphs and Networks

10.1 Introduction to Graphs

The Wolfram Language has extensive capabilities to visualize and also characterize graphs and networks. The fields of systems biology have long tackled problems involving networks, particularly in the context of biochemical reactions, and more recently regulatory networks and gene/protein interaction networks are being used to analyze large datasets (see for example [8]).

10.1.1 Vertices and Edges

We can think of a graph, G, as comprised of two sets of two sets: (i) Vertices, V, which are the nodes of the graph, and (ii) Edges, E, which are the links or connections between vertices [5, 9, 10]. We can represent graphs in the Wolfram Language by providing a list of connectors between vertices to represent edges to the function **Graph**.

The simplest of examples with an edge, where we have two vertices connected, vertex 1 and vertex 2, is shown in Fig. 10.1.

This is an example of an undirected edge. There are multiple ways to input this in the Wolfram Language. You may find the edge symbol a bit tricky to enter, you can type:

- vertex$_1$ esc ue esc vertex$_2$, i.e. use the escape key (esc) before and after typing ue for an undirected edge.
- vertex$_1$ \[UndirectedEdge]vertex$_2$
- Use the input form directly **UndirectedEdge**[vertex$_1$, vertex$_2$].

Electronic supplementary material The online version of this chapter (https://doi.org/10.1007/978-3-319-72377-8_10) contains supplementary material, which is available to authorized users.

In[1]:= `Graph[{1 ⟶ 2}]`

Out[1]= ○─────────────────○

Fig. 10.1 Out[1]= Graph, two vertices, one edge

In[2]:= `Graph[{1 ⟶ 2}, PlotTheme → "ClassicDiagram"]`

Out[2]= 2 ───────────────── 1

Fig. 10.2 Out[2]= Graph, two vertices, one edge, "ClassicDiagram" theme

In[3]:= `Graph[{1 ⟵ 2}, PlotTheme → "ClassicDiagram"]`

Out[3]= 2 ◀───────────────── 1

Fig. 10.3 Out[3]= Graph, two vertices, one directed edge

In[4]:= `simpleGraphExample = Graph[`
`  {1 ⟶ 2, 2 ⟶ 3, 3 ⟶ 4, 2 ⟶ 4, 2 ⟶ 5,`
`   3 ⟶ 5, 5 ⟶ 1}, PlotTheme → "ClassicDiagram"]`

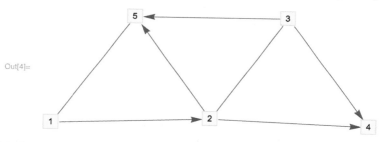

Fig. 10.4 Out[4]= Simple graph with multiple edges and vertices

Graphs have various options and themes. We can use the classic theme that displays the labels, Fig. 10.2.

The edges can also be directed, i.e. have a directionality from one vertex to another. These directed edges are provided in a different format: For example we have two vertices 1,2 where we have a directed edge from 1 to 2, Fig. 10.3.

To enter directed edges as symbols, you can type:

- vertex$_1$ esc de esc vertex$_2$, i.e. use the escape key (esc) before and after typing de for a directed edge.
- vertex$_1$ \[DirectedEdge] vertex$_2$
- Use the input form directly **DirectedEdge**[vertex$_1$,vertex$_2$].

We can create a bigger graph by putting together more vertices and edges, Fig. 10.4.

10.1 Introduction to Graphs

```
In[5]:= simpleGraphExample7 = Graph[{1, 2, 3, 4, 5, 6, 7},
         {1 ↔ 2, 2 ↔ 3, 3 ↔ 4, 2 ↔ 4, 2 ↔ 5, 3 ↔ 5, 5 ↔ 1},
         PlotTheme → "ClassicDiagram"]
```

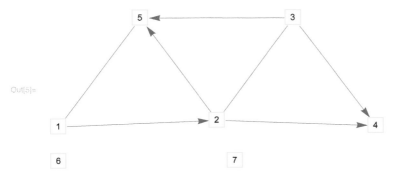

Fig. 10.5 Out[5]= Simple graph with disconnected vertices

Additionally, we can provide the list of vertices if we want separately as the first input, Fig. 10.5. It could be that some vertices are not connected with edges, so this would be a way to enter them.

The vertices and edges have so far been rather abstract. But there are many examples where we can construct graphs. For example we can think of neurons as vertices and synapses as connecting edges, we can construct friend networks where connections indicate friendship between people, and we can think of the internet as a giant graph with internet service providers as vertices connected, or metabolic networks representing biochemical interactions (edges) between molecular species (vertices) [2, 4, 8, 10, 11]. The abstraction for the information from such systems into graphs provides a common framework to think of similarities in the systems considered and can facilitate the analysis.

The study of networks began with the seminal work of the famous mathematician Leonhard Euler [6]. Euler addressed the Seven Bridges of Koningsberg problem (now Kaliningrad): The city of Koningsberg had seven bridges connecting different parts of the city, including an island and the question was whether or not one could walk around the whole city but only cross each bridge once.

```
In[6]:= SystemOpen["paclet:example/VisualizeEulerianCycles"]
         (* the figure is available in the Documentation Center *)
```

We can represent the city as a graph, Fig. 10.6, where the edges represent the bridges, and the vertices essentially dry land.

The problem is then reduced to trying to find a continuous path which passes through every edge once and only once (imagine tracing such a path with a pencil and trying not to go over the same line twice). The problem and Euler's solution was the beginning of graph theory, and a graph where one can find such a path is termed

```
In[7]:= sevenBridges = Graph[{UndirectedEdge[1, 2],
         UndirectedEdge[1, 2], UndirectedEdge[2, 3],
         UndirectedEdge[2, 3], UndirectedEdge[2, 4],
         UndirectedEdge[3, 4], UndirectedEdge[1, 4]},
        PlotTheme → "ClassicDiagram"]
```

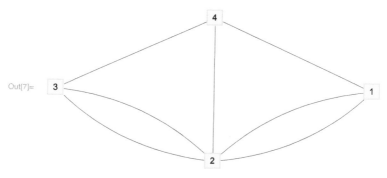

Fig. 10.6 Out[5]= The bridges of Koningsberg graph [6]

Eulerian. The Wolfram Language has a function that can check if a graph is Eulerian, **EulerianGraphQ**. We can thus check the original graph above, and see that the graph is actually not Eulerian.

```
In[8]:= EulerianGraphQ[sevenBridges]
Out[8]= False
```

10.2 Basic Graph Construction

There are many ways to put together and manipulate graphs. Here we show some simple examples:

10.2.1 Entering a Graph

As we have seen above we can enter a graph as lists of vertices and edges connecting them. Let us go back to our example before and now add labels by providing rules for vertex labels (VertexLabels is the option in **Graph**). We first create the labels:

```
In[9]:= vertexLabels=#-> Style[Subscript[ "v",
         ToString[#]], FontFamily->"Arial", Bold]&/@
        Range[7]
```

10.2 Basic Graph Construction   301

Out[9]= $\{1 \to V_1, 2 \to V_2, 3 \to V_3, 4 \to V_4, 5 \to V_5, 6 \to V_6, 7 \to V_7\}$

For the EdgeLabels option we will use "Name" which uses the edge name as the label. Then we can generate a labeled graph, Fig. 10.7.

We can highlight any part of a graph, for example vertices, using **HighlightGraph**[graph, list of vertices], Fig. 10.8 .

Graph vertices can actually be any object, and they do not have to be numbers. Let us draw a simple representation of the central dogma of molecular biology.

In[12]:= verticesCentralDogma={"DNA","mRNA","Protein"};
edgesCentralDogma={**DirectedEdge**["DNA","mRNA"],
 DirectedEdge["mRNA","Protein"]};

In[10]:= **simpleGraphExample7Labeled = Graph[Range[7],**
 {1 ↔ 2, 2 ↔ 3, 3 ↔ 4, 2 ↔ 4, 2 ↔ 5, 3 ↔ 5, 5 ↔ 1},
 VertexLabels → vertexLabels,
 EdgeLabels → "Name"]

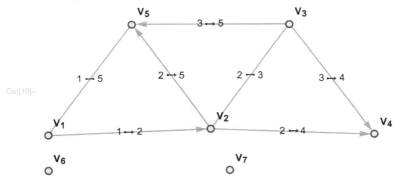

Fig. 10.7 Out[10]= Labeled graph

In[11]:= **HighlightGraph[simpleGraphExample7Labeled, {2, 3}]**

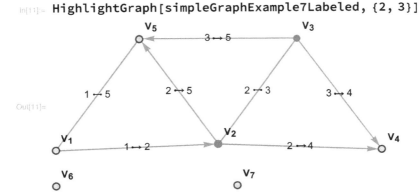

Fig. 10.8 Out[11]= Highlighted graph

```
In[15]:= centralDogmaGraph = Graph[verticesCentralDogma,
            edgesCentralDogma,
            VertexSize -> 0.3,
            PlotTheme -> "ClassicDiagram",
            EdgeLabels -> edgeLabelsCentralDogma]
```

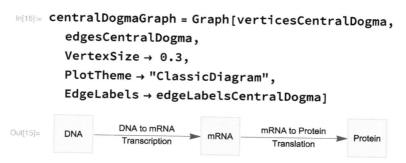

Fig. 10.9 Out[15]= Simple graph of the central dogma of molecular biology

Fig. 10.10 Out[18]= List of edges for a graph

```
In[18]:= EdgeList[centralDogmaGraph]
Out[18]= {DNA -> mRNA, mRNA -> Protein}
```

```
edgeLabelsCentralDogma=
  {DirectedEdge["DNA","mRNA"]->
    Labeled["DNA to mRNA","Transcription"],
   DirectedEdge["mRNA","Protein"]->
    Labeled["mRNA to Protein", "Translation"]};
```

After defining the vertices and edges, we can draw a graph, Fig. 10.9.
We can get a list of vertices, which we see are called as we defined them:

```
In[16]:= VertexList[centralDogmaGraph]
Out[16]= {DNA,mRNA,Protein}
```

We may want to get the exact index of a vertex as it appears in the above list, particularly if we are using named vertices. We use **VertexIndex**[graph, index]:

```
In[17]:= VertexIndex[centralDogmaGraph,"Protein"]
Out[17]=  3
```

Similarly we can get a list the edges, using **EdgeList** Fig. 10.10.
And once more we can extract the index in the above list of a particular edge,

```
In[19]:= EdgeIndex[centralDogmaGraph,
          DirectedEdge["mRNA","Protein"]]
Out[19]= 2
```

Similarly, we can get any vertex index:

```
In[20]:= VertexIndex[centralDogmaGraph,"DNA"]
Out[20]= 1
```

We can delete specific edges on a graph using **EdgeDelete**, Fig. 10.11. We can also add edges using **EdgeAdd**, Fig. 10.12.
And similarly we can delete vertices with **VertexDelete**, Fig. 10.13. We can also add vertices with **VertexAdd**, Fig. 10.14.

10.2 Basic Graph Construction

In[21]:= **EdgeDelete[simpleGraphExample7, 2 ⟷ 5]**
(*Can also input as
EdgeDelete[simpleGraphExample,DirectedEdge[2,5]]*)

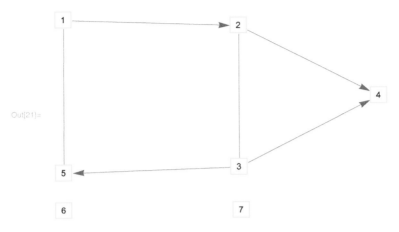

Fig. 10.11 Out[21]= Delete an edge from a graph

In[22]:= **EdgeAdd[simpleGraphExample7, 1 ⟷ 3]**

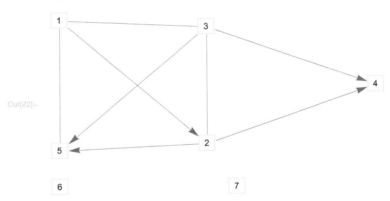

Fig. 10.12 Out[22]= Add an edge to a graph

GraphPlot can generate a plot of our graph, Fig. 10.15.
Note that this is different than the graph, as can be seen from the headers:

In[26]:= **Head[graphPlot]**
Out[26]= **Graphics**

In[23]:= `VertexDelete[simpleGraphExample7, 1]`

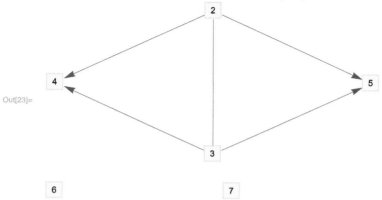

Fig. 10.13 Out[23]= Delete a vertex from a graph

In[24]:= `VertexAdd[simpleGraphExample7, 8]`

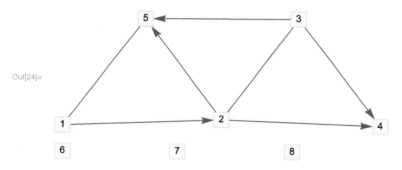

Fig. 10.14 Out[24]= Add a vertex to a graph

In[25]:= `graphPlot = GraphPlot[simpleGraphExample7]`

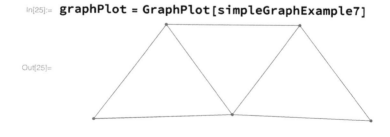

Fig. 10.15 Out[25]= Graph plot

10.2 Basic Graph Construction

```
In[29]:= simpleGraphExampleWeightedEdges =
    Graph[{1 ↔ 2, 2 ↔ 3, 3 ↔ 4, 2 ↔ 4,
      2 ↔ 5, 3 ↔ 5, 5 ↔ 1},
      EdgeWeight → edgeWeights,
      PlotTheme → "ClassicDiagram"];
    simpleGraphExampleWeightedEdges
```

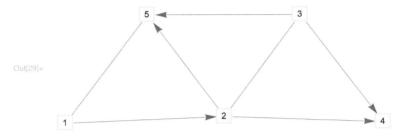

Fig. 10.16 Out[29]= Graph with EdgeWeight defined

```
In[27]:= Head[simpleGraphExample]
Out[27]= Graph
```

10.2.2 Defining Weighted Graphs

A graph's edges and vertices may be assigned weights. For example in biomolecular interaction networks, this could correspond to the strength of the association between the vertices, i.e. the molecular components. Additionally, vertices themselves may be assigned weights. For example the weight of a vertex in a gene network may correspond to the expression level of the gene product as measured in an experiment.

We can assign EdgeWeight as an option in a **Graph**. We can first define weights to include in a graph:

```
In[28]:= edgeWeights = {2, 3, 1, 6, 1, 2, 1};
```

We then assign these weights to the graph and see that the resulting diagram actually looks the same as before, Fig. 10.16.

We can display the weight in labels using "EdgeWeight" as the EdgeLabel, Fig. 10.17.

Similarly, we assign weight to vertices using VertexWeight, and also display these using "VertexWeight" as the value for the VertexLabels option, Fig. 10.18.

In[30]:= Graph[{1 ⇾ 2, 2 ⇾ 3, 3 ⇾ 4, 2 ⇾ 4, 2 ⇾ 5,
    3 ⇾ 5, 5 ⇾ 1},
   EdgeWeight → edgeWeights,
   PlotTheme → "ClassicDiagram",
   EdgeLabels → "EdgeWeight"]

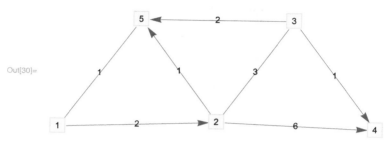

Fig. 10.17 Out[30]= Graph with edges labeled with EdgeWeight definition

In[31]:= simpleGraphExampleWeightedVertices =
   Graph[{1 ⇾ 2, 2 ⇾ 3, 3 ⇾ 4, 2 ⇾ 4,
    2 ⇾ 5, 3 ⇾ 5, 5 ⇾ 1},
   VertexWeight → {12, 31, 32, 10, 0},
   VertexLabels → "VertexWeight"]

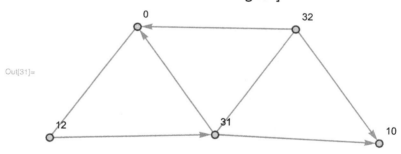

Fig. 10.18 Out[31]= Graph with VertexWeight defined

10.3 Basic Graph Properties

There are different properties that can characterize a graph. Here we review how to obtain certain standard measures.

10.3 Basic Graph Properties

```
In[32]:= graphExample7VerticesEdges =
          {1 -> 2, 2 -> 3, 3 -> 1, 4 -> 1, 5 -> 1, 6 -> 1,
           7 -> 2, 7 -> 3};
        graphExample7Vertices =
         Graph[graphExample7VerticesEdges,
          VertexLabels -> "Name",
          EdgeLabels -> "Name"]
```

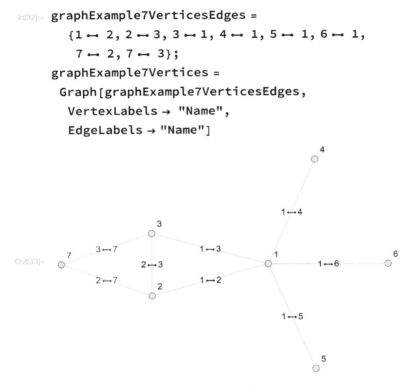

Fig. 10.19 Out[33]= Example graph with varying vertex degree

10.3.1 Degree

We can calculate the degree for any vertex in a graph, which corresponds to the number of incident edges. We define a simple graph, Fig. 10.19.

We can then use **VertexDegree** to obtain all degrees for all vertices in the graph as a list:

```
In[34]:= VertexDegree[graphExample7Vertices]
Out[34]= {5,3,3,1,1,1,2}
```

Now let us consider a directed graph, Fig. 10.20. The vertices can now have in-degree, which is the number of inbound incident edges, or out degree which is the number of outbound incident edges.

VertexOutDegree gives the out-degree. For example the last vertex, 7, does not have any outbound edges.

```
In[37]:= VertexOutDegree[graphExampleLoopsDirected]
Out[37]= {3,3,3,1,3,0,0}
```

Similarly, we can calculate the in-degree for the same graph using **VertexInDegree**:

```
In[35]:= graphExampleLoopsDirectedEdges =
    {1 ↔ 2, 2 ↔ 3, 2 ↔ 2, 3 ↔ 1, 3 ↔ 4, 4 ↔ 4, 2 ↔ 5,
     1 ↔ 6, 3 ↔ 6, 5 ↔ 6, 1 ↔ 7, 5 ↔ 7, 5 ↔ 5};
  graphExampleLoopsDirected =
    Graph[graphExampleLoopsDirectedEdges,
     VertexLabels → "Name"]
```

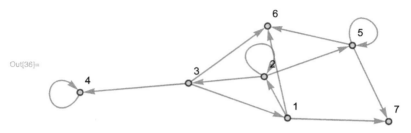

Fig. 10.20 Out[36]= Example directed graph with varying in and out vertex degree

```
In[38]:= VertexInDegree[graphExampleLoopsDirected]
Out[38]= {1,2,1,2,2,3,2}
```

Instead, **VertexDegree** still gives the total number of incident vertices, which is the sum of the in- and out-degrees.

```
In[39]:= VertexDegree[graphExampleLoopsDirected]
Out[39]= {4,5,4,3,5,3,2}
```

10.3.2 Adjacency Matrix

The adjacency matrix is a way to represent a graph in an array. In such a matrix with entries a_{ij} the rows and columns represent the vertices. Hence the matrix is square. If a graph is simple, which means it has no loops or multiple edges then the adjacency matrix entries take values 0 and 1, and the matrix is symmetric. A non-zero entry indicates that the vertices are adjacent (i.e. connected).

In the Wolfram Language, the **AdjacencyMatrix** function is used to produce the matrices. Here we consider a simple undirected graph, Fig. 10.21, and calculate the adjacency matrix:

```
In[41]:= adjGraphSimple=AdjacencyMatrix[graphSimple];
```

We notice that the adjacency matrix is represented as a **SparseArray** object (i.e. representing a matrix which is sparsely populated):

```
In[42]:= Head[adjGraphSimple]
Out[42]= SparseArray
```

10.3 Basic Graph Properties

```
In[40]:= graphSimple = Graph[{1, 2, 3, 4},
         {UndirectedEdge[1, 2], UndirectedEdge[2, 3],
          UndirectedEdge[2, 4], UndirectedEdge[3, 4],
          UndirectedEdge[1, 4]},
         VertexLabels → "Name",
         EdgeLabels → "Name"]
```

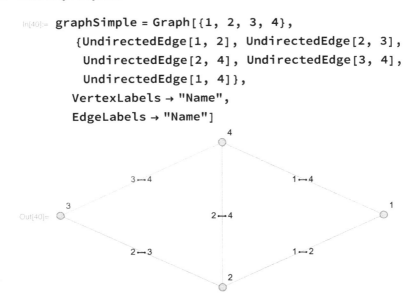

Fig. 10.21 Out[40]= Example undirected simple graph

We can get the "Properties" of the sparse array:

In[43]:= adjGraphSimple["Properties"]
Out[43]= {AdjacencyLists, Background, ColumnIndices,
          Density, MatrixColumns, NonzeroValues,
          PatternArray, Properties, RowPointers}

To see the matrix, we can print to output the **MatrixForm**:

In[44]:= adjGraphSimple // **MatrixForm**

Out[44]//**MatrixForm**= $\begin{pmatrix} 0 & 1 & 0 & 1 \\ 1 & 0 & 1 & 1 \\ 0 & 1 & 0 & 1 \\ 1 & 1 & 1 & 0 \end{pmatrix}$

We can see for example taking row 2 that there are entries in columns 1, 3 and 4. This means there are edges between vertex 2 and each of vertices 1, 3 and 4.

For another example, if we use our seven Bridges of Koningsburg example from above we have for the adjacency matrix:

In[45]:= adjSevenBridges=**AdjacencyMatrix**[sevenBridges];
         adjSevenBridges // **MatrixForm**

Out[46]//**MatrixForm**= $\begin{pmatrix} 0 & 2 & 0 & 1 \\ 2 & 0 & 2 & 1 \\ 0 & 2 & 0 & 1 \\ 1 & 1 & 1 & 0 \end{pmatrix}$

Notice that our matrix has multiple entries that are 2, because there are multiple edges connecting the same adjacent vertices.

```
In[48]:= AdjacencyGraph[{{0, 1, 0, 1}, {1, 0, 1, 1},
           {0, 1, 0, 0}, {1, 1, 0, 0}}]
```

Fig. 10.22 Out[48]= Constructing a graph from an adjacency matrix

If the graph is directed, the adjacency matrix will be asymmetric. For example, we use the graphExampleLoopsDirected graph defined above and see the **MatrixForm**:

In[47]:= AdjacencyMatrix[Graph[graphExampleLoopsDirected]]//
       MatrixForm

$$\text{Out[47]//MatrixForm=} \begin{pmatrix} 0 & 1 & 0 & 0 & 0 & 1 & 1 \\ 0 & 1 & 1 & 0 & 1 & 0 & 0 \\ 1 & 0 & 0 & 1 & 0 & 1 & 0 \\ 0 & 0 & 0 & 1 & 0 & 0 & 0 \\ 0 & 0 & 0 & 0 & 1 & 1 & 1 \\ 0 & 0 & 0 & 0 & 0 & 0 & 0 \\ 0 & 0 & 0 & 0 & 0 & 0 & 0 \end{pmatrix}$$

Conversely, we can use an adjacency matrix to construct a graph directly using **AdjacencyGraph**, Fig. 10.22.

10.3.3 Weighted Adjacency Matrix

For a weighted network, we use a special form of the adjacency matrix, the **WeightedAdjacencyMatrix**. For example on our weighted graph from the previous sections we have:

In[49]:= exampleWeightedAdjacencyMatrix=
       WeightedAdjacencyMatrix[
         simpleGraphExampleWeightedEdges]// MatrixForm

$$\text{Out[49]//MatrixForm=} \begin{pmatrix} 0 & 2 & 0 & 0 & 1 \\ 0 & 0 & 3 & 6 & 1 \\ 0 & 3 & 0 & 1 & 2 \\ 0 & 0 & 0 & 0 & 0 \\ 1 & 0 & 0 & 0 & 0 \end{pmatrix}$$

Note that if we had used adjacency matrix instead we get just ones and zeros, which simply count the connecting edges, but do not take into account the weight:

10.3 Basic Graph Properties

```
In[50]:= AdjacencyMatrix[
          simpleGraphExampleWeightedEdges]//
        MatrixForm
```

$$\text{Out[50]//MatrixForm=} \begin{pmatrix} 0 & 1 & 0 & 0 & 1 \\ 0 & 0 & 1 & 1 & 1 \\ 0 & 1 & 0 & 1 & 1 \\ 0 & 0 & 0 & 0 & 0 \\ 1 & 0 & 0 & 0 & 0 \end{pmatrix}$$

10.4 Some More Definitions

In this section, let us go through some additional definitions of standard terms used in Graph Theory.

10.4.1 Walks and Paths

An important aspect of Graph Theory is the characterization of the different vertices and their connections, and in terms of distances between them. Here are a few relevant definitions [5, 7, 10]:

- **Walk**. A walk on a graph is a sequence of vertices, $\{v_1, v_2, ..., v_n\}$ so that all edges connecting sequential elements in the list are included.
- **Path**. A path is a walk in which the list of vertices does not include duplicates - i.e. no vertex is visited twice. In a **connected** graph any vertex in the graph has a path to any other vertex in the graph. A **disconnected** graph is a graph that is not connected (and will thus have graph parts disconnected from the rest).
- **Trail**. A trail is a walk in which the edges are distinct (but the vertices do not have to be).
- **Cycle**. A cycle is a path with an additional edge connecting the last to the first vertex (think of a circle closing on itself).
- **Geodesic**. A geodesic is the shortest path that can be found connecting two vertices.
- **Length**. The length is defined in terms of the number of edges in any of the above.

We can calculate the shortest distance between vertices in a graph. **FindShortestPath** will give us the vertex sequence to go from a starting vertex to an end vertex. For example going from vertex 1 to vertex 7:

```
In[51]:= shortestPath7Vertices=
         FindShortestPath[graphExample7Vertices,1,7]
Out[51]= {1,2,7}
```

We can use **PathGraph** to construct a path representation, which we then use to highlight the original graph for visualization, Fig. 10.23.

```
In[52]:= shortestPathGraph=PathGraph[shortestPath7Vertices];
```

In[53]:= **HighlightGraph[graphExample7Vertices,
shortestPathGraph]**

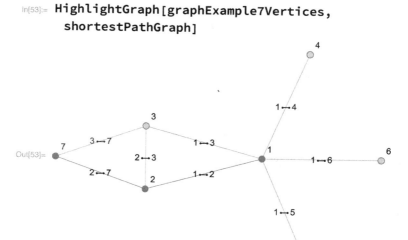

Fig. 10.23 Out[53]= Highlighted shortest path between vertices 1 and 7

10.4.2 Graph Geometry

Having defined walks and paths we can now use them for other network characteristics. For example:

- **Eccentricity**. The eccentricity, ε of a vertex is the maximum shortest path between the given vertex and any other vertex. For any graph we can use the function **VertexEccentricity**[graph, vertex$_i$] to get he eccentricity. For example we extract all the eccentricities for the sevenBridges and simpleGraphExample graphs from above:

In[54]:= **VertexEccentricity[sevenBridges,#]&/@
VertexList[sevenBridges]**
Out[54]= {2,1,2,1}

In[55]:= **VertexEccentricity[graphExample7Vertices,#]&/@
VertexList[graphExample7Vertices]**
Out[55]= {2,2,2,3,3,3,3}

- **Graph Diameter**. The graph diameter, d, is the maximum eccentricity measured across all vertices of the graph. The set of all vertices with eccentricity equal to the diameter d is the **periphery** of the graph.

In[56]:= **GraphDiameter[sevenBridges]**
Out[56]= 2

In[57]:= **GraphDiameter[graphExample7Vertices]**
Out[57]= 3

10.4 Some More Definitions

- **Graph Radius**. The graph radius, r, is the minimum eccentricity measured across all vertices of the graph. Any vertex is called central if its eccentricity matches r, and the set of these central vertices is the center of the graph.

In[58]:= **GraphRadius[sevenBridges]**
Out[58]= 1

In[59]:= **GraphRadius[graphExample7Vertices]**
Out[59]= 2

10.4.3 Centrality

We may be interested in identifying what the important vertices in a graph are. There are many measures of centrality:

- **Degree Centrality**. For a given vertex, we can think of it as being important the higher its degree. **DegreeCentrality** can calculate the measure for all vertices in a graph:

In[60]:= **DegreeCentrality[graphExample7Vertices]**
Out[60]= {5,3,3,1,1,1,2}

- **Betweenness Centrality**. We can calculate the betweenness centrality of an edge in a graph. This is defined as the number of shortest paths between the vertex pairs that include the edge. It measures the numbers of vertex pair sets that essentially go through the edge to connect to each other. For our previous example we get:

In[61]:= **BetweennessCentrality[graphExample7Vertices]**
Out[61]= {12.,2.,2.,0.,0.,0.,0.}

We can use **HighlightGraph** to get the betweenness centrality information highlighted. First we use the **Rescale** function to set the values in the list to range from 0 to 1.

In[62]:= scaledBetweenness=
   **Rescale[BetweennessCentrality[
      graphExample7Vertices]]**
Out[62]= {1.,0.166667,0.166667,0.,0.,0.,0.}

We will highlight all the vertices of the graph, which we next extract with **VertexList**.

In[63]:= vertexListExample7=
   VertexList[graphExample7Vertices];

Finally, we plot the highlighted information in Fig. 10.24. Here we have highlighted the vertices using **HighlightGraph**[graph,Vertex List], and assigned the VertexSize option to correspond to the scaledBetweeness (assigned by a rule using **Thread**).

```
In[64]:= HighlightGraph[graphExample7Vertices,
        vertexListExample7,
        VertexSize →
          Thread[vertexListExample7 → scaledBetweenness]]
```

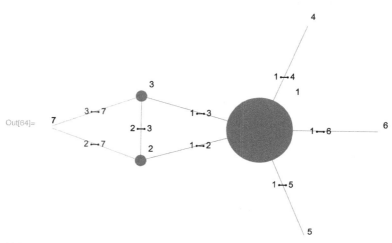

Fig. 10.24 Out[54]= Highlighted betweenness centrality

10.4.4 Clustering Coefficient

10.4.4.1 Local Clustering Coefficient

We may want to measure the interconnectedness of neighbors of a vertex. The local clustering coefficient for a vertex is the probability that a given pair of neighbors to a vertex are themselves neighbors (i.e. connected). It is measured as the ratio (pairs of connected neighbors of the vertex)/(all neighbor vertex pairs for the vertex).

```
In[65]:= localClusteringExample7=
        LocalClusteringCoefficient[graphExample7Vertices]
Out[65]= {1/10,2/3,2/3,0,0,0,1}
```

We can visualize this in a highlighted graph, Fig. 10.25.

10.4.4.2 Global Clustering Coefficient

The **GlobalClusteringCoefficient** function gives us the global clustering coefficient, measuring the fraction (paths of length two that are closed)/(all paths of length two).

```
In[67]:= GlobalClusteringCoefficient[graphExample7Vertices]
Out[67]= 6/17
```

```
In[66]:= HighlightGraph[graphExample7Vertices,
        vertexListExample7,
        VertexSize → Thread[vertexListExample7 →
            localClusteringExample7]]
```

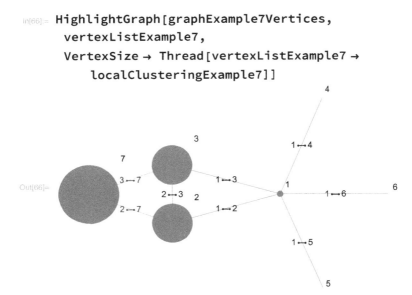

Fig. 10.25 Out[66]= Highlighted local clustering coefficient

10.5 Graph Examples

There are many graphs that can be created through named commands in the Wolfram Language. See in the documentation guide/GraphConstructionAndRepresentation. The Wolfram Language additionally has many available theoretical and empirical graphs that are available through **GraphData** and **ExampleData** respectively.

In the next sections, we will first go through a few definitions of named graph categories,, and then look at a few examples for each category. The graphs considered are often used in graph analysis.

10.5.1 Empty Graphs

An empty graph, E_n has n vertices but no edges [7]. These are available through **GraphData**[{"Empty",n}].

```
In[68]:= emptyGraphs=
        Graph[GraphData[{"Empty",#}],VertexSize -> Medium,
        PlotLabel->Style[Subscript["E",#],Italic,
            Bold]] &/@ Range[5];
```

We will use the same grid function in this section to plot these graphs, so we define it as a function for convenience:

In[70]:= `gridGraphs[emptyGraphs]`

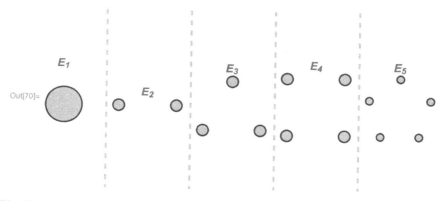

Fig. 10.26 Out[70]= Empty graphs

In[69]:= gridGraphs[x_]:=
    Grid[{x},Dividers->Center,Frame-> True,
    FrameStyle->Directive[Dashing[4],Thickness[2],Opacity
    [0.3]]]

Now we can plot the empty graphs in Fig. 10.26.

10.5.2 Complete Graphs

A Complete Graph K_n has n vertices that are all connected to each other. The function **CompleteGraph** generates the nth graph, Fig. 10.27.

We can look at the **AdjacencyMatrix** of all these graphs:

In[74]:= MatrixForm @ AdjacencyMatrix[#]&/@completeGraphs
Out[74]=

$$\left\{(0), \begin{pmatrix} 0 & 1 \\ 1 & 0 \end{pmatrix}, \begin{pmatrix} 0 & 1 & 1 \\ 1 & 0 & 1 \\ 1 & 1 & 0 \end{pmatrix}, \begin{pmatrix} 0 & 1 & 1 & 1 \\ 1 & 0 & 1 & 1 \\ 1 & 1 & 0 & 1 \\ 1 & 1 & 1 & 0 \end{pmatrix}, \begin{pmatrix} 0 & 1 & 1 & 1 & 1 \\ 1 & 0 & 1 & 1 & 1 \\ 1 & 1 & 0 & 1 & 1 \\ 1 & 1 & 1 & 0 & 1 \\ 1 & 1 & 1 & 1 & 0 \end{pmatrix}, \begin{pmatrix} 0 & 1 & 1 & 1 & 1 & 1 \\ 1 & 0 & 1 & 1 & 1 & 1 \\ 1 & 1 & 0 & 1 & 1 & 1 \\ 1 & 1 & 1 & 0 & 1 & 1 \\ 1 & 1 & 1 & 1 & 0 & 1 \\ 1 & 1 & 1 & 1 & 1 & 0 \end{pmatrix}, \begin{pmatrix} 0 & 1 & 1 & 1 & 1 & 1 & 1 \\ 1 & 0 & 1 & 1 & 1 & 1 & 1 \\ 1 & 1 & 0 & 1 & 1 & 1 & 1 \\ 1 & 1 & 1 & 0 & 1 & 1 & 1 \\ 1 & 1 & 1 & 1 & 0 & 1 & 1 \\ 1 & 1 & 1 & 1 & 1 & 0 & 1 \\ 1 & 1 & 1 & 1 & 1 & 1 & 0 \end{pmatrix}\right\}$$

We notice, as expected, that the adjacency matrices for the complete graphs have a simple form with all 1s everywhere but the diagonal entries (no self-connections).

In these graphs all vertices are equally important:

In[75]:= BetweennessCentrality[#]&/@completeGraphs
Out[75]= {{0.},{0.,0.},{0.,0.,0.},{0.,0.,0.,0.},
    {0.,0.,0.,0.,0.},{0.,0.,0.,0.,0.,0.},
    {0.,0.,0.,0.,0.,0.,0.}}

10.5 Graph Examples

```
In[71]:= completeGraphs = CompleteGraph[#,
           PlotLabel →
             Style[Subscript["K", #], Italic, Bold],
           VertexSize → Medium,
           EdgeStyle → Automatic] & /@ Range[7];
        gridGraphs[completeGraphs[[1 ;; 3]]]
        gridGraphs[completeGraphs[[4 ;; 6]]]
```

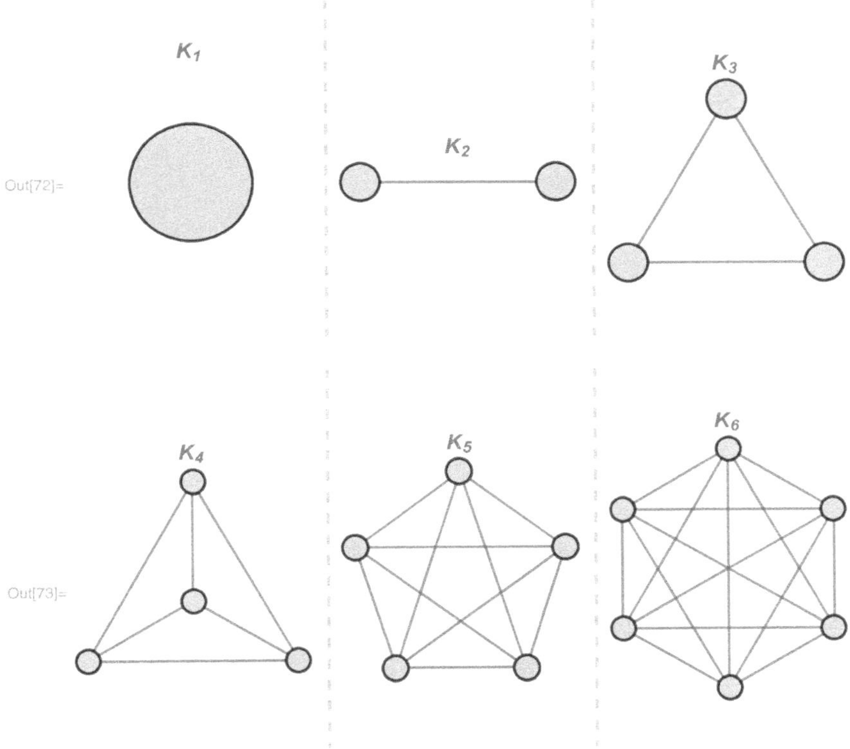

Out[72]=

Out[73]=

Fig. 10.27 **Out**[72]= and **Out**[73]= Complete graphs

```
In[76]:= DegreeCentrality[#]&/@completeGraphs
Out[76]= {{0},{1,1},{2,2,2},
         {3,3,3,3},{4,4,4,4,4},
         {5,5,5,5,5,5},{6,6,6,6,6,6,6}}
```

```
In[78]:= regularGraphs = Graph[GraphData[{"Antiprism", #}],
            VertexSize → Medium,
            PlotLabel →
              TextCell[GraphData[{"Antiprism", #},
                "Name"]]] & /@ Range[5, 6];
         gridGraphs[regularGraphs]
```

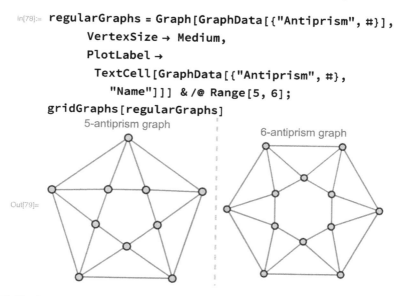

Fig. 10.28 Out[79]= Regular graph examples

10.5.3 Regular Graphs

A *regular* graph has every vertex having the same degree. There are many named examples in **GraphData**:

```
In[77]:= Length @ GraphData["Regular"]
Out[77]= 5353
```

We plot some examples in Figs. 10.28 and 10.29.

The **GraphData** has a lot of "Properties":

```
In[82]:= GraphData[{"Andrasfai",2},"Properties"]//
           Short
Out[82]//Short= {Acyclic, AdjacencyMatrix,<<457>>,
           ZeroSymmetric, ZeroTwo}
```

We can extract vertex and edge counts, centralities and degree sequence for the graph:

```
In[83]:= GraphData[{"Andrasfai",2},
           {"VertexCount","EdgeCount",
            "EdgeBetweennessCentralities","DegreeSequence"}]
Out[83]= {5,5,{6,6,6,6,6},{2,2,2,2,2}}
```

We can also get the AdjacencyMatrix:

```
In[84]:= MatrixForm[#]&/@
           GraphData[{"Andrasfai",3},{"AdjacencyMatrix"}]
```

10.5 Graph Examples

```
In[80]:= regularGraphs2 = Graph[GraphData[{"Andrasfai", #}],
           VertexSize -> Medium,
           PlotLabel -> TextCell[GraphData[{"Andrasfai", #},
             "Name"]]] & /@ Range[2, 4];
         gridGraphs[regularGraphs2]
```

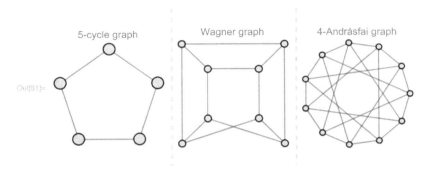

Fig. 10.29 Out[80]= Regular graph examples

$$\text{Out[84]}= \left\{ \begin{pmatrix} 0 & 1 & 0 & 0 & 1 & 0 & 0 & 1 \\ 1 & 0 & 1 & 0 & 0 & 1 & 0 & 0 \\ 0 & 1 & 0 & 1 & 0 & 0 & 1 & 0 \\ 0 & 0 & 1 & 0 & 1 & 0 & 0 & 1 \\ 1 & 0 & 0 & 1 & 0 & 1 & 0 & 0 \\ 0 & 1 & 0 & 0 & 1 & 0 & 1 & 0 \\ 0 & 0 & 1 & 0 & 0 & 1 & 0 & 1 \\ 1 & 0 & 0 & 1 & 0 & 0 & 1 & 0 \end{pmatrix} \right\}$$

10.5.4 Cycle Graphs

We can look at various cycle graphs, C_i using **CycleGraph**. The **CycleGraph** takes as input the number of vertices. Also, we can define whether or not we want directed versions, by setting the option DirectedEdges to **True**. We plot some examples in Fig. 10.30.

The adjacency matrices for cycle graphs have a very simple structure as well:

```
In[88]:= (Subscript["C",#]->
           MatrixForm[ AdjacencyMatrix[CycleGraph[#]]]&/@
           {3,5,10})
```

```
In[85]:= cycleGraphs = {CycleGraph[#, VertexSize → Medium,
           PlotLabel →
             Style[Subscript["C", #], Italic, Bold]] & /@
           Range[3, 5], CycleGraph[#, VertexSize → Medium,
           PlotLabel →
             Style[Subscript["Directed C", #], Italic, Bold],
           DirectedEdges → True] & /@ {3, 12, 16} };
       gridGraphs[cycleGraphs[[1]]]
       gridGraphs[cycleGraphs[[2]]]
```

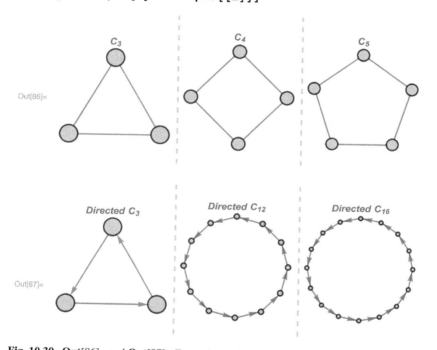

Fig. 10.30 Out[86]= and Out[87]= Examples of **CycleGraph**

$$\text{Out[88]= } \{C_3 \to \begin{pmatrix} 0 & 1 & 1 \\ 1 & 0 & 1 \\ 1 & 1 & 0 \end{pmatrix}, C_5 \to \begin{pmatrix} 0 & 1 & 0 & 0 & 1 \\ 1 & 0 & 1 & 0 & 0 \\ 0 & 1 & 0 & 1 & 0 \\ 0 & 0 & 1 & 0 & 1 \\ 1 & 0 & 0 & 1 & 0 \end{pmatrix}, C_{10} \to \begin{pmatrix} 0 & 1 & 0 & 0 & 0 & 0 & 0 & 0 & 0 & 1 \\ 1 & 0 & 1 & 0 & 0 & 0 & 0 & 0 & 0 & 0 \\ 0 & 1 & 0 & 1 & 0 & 0 & 0 & 0 & 0 & 0 \\ 0 & 0 & 1 & 0 & 1 & 0 & 0 & 0 & 0 & 0 \\ 0 & 0 & 0 & 1 & 0 & 1 & 0 & 0 & 0 & 0 \\ 0 & 0 & 0 & 0 & 1 & 0 & 1 & 0 & 0 & 0 \\ 0 & 0 & 0 & 0 & 0 & 1 & 0 & 1 & 0 & 0 \\ 0 & 0 & 0 & 0 & 0 & 0 & 1 & 0 & 1 & 0 \\ 0 & 0 & 0 & 0 & 0 & 0 & 0 & 1 & 0 & 1 \\ 1 & 0 & 0 & 0 & 0 & 0 & 0 & 0 & 1 & 0 \end{pmatrix}\}$$

10.5 Graph Examples

```
In[89]:= treeGraphs = Graph[
          GraphData[{"BananaTree", {2, #}}],
          VertexSize → Medium, PlotLabel →
            TextCell[GraphData[{"BananaTree", {2, #}},
              "Name"]]] & /@ Range[4, 6];
      gridGraphs[treeGraphs]
```

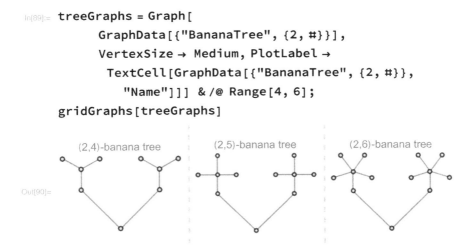

Fig. 10.31 Out[90]= and Out[90]= Examples of tree graphs

There are 1s in the off diagonals of the adjacency matrices (i.e. either side of the diagonal entry), representing an entry's connection to its neighbor, and there is a one in the right and left corner, representing the connection between the start and end vertices, completing the cycle.

10.5.5 Trees

A tree is a connected graph that has no cycles. Vertices of the tree with degree one are called leaves. There are for example "BananaTree" graphs defined through **GraphData**, Fig. 10.31.

We can highlight the leaves that are essentially at the periphery. We use **GraphPeriphery** to extract the vertices and highlight them in red in Fig. 10.32.

10.5.6 Bipartite Graphs

In a *bipartite* graph the vertices can be split into two sets, so that every edge joins a vertex from one set to the other. We plot some examples of complete bipartite graphs that have 2 vertices in one set and varying numbers in the second set (4, 5 and 6) in Fig. 10.33.

In[91]:= `gridGraphs[HighlightGraph[#, GraphPeriphery[#]] & /@
           (treeGraphs)]`

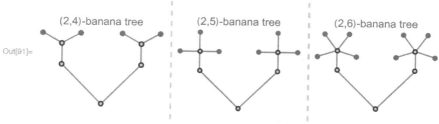

Out[91]=

Fig. 10.32 Out[91]= and Out[91]= Highlighted tree graph periphery

In[92]:= `bipartiteGraphs = Graph[
          GraphData[{"CompleteBipartite", {2, #}}],
          VertexSize → Medium,
          PlotLabel →
           TextCell[GraphData[{"CompleteBipartite",
             {2, #}}, "Name"]]] & /@ Range[4, 6];`
        `gridGraphs[bipartiteGraphs]`

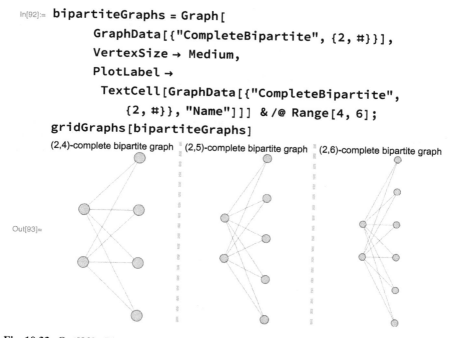

Out[93]=

Fig. 10.33 Out[93]= Bipartite graphs

10.6 Isomorphisms

Often we may have a different representation of two graphs, but they may in fact be in one-to-one correspondence, in that we can relabel the vertices and maintain the correspondence of edges. We can write this as: there exists f: V(G) →V(H), such that the edge v_i-v_j ∈ E(G) if and only if the edge f(v_i)-f(v_j) ∈ E(H). Here is a simple example. Consider the two graphs, 1 and 2 in Fig. 10.34.

10.6 Isomorphisms

```
In[94]:= graph1 = Graph[
          {"a", "b", "c", "d", "e"},
          {"a" ↔ "c", "c" ↔ "e", "e" ↔ "b", "b" ↔ "d",
           "d" ↔ "a"},
          VertexSize → 0.5,
          VertexLabels → Placed["Name", Center],
          VertexLabelStyle -> Directive[Bold, White, 12],
          VertexStyle → Thread[{"a", "b", "c", "d", "e"} →
             {Red, Blue, Magenta, Orange, Purple}]];
        graph2 = CycleGraph[5, VertexSize → 0.5,
          VertexLabels → Placed["Name", Center],
          VertexLabelStyle -> Directive[Bold, White, 12]];
        Row[{"Graph1:", graph1, Spacer[40],
          "Graph2:", graph2}]
```

Out[96]= Graph1: [graph image]   Graph2: [graph image]

Fig. 10.34 Out[96]= Two related graphs

In Fig. 10.34 we used VertexStyle to color Graph1, and used VertexLabels –> **Placed** ["Name",**Center**] to place vertex labels at the center of the vertices. We used a VertexLabelStyle, defined through a directive. Finally, we put the graphs together in a **Row** (and included a **Spacer** for spacing).

Now we can check if these are isomorphic using **IsomorphicGraphQ**:

In[97]:= **IsomorphicGraphQ**[graph1, graph2]
Out[97]= **True**

We can also find the isomorphism itself (this returns a list so we take the first element):

In[98]:= isomorphismG1G2=
             First @ **FindGraphIsomorphism**[graph1, graph2]
Out[98]= <|a→1, b→3, c→5, d→2, e→4|>

We finally put everything together to visualize the isomorphism in Fig. 10.35.

To display the map, we used **PropertyValue** to extract the VertexStyle from graph1 across its vertices, and applied it to highlight the vertices to be graphed.

```
In[99]:= Row@{graph1, Spacer[20],
    Style[Column@Normal[isomorphismG1G2],
     16, FontFamily → "Arial"], Spacer[20],
    HighlightGraph[graph2,
     Style[isomorphismG1G2[#],
       PropertyValue[{graph1, #},
        VertexStyle]] & /@
     VertexList[graph1]]} // TableForm
```

Out[99]//TableForm=

a → 1
b → 3
c → 5
d → 2
e → 4

Fig. 10.35 Out[99]= An isomorphism between the related graphs

10.7 Random Graphs

We often want to generate random graphs. We can use **RandomGraph**[{n,m}] to generate a graph with n vertices and m edges, Fig. 10.36.

If we use an additional argument with **RandomGraph** we can generate multiple random graphs, Fig. 10.37.

```
In[100]:= SeedRandom[7777];
    RandomGraph[{12, 20}]
```

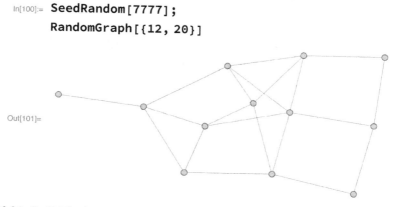

Out[101]=

Fig. 10.36 Out[101]= A random graph

10.7 Random Graphs

In[102]:= **SeedRandom[7777];**
RandomGraph[{12, 20}, 5]

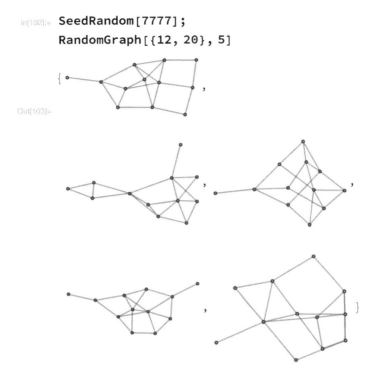

Out[103]=

Fig. 10.37 Out[103]= A collection of random graphs

10.7.1 Barabasi Albert Distribution

There are many random graph distributions that can be used to generate samples. For example, we can generate a Barabasi–Albert [1, 3] distribution for graphs with n vertices, in which an additional vertex with k edges is added at each step:

In[104]:= **SeedRandom[7777];**
barabasiAlbertRandomGraph=**RandomGraph[
BarabasiAlbertGraphDistribution[10^4,3]];**

We create an **EmpiricalDistribution** for the **VertexDegree**:

In[106]:= empiricalDistributionBarabasiAlbert=
**EmpiricalDistribution[
VertexDegree[** barabasiAlbertRandomGraph **]];**

Now we can plot the empirical distribution and see the characteristic behavior of the vertex degree in Fig. 10.38.

In[107]:= **DiscretePlot[**
  PDF[empiricalDistributionBarabasiAlbert, i],
  {i, 4, 20},
  PlotRange →
    All, PlotTheme → "Scientific", PlotLabel →
    "PDF of Empirical Distribution
    of Vertex Degree for Random Graphs
    from BarabasiAlbertGraphDistribution[10^4,3]]"]

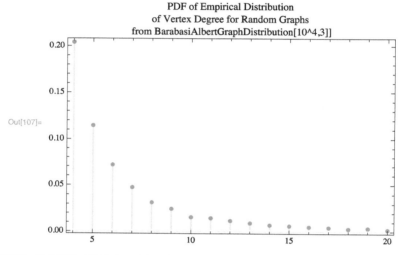

Fig. 10.38 Out[107]= PDF of empirical distribution of vertex degree for a random graph Barabasi–Albert graph distribution [1, 3]

10.7.2 Watts Strogatz

In the Watts–Strogatz paper [11], the authors proposed a rewiring of a ring lattice (having n vertices and k edges per vertex) and randomly rewiring each edge with probability p. In this way p can be tuned to lie between values of zero (which is order) and 1 (disorder).

In[108]:= **?WattsStrogatzGraphDistribution**
  WattsStrogatzGraphDistribution[n,p]
    represents the Watts–Strogatz graph distribution
    for n–vertex graphs with rewiring probability p.
  WattsStrogatzGraphDistribution[n,p,k] represents the
    Watts–Strogatz graph distribution for n–vertex
    graphs with rewiring probability p starting from
    a 2k–regular graph. >>

10.7 Random Graphs

```
In[109]:= SeedRandom[2222]; randomGraphsWattsStrogatz =
  RandomGraph[
    WattsStrogatzGraphDistribution[40, #, 6]] & /@
  {0.01, 0.02, 0.05, 0.1, 0.5, 1}
```

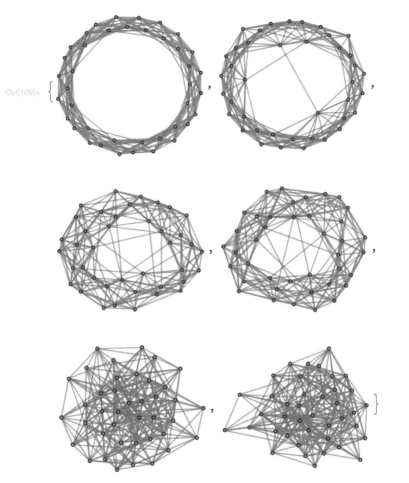

Fig. 10.39 Out[109]= Random graphs with Watts–Strogatz distribution and increasing disorder [11]

We can see in Fig. 10.39 how if we increase the probability p we are increasing the disorder.

Finally, we can get again an emprical distribution of the vertex degree, and plot it for each case in Fig. 10.40.

In[110]:= empiricalDistributionsWattsStrogatz=

```
In[111]:= DiscretePlot[PDF[#, i], {i, 4, 20},
          PlotRange → All, PlotTheme → "Scientific"] & /@
          empiricalDistributionsWattsStrogatz
```

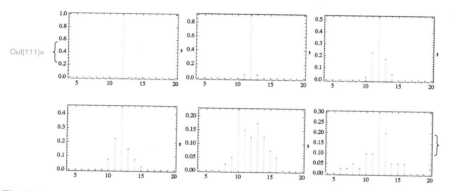

Fig. 10.40 Out[111]= PDF of empirical distribution of vertex degree from random Watts–Strogatz [11] graphs

EmpiricalDistribution[**VertexDegree**[#]] & /
@randomGraphsWattsStrogatz ;

References

1. Albert, R., Barabasi, A.L.: Statistical mechanics of complex networks. Rev. Mod. Phys. **74**(1), 47–97 (2002)
2. Alon, U.: Biological networks: the tinkerer as an engineer. Science (New York, NY) **301**(5641), 1866–1867 (2003)
3. Barabási, A.L., Albert, R.: Emergence of scaling in random networks. Science **286**(5439), 509 (1999)
4. Barabási, A.L., Oltvai, Z.N.: Network biology: understanding the cell's functional organization. Nat. Rev. Genet. **5**(2), 101–113 (2004)
5. Dorogovtsev, S.N.: Lectures on Complex Networks, vol. 24. Oxford University Press, Oxford (2010)
6. Euler, L.: Solutio problematis ad geometriam situs pertinentis. Commentarii academiae scientiarum Petropolitanae **8**, 128–140 (1741)
7. Harris, J.M., Hirst, J.L., Mossinghoff, M.J.: Combinatorics and Graph Theory, vol. 2. Springer, Berlin (2008)
8. Junker, B.H., Schreiber, F.: Analysis of Biological Networks, vol. 2. Wiley, New York (2011)
9. Motter, A.E., Albert, R.: Networks in motion (2012). arXiv:1206.2369
10. Newman, M.: Networks: An Introduction. Oxford University Press, Oxford (2010)
11. Watts, D.J., Strogatz, S.H.: Collective dynamics of 'small-world' networks. Nature **393**(6684), 440–442 (1998)

Chapter 11
Time Series Analysis

11.1 Time Series

Biological processes are inherently dynamic. From the tiniest molecule in our cells, to cells themselves, tissues, organisms and systems, nothing is essentially static. This includes the study of development, disease, and the action of drugs and aging. Continuous dynamic interactions at different spatial and temporal scales are ubiquitous and only through modeling the dynamics can we achieve a systems level knowledge in biology and genetics [3]. Time series analysis is available in the Wolfram Language, and in this chapter we introduce the basic capabilities available through a variety of examples.

We can think of a time series as a collection of values in a sequence, as a list: $X[t] = \{X(t_1), X(t_2), X(t_3), \ldots, X(t_n)\}$ [6, 7, 10, 13]. The series of values represent intensities sequentially changing as time progresses from $t_i \to t_{i+1}$. For a regularly sampled time series, the interval between successive time points, $\Delta t = t_{i+1} - t_i$ is the same. We can conceptualize the time series as an ordered n dimensional vector or list of values. If the variable $X[t]$ is a random variable, then the sequence is a realization of X sampled at the particular time points.

11.1.1 The TimeSeries Function

In the Wolfram Language [22, 23], we can represent time series realizations using the **TimeSeries** function. For our first example, we will construct a random series over 20 timepoints:

In[1]:= **SeedRandom**[123];
(∗set the pseudorandom generator for reproducibility

Electronic supplementary material The online version of this chapter (https://doi.org/10.1007/978-3-319-72377-8_11) contains supplementary material, which is available to authorized users.

© Springer International Publishing AG 2018
G. Mias, *Mathematica for Bioinformatics*,
https://doi.org/10.1007/978-3-319-72377-8_11

— you can remove this input *)

```
        intensities = RandomReal[20,20];
        times = Range[2,40,2]
Out[3]= {2,4,6,8,10,12,14,16,18,20,
        22,24,26,28,30,32,34,36,38,40}
```

We can pair $\{t_i, X[t_i]\}$ values by taking the transpose of the times and intensities lists (basically turning a 2×20 matrix into a 20×2 matrix):

```
In[4]:= pairedValues=Transpose[{times,intensities}]
Out[4]= {{2,9.11438},{4,19.5565},
        {6,18.8643},{8,19.2443},{10,6.04696},
        {12,9.33417},{14,1.23277},{16,7.71289},
        {18,8.59677},{20,15.5749},{22,0.971811},
        {24,12.5653},{26,5.55974},{28,1.80435},
        {30,17.5317},{32,2.18214},{34,5.31515},
        {36,18.3722},{38,3.39832},{40,1.99157}}
```

We can use either the paired values or the intensities and times to define the same time series:

```
In[5]:= ?TimeSeries
        TimeSeries[{{t₁,v₁},…{t₂,v₂}}] represents
        a time series specified by time-value pairs {tᵢ,vᵢ}.
        TimeSeries[{v₁,v₂,…},tspec] represents a time series
        with values vᵢ at times specified by tspec. >>
```

In the case where a time specification is given, the *tspec* specification can take various forms: **Automatic** assumes a uniformly sampled time series beginning at time zero. Alternative *tspec* specifications always take the form of a list. This list can be a single value representing the starting time for a uniformly sampled dataset, t_{min}, or a specification of start and end times, $\{t_{min}, t_{max}\}$ that can include an optional time step size, Δt, as $\{t_{min}, t_{max}, \Delta t\}$. Finally, explicit times can be entered, as we have defined above as $\{\{t_1, \ldots, t_n\}\}$.

A time series generates a **TemporalData** object along a single time path, Fig. 11.1. Alternatively, we use the paired values we generated, Fig. 11.2.

If we press the + sign on the **TemporalData** output we can get some additional information, Fig. 11.3.

We can get the **Head** of the object:

```
In[8]:= Head[timeSeriesObject]
Out[8]= TemporalData
```

```
In[6]:= timeSeriesObject = TimeSeries[intensities, {times}]

Out[6]= TimeSeries[ ▦ ⋀  Time: 2 to 40
                          Data points: 20  ]
```

Fig. 11.1 Out[1]= A **TimeSeries** with a time specification

11.1 Time Series

In[7]:= `timeSeriesObject2 = TimeSeries[pairedValues]`

Out[7]= TimeSeries[▦ ⩔ Time: 2 to 40
                            Data points: 20   ]

Fig. 11.2 Out[7]= A **TimeSeries** as paired values

In[7]:= `timeSeriesObject2 = TimeSeries[pairedValues]`

Out[7]= TimeSeries[▦ ⩔ Time: 2 to 40
                            Data points: 20
                            Regular: **True**   Output dimension: **1**
                            Metadata: **None**   Minimum increment: **2**
                                                      Resampling: **LinearInterpolation**   ]

Fig. 11.3 Out[7]= A **TimeSeries** as paired values, with expanded output, by pressing the + sign on the output

TimeSeries has multiple options, including ways to handle missing data as well as resampling methods:

In[9]:= **Options[TimeSeries]**
Out[9]= {CalendarType→Gregorian, DateFunction→**Automatic**,
         HolidayCalendar→{UnitedStates, **Default**},
         MetaInformation→**None**, MissingDataMethod→**None**,
         ResamplingMethod→**Automatic**,
         TemporalRegularity→**Automatic**,
         ValueDimensions→**Automatic**, **TimeZone**:>$TimeZone}

We can plot the **TemporalData** object directly using a **ListLinePlot**, Fig. 11.4.

For example let us assume we have measurements from time series corresponding to normalized time expression from a gene:

In[11]:= expressionSeries={0.937,1.003,0.966,0.982,1.024,
          1.073,0.921,1.007,0.913,0.937,1.059,1.479,
          1.455,1.689,1.312};

We may have specific times we sampled, and let us say this was done in September of 2017 daily, starting on the 15th of the month. We can form a time series from the data. We can specify a start date, an end date (set to **Automatic** here) and the unit of increment, a "Day":

In[12]:= expressionTemporalData=
          TimeSeries[expressionSeries,
          {"September 15 2017",**Automatic**,"Day"}];

The temporal data have various properties we can extract:

In[13]:= expressionTemporalData["Properties"]
Out[13]= {DatePath, Dates, FirstDate, FirstTime, FirstValue,
          LastDate, LastTime, LastValue, Path, PathFunction,
          PathLength, Times, ValueDimensions, Values}

In[10]:= `ListLinePlot[timeSeriesObject,`
  `FrameLabel → {"Time", "Value"},`
  `Frame → True,`
  `PlotLabel → "A time series plot."]`

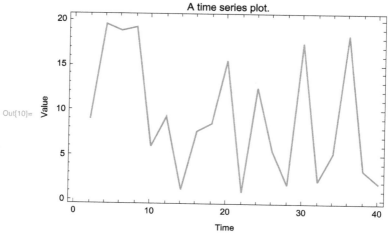

Fig. 11.4 Out[10]= A **TimeSeries** plot using **ListLinePlot**

Fig. 11.5 Out[14]= Extracting the last date in a **TemporalData**

In[14]:= `expressionTemporalData["LastDate"]`

We can get the last date, Fig. 11.5. Notice that the date is output as a DateObject. We may want to extract the values from the time series:

In[15]:= `expressionTemporalData["Values"]`
Out[15]= {0.937, 1.003, 0.966, 0.982, 1.024, 1.073, 0.921, 1.007,
  0.913, 0.937, 1.059, 1.479, 1.455, 1.689, 1.312}

We can get the data value on any date:

In[16]:= `expressionTemporalData["September 19 2017"]`
Out[16]= 1.024

Or we can calculate statistics on the series:

In[17]:= `{Max[#], Min[#], StandardDeviation[#], Median[#]} &@`
  `expressionTemporalData`
Out[17]= {1.689, 0.913, 0.244277, 1.007}

Finally, we can plot the temporal data directly, using **DateListPlot**, Fig. 11.6.
We may also want to calculate the **Differences** between successive timepoints as a **TimeSeries**:

11.1 Time Series

```
In[18]:= DateListPlot[expressionTemporalData,
    PlotLabel →
      "DateList Plot for Expression Change Vs. Date",
    FrameLabel → {"Date", "Normalized Intensity"}]
```

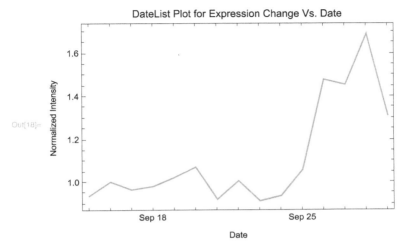

Fig. 11.6 Out[18]= **DateListPlot** of temporal data

In[19]:= ?**Differences**

> **Differences**[*list*] gives the successive differences of elements in *list*.
> **Differences**[*list*,*n*] gives the nth differences of *list*.
> **Differences**[*list*,*n*,*s*] gives the differences of elements step *s* apart.
> **Differences**[*list*, {$n_1, n_2, \ldots$}] gives the successive n_k^{th} differences at level *k* in a nested list. ≫

We calculate the differences in our dataset below, and then plot them, in Fig. 11.7.

In[20]:= diffDaily=**Differences**[expressionTemporalData];

11.1.2 Multiple Time Series Example

We will use the MathIOmica iPOP [8, 14–16] data as another example. First we load the package:

In[22]:= <<**MathIOmica`**

```
In[21]:= DateListPlot[diffDaily,
    PlotLabel →
    "DateList Plot for Daily Expression Change Vs. Date",
    FrameLabel → {"Date", "Normalized Intensity"}]
```

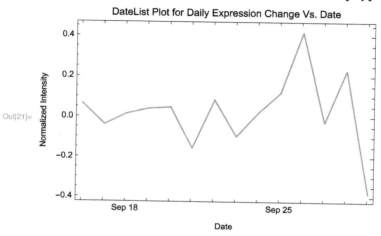

Fig. 11.7 Out[21]= **DateListPlot** of successive time point differences in temporal data

We can import an example for RNA-sequencing data, from the iPOP study. The data is stored in the directory **ConstantMathIOmicaExamplesDirectory**. The following command should be operating system independent and will load the RNA data:

```
In[23]:= rnaExample=Get[FileNameJoin[{
    ConstantMathIOmicaExamplesDirectory,"rnaExample"}]];
```

We take a quick look at the data:

```
In[24]:= rnaExample//Short[#,3]&

Out[24]//Short=  <|7→ <|{FAM138A,RNA}→{{0},{OK}},
        {OR4F5,RNA}→{{0},{OK}},<<25265>>,
        {DUX4L,RNA}→{{0},{OK}}|>,<<13>>,
        21→ <|<<1>>|>|>
```

The outer keys in the rnaExample nested association, which is an **OmicsObject** in MathIOmica, correspond to a sample number. Each sample corresponds in turn to a day in the study (e.g. sample 7 is day 186, sample 20 is day 380). We make an association of the sample number to the study day:

```
In[25]:= sampleToDays= <|"7"->"186","8"->"255","9"->"289",
    "10"->"290","11"->"292","12"->"294","13"->"297",
    "14"->"301","15"->"307","16"->"311","17"->"322",
    "18"->"329","19"->"369","20"->"380","21"->"400"|>;
```

11.1 Time Series

The inner keys in the rnaExample nested association, have the form {"Official Gene Symbol","RNA"} , to indicate that this is transcriptome information for the given gene. The values correspond to {{measurements},{metadata}} are {{"FPKM" values},{"FPKM status"}} as obtained from mapped RNA-sequencing data. FPKM stands for Fragments Per Kilobase of transcript per Million mapped reads.

We can first pretend the samples are sequential states and extract each sample's number as its token time:

In[26]:= sampleList=**TimeExtractor**[rnaExample]
Out[26]= {7,8,9,10,11,12,13,14,15,16,17,18,19,20,21}

We can extract the series FPKM data (N.B. **MathIOmica** orders the data based on the sorted order of the outer keys):

In[27]:= exampleRNASampleSeries=
    CreateTimeSeries[rnaExample];

In[28]:= exampleRNASampleSeries[[1;;3]]
Out[28]= <|{FAM138A,RNA}→{0,0,0.00257233,0.00206039,0,0,0,
    0.00429509,0,0,0,0,0,0,0},
    {OR4F5,RNA}→{0,0,0,0,0,0,0,0,0,0,0,0,0,0,0},
    {LOC729737,RNA}→{2.73998,0.555218,5.15563,4.53362,
    5.71829,0.889102,1.413,2.38838,0.83811,0.706081,
    2.2049,2.18935,4.05165,2.70102,1.22675}|>

We are not interested in constant series, so we can remove those using the function **ConstantSeriesClean**:

In[29]:= exampleRNASampleSeriesNonConstant=
    ConstantSeriesClean[exampleRNASampleSeries];

(* the lines below are printed during the evaluation *)
Removed series and returning filtered list. If you
    would like a list of removed keys run the command
    ConstantSeriesClean[data,ReturnDropped] → True.

Let us look at a few of these series, and verify that constant series have been removed:

In[30]:= exampleRNASampleSeriesNonConstant//**Short**[#,4]&
Out[30]//**Short**= <|{FAM138A,RNA}→{0,0,0.00257233,0.00206039,
    0,0,0,0.00429509,0,0,0,0,0,0,0},
    <<20711>>,{UTY,RNA}→{3.16532,3.28427,
    3.06644,<<9>>,1.59413,1.13702,0.98188}|>

We can get a list of all the Official Gene Symbols, say the first 10:

In[31]:= genesTimeSeriesExample=
    Query[**Keys**,1]@
    exampleRNASampleSeriesNonConstant[[1;;10]]
Out[31]= {FAM138A,LOC729737,DDX11L1,WASH7P,MIR6859-1,
    OR4F16,LOC100132287,LOC100133331,MIR6723,OR4F29}

These all have 15 times associated with them:

```
In[36]:= ListLinePlot[exampleRNASampleTemporalData,
    PlotLabel →
      "RAW FPKM Data over iPOP study samples 7-21",
    FrameLabel → {"Ordered Samples",
      "Normalized Intensity"},
    Frame → True,
    PlotLegends → genesTimeSeriesExample]
```

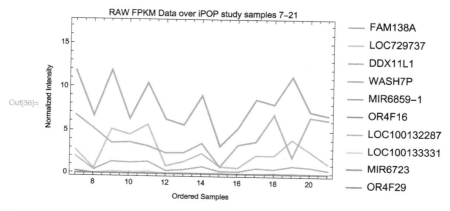

Fig. 11.8 Out[36]= **ListLinePlot** of temporal data

```
In[32]:= Length[#]&/@
          Values @ exampleRNASampleSeriesNonConstant[[1;;10]]
Out[32]= {15,15,15,15,15,15,15,15,15,15}
```

We need to create a list of points at every timepoint to use **TimeSeries**:

```
In[33]:= dataTimeSeriesExample=
          Transpose[
            Values @ exampleRNASampleSeriesNonConstant[[1;;10]]];
In[34]:= dataTimeSeriesExample//Short[#,5]&
Out[34]//Short= {{0,2.73998,6.75461,11.8883,0.256318,0.0053034,
            1.97164,0.0000580254,20.3672,0.0053034},
            <<13>>,{0,1.22675,6.31519,6.8114,0,
            0,0.462767,0.000176344,50.1361,0}}
```

Now we can pass the information to **TimeSeries** with a time specification given by sampleList:

```
In[35]:= exampleRNASampleTemporalData=TimeSeries [
    dataTimeSeriesExample,{ sampleList }];
```

We can plot these un-normalized data in a **ListLinePlot**, Fig. 11.8. The lines are essentially linearly interpolating between adjacent sample points.

11.1 Time Series

```
In[37]:= ListStepPlot[exampleRNASampleTemporalData,
    PlotLabel →
      "RAW FPKM Data over iPOP study samples 7-21",
    FrameLabel → {"Time", "Normalized Intensity"},
    Frame → True, Joined → False,
    PlotLegends →
      PointLegend[Automatic, genesTimeSeriesExample,
        LegendFunction → Panel, LegendMargins → 5]]
```

Out[37]=

Fig. 11.9 Out[37]= **ListStepPlot** of temporal data, without interpolation

We can instead use a **ListStepPlot** if we do not want to interpolate, Fig. 11.9. Also, notice that we have used a **PointLegend** function to define our legends in the figure.

In reality, the time series is unevenly sampled, and adjacent time points are not equidistant. We will now process the same data using real times:

```
In[38]:= daysList=Values @ sampleToDays
Out[38]= {186,255,289,290,292,294,297,301,
         307,311,322,329,369,380,400}
```

Again, we use **TimeSeries** and this time we pass the daysList as the time specification, and then plot the data in Fig. 11.10.

```
In[39]:= exampleRNADatesTemporalData=TimeSeries[
    dataTimeSeriesExample,{daysList}];
```

The plotted data in Fig. 11.10 now look very different. Unless we are dealing with pre-defined states, the sampling interval is very important, and the plot interpolates linearly between time points. Given the large variation in the intervals between time points, it is unclear whether a linear interpolation is actually the correct approach, but it is a usual first approximation.

```
In[40]:= DateListPlot[exampleRNADatesTemporalData,
    PlotLabel → "RAW FPKM Data over iPOP study Dates",
    FrameLabel → {"Ordered Samples",
      "Normalized Intensity"}, Frame → True,
    PlotLegends → PointLegend[Automatic,
      genesTimeSeriesExample,
      LegendFunction → Panel, LegendMargins → 5]]
```

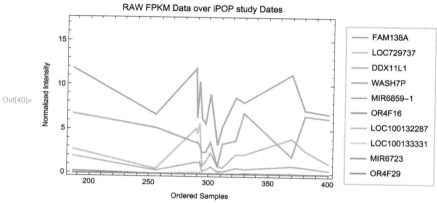

Fig. 11.10 Out[40]= **ListLinePlot** of data with actual time differences shown

11.2 Vignette: FinancialData

In the Wolfram Language we can visualize easily time series from entity data or from Wolfram|Alpha. As an example, we can get financial data on companies over time using **FinancialData**. For example, let us look at Illumina Inc., using the symbol "ILMN":

In[41]:= **FinancialData**["ILMN","Name"]
Out[41]= Illumina

We can check which sector the company is defined to be in:

In[42]:= **FinancialData**["ILMN","Sector"]
Out[42]= Biotechnology

We can now construct a time series from the financial data, since the year 2000, setting {2000} as the time specification for the date minimum. Then we use **DateListPlot** to plot the information in Fig. 11.11.

In[43]:= ilm=**TimeSeries**[**FinancialData**["ILMN",{2000}]];

We can also get information on major drug manufacturers,

In[45]:= manufacturers=**FinancialData**["DrugManufacturers-Major",
      "Members"];

11.2 Vignette: FinancialData

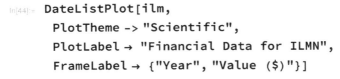

```
In[44]:= DateListPlot[ilm,
    PlotTheme -> "Scientific",
    PlotLabel → "Financial Data for ILMN",
    FrameLabel → {"Year", "Value ($)"}]
```

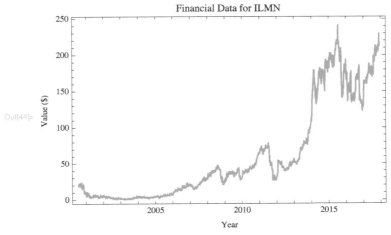

Fig. 11.11 Out[44]= **DateListPlot** of **FinancialData** for Illumina (symbol "ILMN") since 2000

```
In[46]:= manufacturers // Short
Out[46]//Short= {AMEX:AXN,AMEX:CPHI,<<74>>,SI:A61,SI:E02}
```

And additional biotechnology companies,

```
In[47]:= biotechnology=FinancialData["Biotechnology"];
```

```
In[48]:= biotechnology // Short
Out[48]//Short= {AMEX:BTX,AMEX:CRMD,<<235>>,V:SSS,V:VPI}
```

11.3 Time Series Model Fitting

The function **TimeSeriesModelFit** can construct a model for input time series data from various model families. The Wolfram Language can fit to a variety of model families, with the defined parameters as briefly shown in Table 11.1. We will not go into details of the various time series models as these are extensively discussed in various time series theory books [6, 7, 10, 13].

In this section we use an example utilizing Body Mass Index data from 1975 to 2014. The data is a **ResourceObject** which can be obtained under the name "BMI by Country".

Table 11.1 Out[49]= Models and Parameters for **TimeSeriesModelFit**

Model	Parameter	Family
AR	{p}	autoregressive model family
MA	{q}	moving-average model family
ARMA	{p, q}	autoregressive moving-average model family
ARIMA	{p, d, q}	integrated ARMA model family
SARMA	{{p, q},{sp, sq}, s}	seasonal ARMA model family
SARIMA	{{p, d, q},{sp, sd, sq}, s}	seasonal ARIMA model family
ARCH	{q}	ARCH model family
GARCH	{q, p}	GARCH model family

In[50]:= roBMI=**ResourceObject**["BMI by Country"];

We can get a "Description" of the **ResourceObject**,

In[51]:= roBMI["Description"]
Out[51]= Estimated mean Body Mass Index (BMI) trend by country

And we can get the "Details":

In[52]:= roBMI["Details"]
Out[52]= Estimated mean Body Mass Index (BMI)
        trend between 1975 and 2014 by country.
        BMI (kg/m2) estimates are from male/female
        populations over the age of 18.

There are 382 elements:

In[53]:= roBMI["InformationElements"]
Out[53]= <|**Length**→382,**Dimensions**→{382,3}|>

We can also get the last release date, which was June 16th 2017:

In[54]:= **FullForm** @ roBMI["ReleaseDate"]
Out[54]//**FullForm**=
        DateObject[List[2017,6,16],"Day","Gregorian",-4.`]

Let us now obtain the data by using **ResourceData**:

In[55]:= dataBMI=**ResourceData**["BMI by Country"];

The data is in the form of a dataset:

In[56]:= **Head**[dataBMI]
Out[56]= **Dataset**

There are three keys, corresponding to Country, Gender and BMI:

In[57]:= **Normal[Query[1,Keys]@dataBMI]**
Out[57]= {Country, Gender, BMI}

These are respectively an **Entity**, a **String** and **TemporalData**:

In[58]:= **Normal[Query[1,Values,Head[#]&]@dataBMI]**
Out[58]= {Entity, String, TemporalData}

11.3.1 USA BMI Data Modeling

Now let us carry out an analysis on the specific BMI data from the United States. We carry out a **Query** in which we select the elements in which the named pattern Country matches an **Entity** of type "Country" and "UnitedStates":

In[59]:= usaBMI=
        Query[
          Select[
            MatchQ[#Country, Entity["Country",
              "UnitedStates"]]&]]@dataBMI;

This is a two part list that contains both male and female time data for BMI:

In[60]:= **Head[#]& /@ #& /@ (Normal @ usaBMI)**
Out[60]= {<|Country→**Entity**, Gender→**String**,
         BMI→**TemporalData**|>, <|Country→**Entity**,
         Gender→**String**, BMI→**TemporalData**|>}

We can extract the time series for male and female from elements 1 and 2 respectively:

In[61]:= timeSeriesUSABMImale=**Query**[1,"BMI"]@usaBMI;

In[62]:= timeSeriesUSABMIfemale=**Query**[2,"BMI"]@usaBMI;

If we want, we can calculate directly on the time series, for example we can get the **Mean** time series, Fig. 11.12.

In[63]:= `Mean[{timeSeriesUSABMImale, timeSeriesUSABMIfemale}]`

Out[63]= `TimeSeries[` ▫ / Time: 01 Jan 1975 to 01 Jan 2014
                          Data points: 40   `]`

Fig. 11.12 Out[63]= **Mean** of two time series

In[65]:= `DateListPlot[{timeSeriesUSABMImale,`
   `timeSeriesUSABMIfemale},`
  `PlotTheme → "Scientific",`
  `PlotLabel →`
   `("Average BMI in the US From 1975 to 2014"),`
  `PlotLegends → {"Male", "Female"},`
  `FrameLabel → {"Year", "BMI"}]`

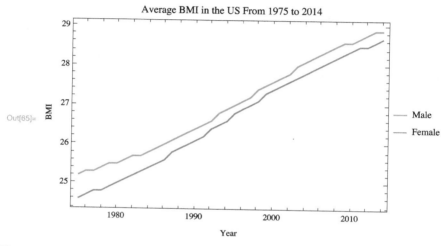

Fig. 11.13 Out[65]= Average BMI in the US from 1975-2014

There are a total 40 time points:

In[64]:= lengthBMI=timeSeriesUSABMIfemale["PathLength"]
Out[64]= 40

We can plot directly the **TemporalData** using a **DateListPlot**, Fig. 11.13. Both female and male BMIs display a rather linear increase consistenlty since 1975.

We will now and try and fit the US BMI data to a model. First let us pick a specific model. We will fit to an "AR", autoregressive model for each series, using **TimeSeriesModelFit**:

In[66]:= modelUSABMIfemale=
   TimeSeriesModelFit[timeSeriesUSABMIfemale,"AR"];

In[67]:= modelUSABMImale=
   TimeSeriesModelFit[timeSeriesUSABMImale,"AR"];

The fitted model has multiple "Properties":

In[68]:= modelUSABMIfemale["Properties"]
Out[68]= {ACFPlot,ACFValues,AIC,AICc,BIC,

11.3 Time Series Model Fitting

BestFit, BestFitParameters, CandidateModels,
CandidateModelSelectionValues,
CandidateSelectionTable,
CandidateSelectionTableEntries, CovarianceMatrix,
ErrorVariance, FitResiduals, ForecastStandardErrors,
InformationMatrix, LjungBoxPlot, LjungBoxValues,
ModelFamily, PACFPlot, PACFValues,
ParameterConfidenceIntervals, ParameterTable,
ParameterTableEntries, PredictionLimits,
Properties, SBC, SelectionCriterion,
StandardizedResiduals, TemporalData}

We can obtain the "ParameterTable" of fitted coefficients:

In[69]:= modelUSABMIfemale["ParameterTable"]

Out[69]=
	Estimate	Standard Error	t-Statistic	P-Value
a_1	0.93382	0.0565642	16.509	9.29499×10^{-20}

We can also get the "BestFit" model:

In[70]:= modelUSABMIfemale["BestFit"]
Out[70]= ARProcess[1.76485, {0.93382}, 0.215031]

Additionally, we can plot the residual autocorrelations, Fig. 11.14.

We can also plot the Ljung-Box [4, 11] residual autocorrelation test p values, testing the null hypothesis that the autocorrelations at a given lag in the residuals are zero, Fig. 11.15.

Instead of selecting a particular model, as we have done above, the Wolfram Language allows us to try to get the best model automatically, by setting the selection of model to **Automatic**:

In[73]:= modelUSABMIfemaleAutomatic=
 TimeSeriesModelFit[timeSeriesUSABMIfemale, **Automatic**];

In[74]:= modelUSABMImaleAutomatic=
 TimeSeriesModelFit[timeSeriesUSABMImale, **Automatic**];

It seems that an ARIMAProcess, an autoregressive integrated moving-average process, is the best fit for both female and male time series:

In[75]:= modelUSABMIfemaleAutomatic["BestFit"]
Out[75]= ARIMAProcess[0.105128, {}, 1, {}, 0.00151216]

In[76]:= modelUSABMImaleAutomatic["BestFit"]
Out[76]= ARIMAProcess[0.0948718, {}, 1, {}, 0.00202498]

We can see what the candidate selection table looks like. These includes the Akaike information criterion (AIC) [1, 6] which estimates the relative quality between the various models that the function uses:

In[77]:= modelUSABMIfemaleAutomatic["CandidateSelectionTable"]

```
In[71]:= TableForm[#, TableHeadings →
            {{"Female", "Male"}, None}] &@
         (#["ACFPlot"] & /@
            {modelUSABMIfemale, modelUSABMImale})
```

Out[71]//TableForm=

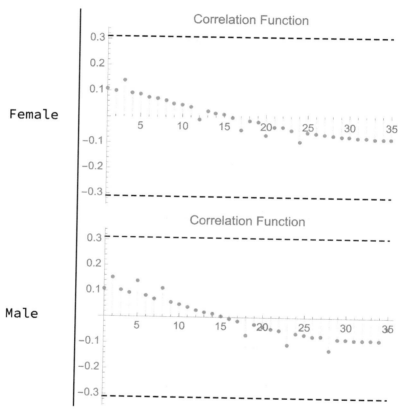

Fig. 11.14 Out[71]= Plot of residual autocorrelations for time series model for BMI

Out[77]=

	Candidate	AIC
1	**ARIMAProcess[0,1,0]**	**−255.769**
2	ARIMAProcess(0, 1, 1)	−253.782
3	ARIMAProcess(1, 1, 0)	−253.781
4	ARIMAProcess(1, 2, 1)	−251.5
5	ARIMAProcess(1, 1, 1)	−251.354
6	ARIMAProcess(1, 2, 2)	−248.683
7	ARIMAProcess(2, 2, 1)	−248.319
8	ARIMAProcess(2, 2, 2)	−246.412
9	ARIMAProcess(0, 2, 1)	−244.195
10	ARIMAProcess(1, 2, 0)	−237.821

11.3 Time Series Model Fitting

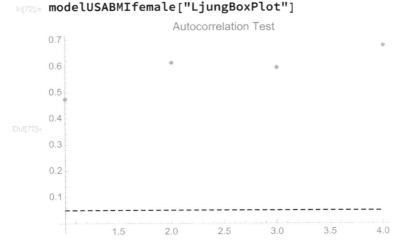

Fig. 11.15 Out[72]= Ljung-Box residual autocorrelation test p-values

In this case we see that all the top models were ARIMA processes with different parameters.

We can again plot the `"ACFPlot"` for the ARIMA process model for both Female and Male, Fig. 11.16.

Now that we have a fitted model, we may want to use our model to make a forecast. For each model for female and male BMI, we will obtain a forecast/prediction for 4 timepoints in the future (in this case 4 years ahead of data to 2018). We use **TimeSeriesForecast** and **Map** to both the female and male models respectively:

```
In[79]:=  forecastsBMIUS=TimeSeriesForecast[#,{4}]&/@
            {modelUSABMIfemaleAutomatic,
             modelUSABMImaleAutomatic};
```

The forecast also has various properties:

```
In[80]:=  forecastsBMIUS[[1]]["Properties"]
Out[80]=  {Components, DateList, DatePath, DatePaths, Dates,
           FirstDates, FirstTimes, FirstValues, LastDates,
           LastTimes, LastValues, MeanSquaredErrors, Part,
           Path, PathCount, PathFunction, PathFunctions,
           PathLength, PathLengths, Paths, PathTimes,
           SliceData, SliceDistribution, TimeList,
           Times, ValueDimensions, ValueList, Values}
```

For each model we are usually interested in either the error or the confidence interval. Here we get the error so we can include it in a plot of the forecast further down:

```
In[81]:=  errorsBMIForecastUS=
            #["MeanSquaredErrors"]^0.5&/@ forecastsBMIUS;
```

```
In[78]:= TableForm[#, TableHeadings →
            {{"Female", "Male"}, None}] &@
         (#["ACFPlot"] & /@ {modelUSABMIfemaleAutomatic,
            modelUSABMImaleAutomatic})
```
Out[78]//TableForm=

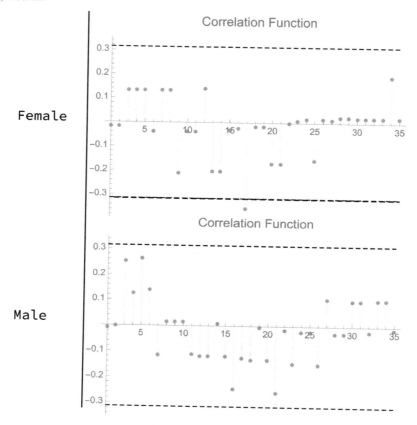

Fig. 11.16 Out[78]= Plot of residual autocorrelations for ARIMA process model for BMI

We would like to define bands to plot for the error, an upper band and a lower band for each of female and male forecast time series. We can use the **TimeSeriesThread** function that allows us to to combine time series by applying a function f of our choosing:

```
In[82]:= ?TimeSeriesThread
         TimeSeriesThread[f, {tseries₁, tseries₂, ...}]
             combines the tseries_i using the function f.  >>
```

11.3 Time Series Model Fitting

```
In[87]:= DateListPlot[
    {modelUSABMIfemaleAutomatic["TemporalData"],
     TimeSeriesForecast[modelUSABMIfemaleAutomatic,
     {4}], upperBandFemale, lowerBandFemale},
    PlotTheme → "Scientific",
    PlotLegends → {"Female", "Female Forecast",
     "Error", None},
    FrameLabel → {"Year", "BMI"},
    PlotLabel → "Average BMI in the US From
    1975 to 2014 and Forecast to 2018",
    PlotStyle → {Automatic, Automatic, Gray, Gray},
    Filling → {3 -> {4}}]
```

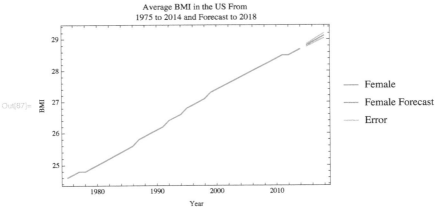

Fig. 11.17 Out[87]= Plot of mean BMI and forecast for females in the US

We use the functions {1,1}.#& and {1,−1}.#& . These correspond to summing and subtracting elements in a paired list respectively for the upper and lower band data (i.e. add the error to the forecast and subtract the error from the forecast):

```
In[83]:= upperBandFemale=TimeSeriesThread[{1,1}.#&,
         {forecastsBMIUS[[1]],errorsBMIForecastUS[[1]]}];
       upperBandMale=TimeSeriesThread[{1,1}.#&,
         {forecastsBMIUS[[2]],errorsBMIForecastUS[[2]]}];
       lowerBandFemale=TimeSeriesThread[{1,−1}.#&,
         {forecastsBMIUS[[1]],errorsBMIForecastUS[[1]]}];
       lowerBandMale=TimeSeriesThread[{1,−1}.#&,
         {forecastsBMIUS[[2]],errorsBMIForecastUS[[2]]}];
```

Finally, we can plot all the information together for female, Fig. 11.17, and male, Fig. 11.18 mean BMI in the US. In each of the Figs. 11.17 and 11.18 we have plotted

```
In[88]:= DateListPlot[
    {modelUSABMImaleAutomatic["TemporalData"],
     TimeSeriesForecast[modelUSABMImaleAutomatic,
     {4}], upperBandMale, lowerBandMale},
    PlotTheme -> "Scientific",
    PlotLegends -> {"Male", "Male Forecast",
     "Error", None},
    FrameLabel -> {"Year", "BMI"},
    PlotLabel -> "Average BMI in the US
   From 1975 to 2014 and Forecast to 2018",
    PlotStyle -> {Automatic, Automatic, Gray, Gray},
    Filling -> {3 -> {4}}]
```

Fig. 11.18 Out[88]= Plot of mean BMI and forecast for males in the US

the model fitted (red), the forecast time series (blue) and the upper and lower bands (gray).

We see that the prediction in both cases is an increase with a fairly small error band.

The methods of fitting time series to models and obtaining forecasts for the data can be extended to any time series - as always, the more data the better the model fitting, and also the predictions.

11.4 Unevenly Sampled Time Series Classification Examples

The **MathIOmica** package has various simple methods for classifying trends in time series. In this example we will use the `"LombScargle"` method for the **TimeSeriesClassification**[data, setTimes] function. We will first discuss the approach a bit more, as found in the **MathIOmica** documentation. Please note that there are many methods for time-series classification, and this method is a fairly elementary approach.

11.4.1 Lomb Scargle Classification in MathIOmica

The `"LombScargle"` method used in MathIOmica has at its basis a transformation of a time signal (may include missing data or be unevenly sampled) to a periodogram/power spectrum [5, 12, 17, 18, 21, 24]. The maximum intensity across all frequencies is then used to filter signals based on whether they display a dominant frequency signal, typically using a cutoff that is user-supplied based on time series simulations. If a frequency is found that is judged to be significant based on the cutoff, the signals displaying such a dominant frequency are selected and separated into classes based on what the frequency is.

Following the discussion in MathIOmica's documentation, for each input series, the data has to be in the order that the setTimes appear. In general the data are lists of time series signal values, $X_j = \{X_j(t_1), X_j(t_2), ..., X_j(t_N)\}$ taken over a list of ordered sampling times, where $t_1 < t_2 < \cdots < t_N$.

Note that $\{t_1, t_2, ..., t_N\}$ is necessarily a subset of the possible setTimes, since any time point during which data was collected is necessarily a possible time point, but some time points may be missing.

For a given signal X_j, we can then calculate the signal's periodogram. The method uses the **LombScargle** function in **MathIOmica** to calculate the periodogram, and to provide $n =$**Floor**$[N/2]$ frequencies, $f_j = \{f_{j1}, f_{j2}, \ldots, f_{jk}, \ldots, f_{jn}\}$, and corresponding n intensities, $I_j = \{I_{j1}, I_{j2}, \ldots, I_{jn}\}$.

The default is for the intensity vector returned I_j to be calculated as a normalized vector. The maximum intensity of this vector, $I_{jk_{\max}} = \max(I_j)$ corresponds to a dominant frequency $f_{jk_{\max}}$, and occurs at some index $k_{\max}$. For each signal $I_{jk_{\max}}$ we can then compare $I_{jk_{\max}}$ to a cutoff intensity I_{cutoff} to see if $I_{jk_{\max}} > I_{\text{cutoff}}$. If so, the signal is placed in class $f_{k_{\max}}$.

A maximum of n classes is possible, corresponding to n possible frequencies that can be dominant, and the possible classes are labeled as $\{f_1, f_2, ..., f_k, ..., f_n\}$. The exact frequency list will depend on n, and hence the length of the input set times N, and is determined automatically by the LombScargle function.

After signals that have dominant frequencies are identified, those that do not show a maximum intensity in frequency space above the cutoff intensity, i.e. signals j for which $I_{jk} \leq I_{\text{cutoff}}$ for all k, are checked for sudden signal spikes at any time point, and if they display such spikes, are classified as spike maxima or minima.

For each signal not showing a maximum periodogram intensity, $\tilde{X}_j$, we can calculate the real signal maximum, $\max_j = \mathbf{Max}[\tilde{X}_j]$, and minimum, $\min_j = \mathbf{Min}[\tilde{X}_j]$, from signal intensities across all time points. We can compare these values against cutoffs {Maximum Spike Cutoff$_n$, Minimum Spike Cutoff$_n$} that are provided by the user.

These Spike cutoffs are dependent on the length of a time series, n, and typically would correspond to the 95th or 99th quantile of distributions of maxima and minima of randomly generated signals. These cutoff values are provided to the **TimeSeriesClassification** by the SpikeCutoffs option, for the value for each length n involved in the computation as part of an association for different lengths,

<|..., $n \to$ {Maximum Spike Cutoff$_n$,
    Minimum Spike Cutoff$_n$},
    ...
length $i \to$ {Maximum Spike Cutoff$_i$,
    Minimum Spike Cutoff$_i$}, ...|> .

If a signal of length n, $\tilde{X}_j$, has $\max_j >$ Maximum Spike Cutoff$_n$ it is classified in the "SpikeMax" class, or otherwise if a signal $\tilde{X}_j$ has $\min_j <$ Minimum Spike Cutoff$_n$ it is classified in the "SpikeMin" class. Signals for which the maximum signal intensity is not above the cutoffs are not reported.

This classification method identifies which signals contain significant dominant frequencies compared to the LombScargleCutoff, and of those signals that do not contain dominant frequencies, which contain significant positive or negative spikes, compared to values provided in the SpikeCutoffs associations. Three groups of classes are reported: (1) Significant intensity at various frequencies, (2) No dominant frequencies, but positive spikes present, (3) No dominant frequencies, but negative spikes present.

The default output for this method is an **Association** with outer keys being the classification classes C, where $C \in \{f_1, f_2, \ldots, f_k, \ldots, f_n, \text{SpikeMax}, \text{SpikeMin}\}$, inner keys being the class members, signals X_j, and each class member value being a list of {{periodogram intensity list for signalX_j}, {input data list for X_j}} .

11.4.2 Classification Simulation Example

On many occasions, data might be missing, or the sampling might be uneven. In the case of even sampling a missing time series creates uneven spacing as well. Let us create some random time series, some that have positive or negative spikes at a time point ("SpikePositive/SpikeNegative), some that are a linear ramp up or down set of intensities (LinearPositive/LinearNegative, or sinusoidal (Sinusoidal1/2/3):

In[89]:= **SeedRandom**[7777];
    classificationExample1 =

11.4 Unevenly Sampled Time Series Classification Examples

```
Normalize[#]&/@
Association[
 Join[Table["SpikePositive_"<>ToString[i]->
   UnitVector[20,1]+RandomReal[{-.1,.1},20],
   {i,1,10}],
  Table["SpikePositive_"<>ToString[i]->
   UnitVector[20,3]+RandomReal[{-.1,.1},20],
   {i,11,20}],
  Table["SpikeNegative_"<>ToString[i]->
   -UnitVector[20,1]+RandomReal[{-.1,.1},20],
   {i,1,10}],
  Table["SpikeNegative_"<>ToString[i]->
   -UnitVector[20,3]+RandomReal[{-.1,.1},20],
   {i,11,20}],
  Table["LinearPositive_"<>ToString[i]->
   Range[20]+RandomReal[{-.1,.1},20],
   {i,1,10}],
  Table["LinearPositive_"<>ToString[i]->
   0.2Range[20]+RandomReal[{-.1,.1},20],
   {i,11,20}],
  Table["LinearNegative_"<>ToString[i]->
   -0.1Range[20]+RandomReal[{-.1,.1},20],
   {i,1,10}],
  Table["LinearNegative_"<>ToString[i]->
   -0.2Range[20]+RandomReal[{-.1,.1},20],
   {i,11,20}],
  Table["Sinusoidal1_"<>ToString[i]->
   Cos[2Pi*0.65Range[20]]+RandomReal[{-.1,.1},20],
   {i,1,10}],
  Table["Sinusoidal2_"<>ToString[i]->
   5Cos[2Pi*0.3Range[20]]+RandomReal[{-.1,.1},20],
   {i,11,20}],
  Table["Sinusoidal3_"<>ToString[i]->
   5Cos[2Pi*0.15*Range[20]]+RandomReal[{-.1,.1},20],
   {i,21,30}]]];
```

We also want to simulate missing data. Here we take the above series, and at random points we introduce missing points. This is done by using **Nest** to apply the removal of numbers randomly:

```
In[90]:= classificationExampleMissing=
    (Query[All,
     Nest[
      ReplacePart[#,RandomInteger[{1,20}]->
       Missing[]]&,#,RandomInteger[{1,2}]]&]@
     classificationExample1);
```

Let us take a short look at the series:

```
In[91]:= classificationExampleMissing//Short[#,6]&
Out[91]//Short= <|SpikePositive_1->{0.966088,-0.0037329,
        Missing[], 0.0610785,0.0398251,<<10>>,
        0.0731667, -0.0481196,-0.0272324,
        0.0690434,0.0282179}, <<108>>,
        Sinusoidal3_30->{<<1>>}|>
```

We set the time points to be 1 through 20:

In[92]:= setTimesExample=**Range**[20];

If we are looking for non-random trends in the data, we can use a simulated set of 10^5 random noisy signals to compare against, as a control background. We use the backgoundExample data below as representative of our null distribution of time series. We also use **SeriesApplier** which can apply a function to a set of data by masking _Missing values:

In[93]:= backgroundExample=
    SeriesApplier[**Normalize**,#]&@
    (**Query**[**All**,
      Nest[**ReplacePart**[#,**RandomInteger**[{1,20}]→
        Missing[]]&,#,**RandomInteger**[{1,2}]]&]@
      RandomReal[{−.1,.1},{10^5,20}]);

We can use the background to determine the cutoffs we would like to assess as significant for our "LombScargle" method that we are using. We use the **QuantileEstimator** function that will generate the 0.95th quantile value cutoffs for different lags, for the method of choice which is here "LombScargle", which is also the default method.

In[94]:= bootstrapQ95=**QuantileEstimator**[backgroundExample,
    setTimesExample]
Out[94]= 0.886511

The same function can estimate the 0.95 quantile cutoffs for positive or negative spikes:

In[95]:= bootstrapQ95Spikes=
    QuantileEstimator[backgroundExample,
    setTimesExample,**Method**→ "Spikes",
    QuantileValue→ .95]
Out[95]= <|19→{0.434204,−0.433724},18→{0.445524,−0.445848}|>

We can now put it all together to classify the data:

In[96]:= classificationLombScargle1=
    TimeSeriesClassification[
    classificationExampleMissing,
    setTimesExample,
    LombScargleCutoff→ bootstrapQ95,
    SpikeCutoffs→bootstrapQ95Spikes];

    (∗ the line below is printed during the evaluation ∗)
    Method → "LombScargle"

We obtain the classification, and also we can see how many members are in each class.

In[97]:= **Query**[**All**,
    Length[#]−>
      Tally[
        StringSplit[**Keys**[#],"_"][[**All**,1]]]&]@

11.4 Unevenly Sampled Time Series Classification Examples

```
                  classificationLombScargle1
Out[97]= <|SpikeMax  → 20 → {{SpikePositive,20}},
          SpikeMin  → 18 → {{SpikeNegative,18}},
          f1        → 40 → {{LinearPositive,20},{LinearNegative,20}},
          f3        → 10 → {{Sinusoidal3,10}},
          f5        →  6 → {{Sinusoidal2,6}},
          f6        → 11 → {{SpikeNegative,1},{Sinusoidal1,10}}|>
```

We have used a string split to obtain the signal name, which is the first member of the split list. The names correspond to our simulation list from above, and by using **Tally** we can see how well the classification is doing. In terms of misclassification, the majority of the signals are classified correctly. A SpikeNegative signal was misclassified in f6 and additionally, 1 SpikeNegative, and 4 Sinusoidal signal 2 were not classified.

We can now plot the data and see the classes. The Spike Minima, Spike Maxima and f1 classes are plotted together in Fig. 11.19, while the f3, f5 and f6 signals are shown in Fig. 11.20.

From the classification, we notice that the classes have correctly captured the trends in the data. In terms of breaking the signals into frequencies, we also notice that the linear signals of positive and negative slope are classified in the same frequency classification, the f1 frequency, which corresponds to the lowest frequency available. To figure out the sense of the signal we can carry out a two-level classification approach as shown in the next section.

Finally, we can also obtain the frequencies corresponding to the various classes:

```
In[99]:= LombScargle[classificationExampleMissing[[1]],
         setTimesExample,FrequenciesOnly -> True]
Out[99]= <|f1→0.0554017,f2→0.110803,
          f3→0.166205,f4→0.221607,f5→0.277008,
          f6→0.33241,f7→0.387812,f8→0.443213,
          f9→0.498615,f10→0.554017|>
```

Our simulated signals did not contain any frequencies that would match f2 and f4, and hence we did not observe any signals classified in these classes.

11.4.3 Classification RNA Sequencing Data Example

In this section we carry out an example of classification of time series data from the iPOP experiment. Again we will use the Lomb Scargle approach as an illustration given its simplicity in interpretation. We will use the RNA Example data we generated above (rnaExample) in the following sections.

11.4.3.1 Time Series Creation

As we saw above, the samples and day intervals are not directly related. We can use a **KeyMap** to change from the samples to Days as outer keys in the **OmicsObject**:

```
In[98]:= Query[All,
          ListLinePlot[#[[All, 2]],
           PlotRange → Full,
           PlotLegends → PointLegend[Automatic,
             LegendFunction → Panel, LegendMargins → 5]] &] @
        classificationLombScargle1
```

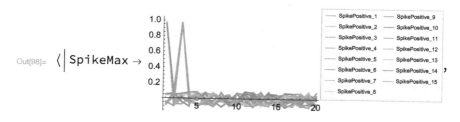

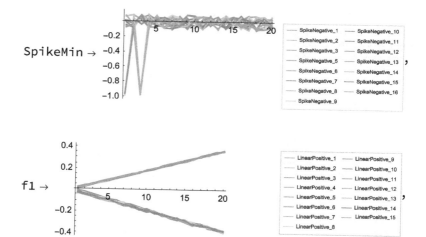

Fig. 11.19 Out[98]= Classification of Spike Maxima, Spike Minima and f1 signals

```
In[100]:= rnaLongitudinal=KeyMap[sampleToDays,rnaExample];
In[101]:= rnaLongitudinal[[1;;3,1;;3]]
Out[101]= <|186→ <|{FAM138A,RNA}→{{0},{OK}},
           {OR4F5,RNA}→{{0},{OK}},
           {LOC729737,RNA}→{{2.73998},{OK}}|>,
         255→ <|{FAM138A,RNA}→{{0},{OK}},
           {OR4F5,RNA}→{{0},{OK}},
           {LOC729737,RNA}→{{0.555218},{OK}}|>,
         289→ <|{FAM138A,RNA}→{{0.00257233},{OK}},
           {OR4F5,RNA}→{{0},{OK}},
```

11.4 Unevenly Sampled Time Series Classification Examples 355

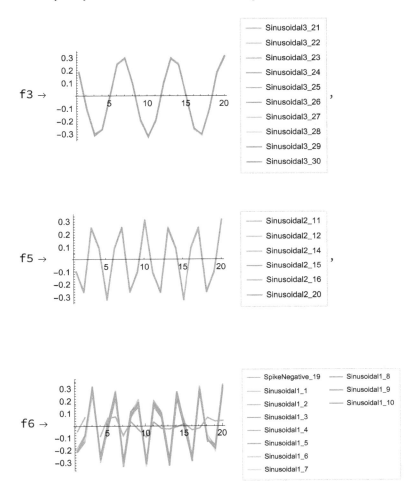

Fig. 11.20 Out[98]= ...continued. Classification of f3, f5 and f6 signals

{LOC729737,RNA} → { { 5.15563 } ,{OK} }|>|>

We use here **QuantileNormalization** to normalize the transcriptome data. Essentially all the distributions are equated across all the samples. The **QuantileNormalization** function can act on an **OmicsObject** and is available in **MathIOmica**. Since we are normalizing the FPKM values which are at location Index 1, Position 1 in each value in the inner association we specify this in the function options:

In[102]:= rnaQuantileNormed=
            QuantileNormalization[rnaLongitudinal,
            ListIndex −>1,ComponentIndex −>1];

We then tag 0 values as **Missing** for this example, using the function **LowValueTag**:

In[103]:= rnaZeroTagged=**LowValueTag**[rnaQuantileNormed,0];

We can compare the first three values to before and see that this worked:

In[104]:= rnaZeroTagged[[1;;3,1;;3]]
Out[104]= <|186→ <|{FAM138A,RNA}→{{**Missing**[]},{OK}},
   {OR4F5,RNA}→{{**Missing**[]},{OK}},
   {LOC729737,RNA}→{{2.2946},{OK}}|>,
   255→ <|{FAM138A,RNA}→{{0.0000155736},{OK}},
   {OR4F5,RNA}→{{0.0000155736},{OK}},
   {LOC729737,RNA}→{{0.494723},{OK}}|>,
   289→ <|{FAM138A,RNA}→{{0.00140424},{OK}},
   {OR4F5,RNA}→{{0.0000175111},{OK}},
   {LOC729737,RNA}→{{4.67694},{OK}}|>|>

Next, we set all FPKM values that are smaller than 1 to 1 using again **LowValueTag** and specifying a value replacement. The main assumption here is that values less than 1 are essentially noise.

In[105]:= rnaNoiseAdjusted=**LowValueTag**[rnaZeroTagged,1,
   ValueReplacement-> 1.];

In[106]:= rnaNoiseAdjusted[[1;;3,1;;3]]
Out[106]= <|186→ <|{FAM138A,RNA}→{{**Missing**[]},{OK}},
   {OR4F5,RNA}→{{**Missing**[]},{OK}},
   {LOC729737,RNA}→{{2.2946},{OK}}|>,
   255→ <|{FAM138A,RNA}→{{1.},{OK}},
   {OR4F5,RNA}→{{1.},{OK}},
   {LOC729737,RNA}→{{1.},{OK}}|>,
   289→ <|{FAM138A,RNA}→{{1.},{OK}},
   {OR4F5,RNA}→{{1.},{OK}},
   {LOC729737,RNA}→{{4.67694},{OK}}|>|>

We now want to keep only data if the reference healthy point "255" is not missing for which at least 3/4 points available. Note that we are using 255 days as a reference, not 186 which is the start of the data. We can do this using the FilterMissing function, which also generates information regarding the **Missing** data, Fig. 11.21.

The Histogram and Pie Chart in Fig. 11.21 give us a visual representation of the counts. The association beneath the histogram also has the number of missing points to number of counts as an association:

<|0→18427,1→6841|>.

We see that close to one third of the data is missing a point only. Of course, we have generated the **Missing** tag above, and different tagging options would change the distribution of **Missing** data in our dataset. In other circumstances, there are inherent missing data, for example mass features in mass spectrometry proteomics or metabolomics, missed samples in clinical trials, experimental failures, etc.

We can extract the times for the filtered RNA data using **TimeExtractor** - note that these are what we expect since we had set the timepoints above. The times are in increasing order:

11.4 Unevenly Sampled Time Series Classification Examples

In[107]:= `rnaFiltered = FilterMissing[rnaNoiseAdjusted, 3/4, Reference → "255"];`

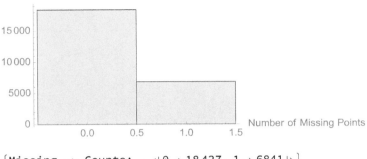

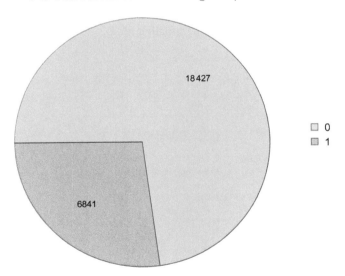

Fig. 11.21 Out[107]= Filtering of **Missing** data

In[108]:= timesRNAiPOP=**TimeExtractor**[rnaFiltered]
Out[108]= {186,255,289,290,292,294,297,
301,307,311,322,329,369,380,400}

358   11 Time Series Analysis

We can then use **CreateTimeSeries**, which pools information for each gene (inner **Key**) across the outer associations corresponding to each time to give a series of values:

In[109]:= timeSeriesRNA=**CreateTimeSeries**[rnaFiltered];

Some values are **Missing**[] and some are constant. We will deal with the constants further below.

In[110]:= timeSeriesRNA[[1;;5]]
Out[110]= <|{FAM138A,RNA}→{**Missing**[],1.,1.,1.,
            1.,1.,1.,1.,1.,1.,1.,1.,1.,1.,1.},
         {OR4F5,RNA}→{**Missing**[],1.,1.,1.,1.,
            1.,1.,1.,1.,1.,1.,1.,1.,1.,1.},
         {LOC729737,RNA}→{2.2946,1.,4.67694,4.48131,
            4.95507,1.,1.25726,2.14767,1.93219,1.,
            2.58217,2.31301,4.10284,3.80929,1.45471},
         {DDX11L1,RNA}→{5.91665,4.32081,3.19599,3.64164,
            2.7327,2.13461,2.17168,3.23429,1.89576,3.0267,
            4.34004,7.27001,2.01132,9.27701,7.54415},
         {WASH7P,RNA}→{10.8594,5.61191,11.1933,
            6.24138,9.24944,5.35752,4.95971,
            8.33658,7.26071,4.63689,9.41982,
            8.54148,11.0612,10.1701,8.13348}|>

We implement a logarithmic transformation on the data, using **SeriesApplier** to apply Log:

In[111]:= ?**SeriesApplier**
         SeriesApplier[function,data] applies a given function
            to data, an association of lists, implementing
            masking for **Missing** values.

In[112]:= timeSeriesRNALog=**SeriesApplier**[Log,timeSeriesRNA];

In[113]:= timeSeriesRNALog[[1;;5]]
Out[113]= <|{FAM138A,RNA}→{**Missing**[],0.,0.,0.,
            0.,0.,0.,0.,0.,0.,0.,0.,0.,0.,0.},
         {OR4F5,RNA}→{**Missing**[],0.,0.,0.,0.,
            0.,0.,0.,0.,0.,0.,0.,0.,0.,0.},
         {LOC729737,RNA}→{0.830556,0.,1.54264,1.49992,
            1.60041,0.,0.228935,0.764385,0.658653,0.,
            0.94863,0.838548,1.41168,1.33744,0.374807},
         {DDX11L1,RNA}→{1.77777,1.46344,1.1619,
            1.29243,1.00529,0.758282,0.775501,
            1.17381,0.639619,1.10747,1.46788,1.98376,
            0.698792,2.22754,2.02077},
         {WASH7P,RNA}→{2.38503,1.72489,2.41532,
            1.8312,2.22456,1.6785,1.60135,2.12065,1.98248,
            1.53404,2.24282,2.14493,2.40345,2.31945,2.09599}|>

We compare every value in each series to the healthy "255" time point, which is the second element in each series, using **SeriesInternalCompare**:

In[114]:= rnaCompared=**SeriesInternalCompare**[timeSeriesRNALog,

11.4 Unevenly Sampled Time Series Classification Examples   359

                     ComparisonIndex −>2];
In[115]:= rnaCompared[[1;;5]]
Out[115]= <|{FAM138A,RNA}→{**Missing**[],0.,0.,0.,
             0.,0.,0.,0.,0.,0.,0.,0.,0.,0.,0.},
           {OR4F5,RNA}→{**Missing**[],0.,0.,0.,0.,
             0.,0.,0.,0.,0.,0.,0.,0.,0.,0.},
           {LOC729737,RNA}→{0.830556,0.,1.54264,1.49992,
             1.60041,0.,0.228935,0.764385,0.658653,0.,
             0.94863,0.838548,1.41168,1.33744,0.374807},
           {DDX11L1,RNA}→{0.314326,0.,−0.301545,
             −0.171011,−0.458154,−0.705162,−0.687943,
             −0.289634,−0.823824,−0.35597,0.00444068,
             0.520314,−0.764652,0.764095,0.557328},
           {WASH7P,RNA}→{0.660143,0.,0.690425,
             0.10631,0.499672,−0.0463898,−0.123544,
             0.395761,0.257587,−0.190848,0.517924,
             0.420043,0.678555,0.594558,0.371097}|>

     Next, we normalize each series to unity across the time points:

In[116]:= normedRNACompared=**SeriesApplier**[**Normalize**,rnaCompared];
In[117]:= normedRNACompared[[1;;5]][[1;;5]]
Out[117]= <|{FAM138A,RNA}−>{**Missing**[],0.,0.,0.,
             0.,0.,0.,0.,0.,0.,0.,0.,0.,0.,0.},
          {OR4F5,RNA}−>{**Missing**[],0.,0.,0.,0.,
             0.,0.,0.,0.,0.,0.,0.,0.,0.,0.},
          {LOC729737,RNA} −>{0.218293,0.,0.40545,0.39422,
             0.420632,0.,0.0601705,0.200902,0.173112,0.,
             0.249326,0.220394,0.371029,0.351517,0.0985097},
          {DDX11L1,RNA} −>{0.156411,0.,−0.150051,
             −0.0850959,−0.22798,−0.350893,−0.342324,
             −0.144124,−0.40994,−0.177133,0.00220971,
             0.258911,−0.380495,0.380218,0.27733},
          {WASH7P,RNA} −>{0.391269,0.,0.409217,
             0.0630104,0.296157,−0.0274954,−0.0732251,
             0.234569,0.152672,−0.113116,0.306975,
             0.248961,0.402181,0.352396,0.21995}|>

     Finally, we remove constant series, as we are interested in temporal patterns that are not constant:

In[118]:= rnaFinalTimeSeries=
            ConstantSeriesClean[normedRNACompared];

        (∗ the lines below are printed during the evaluation ∗)
        Removed series and returning filtered list. If you
          would like a list of removed keys run the command
          ConstantSeriesClean[data,ReturnDropped → True].

     We can do a visual inspection and see that the constant series are now removed:

In[119]:= rnaFinalTimeSeries[[1;;3]]
Out[119]= <|{LOC729737,RNA}→
            {0.218293,0.,0.40545,0.39422,0.420632,0.,
             0.0601705,0.200902,0.173112,0.,0.249326,

                    0.220394,0.371029,0.351517,0.0985097},
       {DDX11L1,RNA}→{0.156411,0.,−0.150051,
         −0.0850959,−0.22798,−0.350893,−0.342324,
         −0.144124,−0.40994,−0.177133,0.00220971,
            0.258911,−0.380495,0.380218,0.27733},
       {WASH7P,RNA}→{0.391269,0.,0.409217,
           0.0630104,0.296157,−0.0274954,−0.0732251,
           0.234569,0.152672,−0.113116,0.306975,
           0.248961,0.402181,0.352396,0.21995}|>

11.4.3.2 Random Distribution Generation

We would next like to identify trends in the data. To do this we first want to create a resampled null distribution for the transcriptome dataset, as we did in the previous section for the simulation. We do this prior to classification and clustering so that we can compare against random time series from the same type of data without any assumptions regarding the underlying distribution.

We repeat the steps in the processing of the original data described above, instead now using a resampled set of measurements. The resampling is done with replacement in the first step. For brevity all nine steps are listed together below. Additionally, note that we seed the pseudorandom generator for reproducibility for this manuscript.

```
In[120]:= (*Bootstrap generation of 100,000
            series using resampling*)
          SeedRandom[12345]
          rnaBootstrap=BootstrapGeneral[rnaLongitudinal,
            100000];
          (*1. Quantile Normalization*)
          rnaBootstrapQuantileNormed=
            QuantileNormalization[rnaBootstrap,ListIndex−> 1,
              ComponentIndex−> 1];
          (*2. Zero values tagged as Missing*)
          rnaBootstrapZeroTagged=
            LowValueTag[rnaBootstrapQuantileNormed,0];
          (*3. Values less than one replaced*)
          rnaBootstrapNoiseAdjusted=
            LowValueTag[rnaBootstrapZeroTagged,1,
              ValueReplacement−> 1];
          (*4. Missing data filtered out*)
          rnaBootstrapFiltered=
            FilterMissing[rnaBootstrapNoiseAdjusted,3/4,
              Reference−> "255",ShowPlots−> False];
          (*5. Create time series*)
          timeSeriesBootstrapRNA=
            CreateTimeSeries[rnaBootstrapFiltered];
          (*6. Apply Log to values*)
          timeSeriesBootstrapRNALog=
            SeriesApplier[Log,timeSeriesBootstrapRNA];
          (*7. Compare against reference in position 2*)
          rnaBootstrapCompared=
```

11.4 Unevenly Sampled Time Series Classification Examples

```
SeriesInternalCompare[timeSeriesBootstrapRNALog,
    ComparisonIndex ->2];
(*8. Normalize to unity*)
normedBootstrapRNACompared=
    SeriesApplier[Normalize, rnaBootstrapCompared];
(*9. Check for and remove constant series*)
rnaBootstrapFinalTimeSeries=
    ConstantSeriesClean[normedBootstrapRNACompared];

(* the lines below are printed during the evaluation *)
Removed series and returning filtered list. If you
    would like a list of removed keys run the command
    ConstantSeriesClean[data,ReturnDropped -> True].
```

11.4.3.3 Classification and Clustering

Now we have generated the random distribution to use for our null statistic distribution creation. We can proceed with classification using the LombScargle method as in the previous section. Before we classify our transcriptome data, we estimate for the "LombScargle" **Method** a 0.95 quantile cutoff from the bootstrap transcriptome data, using **QuantileEstimator**. Please note that depending on your hardware this can be a lengthy computation:

In[131]:= q95RNA=**QuantileEstimator**[
    rnaBootstrapFinalTimeSeries ,timesRNAiPOP]
Out[131]= 0.860294

Next, we estimate the "Spikes" 0.95 quantile cutoff from the bootstrap transcriptome data. We can again use **QuantileEstimator**, and specify "Spikes" as a **Method** option:

In[132]:= q95RNASpikes=
   QuantileEstimator[rnaBootstrapFinalTimeSeries,
   timesRNAiPOP,**Method**-> "Spikes"]
Out[132]= <|15->{0.856796,−0.337722},
   14->{0.884214,−0.348537}|>

Now we can classify the transcriptome time series data based on these cutoffs:

In[133]:= rnaClassification=
   TimeSeriesClassification[rnaFinalTimeSeries,
   timesRNAiPOP, LombScargleCutoff-> q95RNA,
   SpikeCutoffs->q95RNASpikes];

(* the line below is printed during the evaluation *)
Method -> "LombScargle"

As before, the option FrequenciesOnly allows us to obtain the frequencies over which the classification is taking place, corresponding to the class labels f_i:

In[134]:= **LombScargle**[rnaFinalTimeSeries[[1]],timesRNAiPOP,
   FrequenciesOnly-> **True**]

Out[134]= <|f1→0.00500668,f2→0.0104306,
          f3→0.0158545,f4→0.0212784,f5→0.0267023,
          f6→0.0321262,f7→0.0375501|>

The classification has multiple components in the various classes:

In[135]:= **Query[All, Length]** @rnaClassification
Out[135]= <|SpikeMax→829,SpikeMin→5962,f1→116,f2→3,
          f3→30,f4→128,f5→35,f6→13,f7→61|>

We now cluster our RNA-sequencing data. The **TimeSeriesClusters** function uses two tiers of hierarchical clustering of data performed sequentially using two classification vectors for each signal:

{{classification vector$_1$}, {classification vector$_2$}}.

The classification vectors are generated as the output from **TimeSeriesClassification** (or can be created as needed). We can see this from our classification if we look at a few examples:

In[136]:= rnaClassification[[1;;2,1;;2]]
Out[136]= <|SpikeMax→
          <|{ATAD3C,RNA}→{{0.0855374,0.204135,0.219303,
              0.378496,0.5849,0.346012,0.545735},
              {0.,0.,0.,0.,0.,0.,0.,0.,0.997114,
              0.,0.,0.,0.,0.075919,0.}},
          {MMP23B,RNA}→{{0.237345,0.471368,0.335256,
              0.450949,0.568953,0.201913,0.203105},
              {0.,0.,0.,0.,0.,0.,0.,0.,0.,
              1.,0.,0.,0.,0.}}|>, SpikeMin→
          <|{DDX11L1,RNA}→{{0.447167,0.069665,0.842392,
              0.173482,0.0417282,0.112542,0.202636},
              {0.156411,0.,−0.150051,−0.0850959,
              −0.22798,−0.350893,−0.342324,−0.144124,
              −0.40994,−0.177133,0.00220971,
              0.258911,−0.380495,0.380218,0.27733}},
          {LINC01128,RNA}→{{0.0905487,0.406082,0.0297171,
              0.578597,0.239233,0.658612,0.0154399},
              {−0.0991061,0.,0.0187721,−0.270964,
              −0.206247,−0.0847502,−0.145098,−0.0170673,
              −0.671463,−0.103483,−0.12124,−0.0569299,
              −0.170415,−0.173748,−0.553712}}|>|>

Similarities at each clustering tier are then computed using in succession from each time series first {classification vector$_1$}, and subsequently {classification vector$_2$} (which corresponds to the {input data time series}).

The result is that we create groups and sub-groups of the data within each class, based on pairwise similarities using the respective classification vectors. The groups and subgroups are labeled as $GiSj$, where Gi is the group i based on the first clustering, and Sj is the corresponding subsequent subgroup for group Gi. For example G3S4 is the subgroup 4 of the 3rd group.

We now cluster the data:

In[137]:= rnaClusters=**TimeSeriesClusters**[rnaClassification];

11.4 Unevenly Sampled Time Series Classification Examples

You may receive messages during the clustering that Agglomerate has detected ties in the data., for example:

Agglomerate::ties: 426 ties have been detected; reordering input may produce a different result.
...
...
General::stop: Further output of Agglomerate::ties will be suppressed during this calculation.

This is often the case in hierarchical clustering, particularly if you are using digital data that can have equal points, and as the message indicates, reordering the data can change the results.

The clusters have the outer keys corresponding to each class:

In[138]:= **Keys** @ rnaClusters
Out[138]= {SpikeMax, SpikeMin, f1, f2, f3, f4, f5, f6, f7}

Internally, other keys are available for each external key value, with various information.

In[139]:= **Query**[1, **Keys**] @rnaClusters
Out[139]= {Cluster, InitialSplitCluster, IntermediateClusters, SubsplitClusters, Data, GroupAssociations}

For example we can get any of the groups and subgroups for class f1. There are 4 subgroupings:

In[140]:= **Query**["f1", "GroupAssociations", **Length**] @rnaClusters
Out[140]= 4

Within each there are different numbers of results.

In[141]:= **Query**["f1", "GroupAssociations", **All**, **Length**]@ rnaClusters
Out[141]= <|G1S1→80, G1S2→27, G2S1→6, G2S2→3|>

Let's look at the second list as an example:

In[142]:= **Query**["f1", "GroupAssociations", "G1S2"] @rnaClusters
Out[142]= {{POC5,RNA}, {EWSR1,RNA}, {RCCD1,RNA},
           {MBLAC1,RNA}, {FAM173A,RNA}, {CCDC12,RNA},
           {SDHAF1,RNA}, {CTU1,RNA}, {CCDC28B,RNA},
           {EMG1,RNA}, {CDK2AP2,RNA}, {COA5,RNA},
           {LOC102288414,RNA}, {DRAP1,RNA}, {THAP7,RNA},
           {TRMT112,RNA}, {TRIM56,RNA}, {ZRSR2,RNA},
           {STAG3L3,RNA}, {STAG3L1,RNA}, {ZNF529,RNA},
           {TNFRSF13B,RNA}, {GEMIN6,RNA}, {C1orf52,RNA},
           {MTERF,RNA}, {RNF122,RNA}, {LINC00921,RNA}}

Similarly this can be done across all classes:

In[143]:= **Query**[**All**, "GroupAssociations", **All**, **Length**] @rnaClusters
Out[143]= <|SpikeMax→
           <|G1S1→360, G1S2→46, G1S3→38, G1S4→8, G1S5→10,
           G1S6→30, G1S7→5, G1S8→194, G1S9→21, G1S10→23,
           G1S11→21, G1S12→7, G1S13→3, G1S14→63|>,

```
              SpikeMin→ <|G1S1→ 5667,G1S2→ 295|>,
              f1→ <|G1S1→ 80,G1S2→ 27,G2S1→ 6,G2S2→ 3|>,
              f2→ <|G1S1→ 2,G2S1→ 1|>,
              f3→ <|G1S1→ 2,G1S2→ 4,
                  G2S1→ 1,G2S2→ 4,G3S1→ 15,G3S2→ 1,G4S1→ 2,G4S2→ 1|>,
              f4→ <|G1S1→ 79,G1S2→ 48,G2S1→ 1|>,
              f5→ <|G1S1→ 1,G1S2→ 2,G1S3→ 4,
                  G2S1→ 16,G2S2→ 5,G2S3→ 7|>,
              f6→ <|G1S1→ 2,G1S2→ 4,G2S1→ 5,G2S2→ 2|>,
              f7→ <|G1S1→ 31,G1S2→ 24,G2S1→ 4,G2S2→ 2|>|>
```

If we want we can also obtain the Cluster information, which has the form

Cluster[$cluster_1$, $cluster_2$, $dissimilarity$, n_1, n_2]

which is used from the HierarchicalClustering package by MathIOmica, and represents a merger of the clusters 1 and 2 with dissimilarity value, and having n_1 and n_2 data respectively.

```
In[144]:= Query["f1","Cluster"] @ rnaClusters //Short[#,5]&
Out[144]//Short= Cluster[Cluster[Cluster[Cluster[<<1>>],
                  Cluster[<<1>>,Cluster[<<1>>],
                  0.268368,10,8],0.302636,21,18],
                  Cluster[Cluster[Cluster[<<1>>],
                  Cluster[<<1>>],<<20>>,47,20],<<3>>,1],
                  <<19>>,39,68],<<3>>,9]
```

11.4.4 Heatmaps and Dendrograms

In the above clustering we can actually request the dendrograms to be returned instead:

```
In[145]:= rnaClustersDendrogramsReturn=
          TimeSeriesClusters[rnaClassification,
          ReturnDendrograms-> True];
```

Now each group's dendrograms are returned if the group has at least two members. For example, in class "f1" and group "G1" we have two clustering dendrograms corresponding to the 2 subgroups generated, Fig. 11.22.

The dendrogram data is arranged by groups:

```
In[147]:= Query[All,Keys] @rnaClustersDendrogramsReturn
Out[147]= <|SpikeMax→ {G1},SpikeMin→ {G1},f1→ {G1,G2},
            f2→ {G1,G2},f3→ {G1,G2,G3,G4},f4→ {G1,G2},
            f5→ {G1,G2},f6→ {G1,G2},f7→ {G1,G2}|>
```

We can obtain in a similar fashion the clustering for each group.

For each class we may instead want to generate a dendrogram/heatmap plot using **MathIOmica**'s **TimeSeriesDendrogramsHeatmaps** (or **TimeSeriesDendrogramHeatmap** if you have only a single classification).

11.4 Unevenly Sampled Time Series Classification Examples

In[146]:= `Query["f1", "G1"] @ rnaClustersDendrogramsReturn`

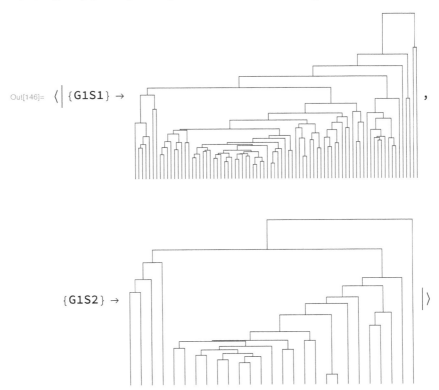

Out[146]= ⟨| {G1S1} → [dendrogram], {G1S2} → [dendrogram] |⟩

Fig. 11.22 Out[146]= Dendrograms representing the clustering within subgroups for group "G1"

In[148]:= `?TimeSeriesDendrogramsHeatmaps`
　　　　 TimeSeriesDendrogramsHeatmaps[data]
　　　　　　generates a dendrogram and heatmap plots for
　　　　　　all classified classes of time series data clusters.

Since we have multiple classes we can use the function directly:

In[149]:= **TimeSeriesDendrogramsHeatmaps[rnaClusters]**

The output is rather large and is shown in Fig. 11.23 for both Spike classes and Fig. 11.24 for f1–f7.

For each class we get a plot titled by the class name. On the left there is a Dendrogram corresponding to the outer classification. The heatmap itself has two bars that correspond to the groupings and subgroupings within the class, and the legend is shown to the right. The G, S, columns represent the groupings and subgroupings as discussed above, with the arrow pointing to the number of components in each group as also stated in the title. The function has multiple options that allow you to change colors and other considerations.

In[149]:= `TimeSeriesDendrogramsHeatmaps[rnaClusters,`
`  FunctionOptions →`
`  {HorizontalAxisName → "Timepoints"}]`

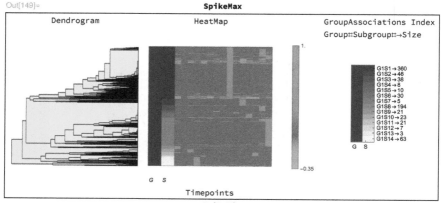

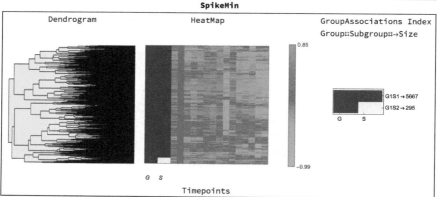

Fig. 11.23 Out[149]= Heatmaps and dendrograms representing the Spike Maxima and Spike Minima classes

In terms of results, the simplified analysis Fig. 11.23 shows the Spike Maxima and Spike Minima classes. These are characterized by a spike in both cases coinciding with the onset of high glucose in the study subject.

Figure 11.24 shows the f1 changes, that appear as linear increases and decreases following a viral infection, and accelerated in the high glucose regime. f2-f7 are shown schmatically in the figure as well.

11.4 Unevenly Sampled Time Series Classification Examples

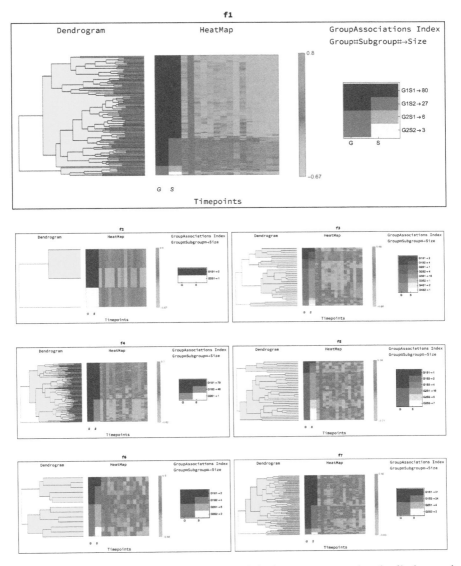

Fig. 11.24 Out[149]= ... continued Heatmaps and dendrograms representing the f1 class, and minuture representations of the f2–f7 remaining classes

11.4.4.1 Annotation and Over Representation Analysis

For each class and subgroup identified in the clustering we will want to carry out over representation analysis for Gene Ontology [2] terms, using **MathIOmica**'s **GOAnalysis** function.

In[150]:= ?GOAnalysis
   GOAnalysis[data] calculates input data
      over-representation analysis for Gene
      Ontology (GO) categories.

The GOAnalysis function uses GO Consortium ontologies and is optimized for human data annotated using UniProt [19, 20] accessions. It has an option to specify the input identifiers, that also is by default set to accept Gene Symbols as well. The function can take results from the **TimeSeriesClusters** directly. Please note that the function might take some time to compute depending on the length of your lists and your hardware configuration.

In[151]:= goAnalysisRNA=**GOAnalysis**[rnaClusters,
   OntologyLengthFilter-> 3,ReportFilter-> 3];

We have selected here the option OntologyLengthFilter->3 to exclude GO terms with fewer than three members. Also the ReportFilter -> 3 option, is another cutoff for membership of our inputs in ontologies in selecting which terms/categories to report back, and will report GO terms with 3 or more members from our test data.

The output has enrichment analysis for each class and group subgrouping:

In[152]:= **Query**[**All**, **All**, **Length**]@goAnalysisRNA
Out[152]= <|SpikeMax→
   <|G1S1→40,G1S2→10,G1S3→1,G1S4→5,G1S5→1,
   G1S6→3,G1S7→1,G1S8→189,G1S9→2,G1S10→1,
   G1S11→1,G1S12→1,G1S13→0,G1S14→17|>,
   SpikeMin→ <|G1S1→2931,G1S2→110|>,
   f1→ <|G1S1→41,G1S2→4,G2S1→0,G2S2→0|>,
   f2→ <|G1S1→0,G2S1→0|>,
   f3→ <|G1S1→0,G1S2→0,G2S1→0,G2S2→0,
   G3S1→3,G3S2→0,G4S1→0,G4S2→0|>,
   f4→ <|G1S1→6,G1S2→24,G2S1→0|>,
   f5→ <|G1S1→0,G1S2→0,G1S3→1,G2S1→2,G2S2→0,G2S3→0|>,
   f6→ <|G1S1→0,G1S2→1,G2S1→0,G2S2→0|>,
   f7→ <|G1S1→10,G1S2→8,G2S1→0,G2S2→0|>|>

We can go ahead and view the results for any of the groups in our clustering corresponding to any trend that we are interested in from our heatmaps as well. For example, we may want class f1, group 1, subgroup 1 (G1S1):

In[153]:= **Query**["f1","G1S1",1;;2]@goAnalysisRNA
Out[153]= <|GO:0005515→{{3.34254×10^{-17},2.7576×10^{-14},**True**},
   {77,8801,47241,48},
   {{protein binding,molecular_function},
   {{{PADI4,RNA}},{{USP25,RNA}},
   {{ZNF207,RNA}},{{ADAM9,RNA}},
   {{HACE1,RNA}},{{AGO2,RNA}},{{JAK1,RNA}},
   {{MBP,RNA}},{{KLF3,RNA}},{{FLI1,RNA}},
   {{PRKDC,RNA}},{{IPO5,RNA}},{{FOCAD,RNA}},
   {{ELK3,RNA}},{{EFCAB4B,RNA}},
   {{ARHGEF6,RNA}},{{GTF3C4,RNA}},
   {{MAPK1,RNA}},{{ANTXR2,RNA}},

11.4 Unevenly Sampled Time Series Classification Examples

```
            {{GSK3B,RNA}},{{AHCYL2,RNA}},
            {{ERAP1,RNA}},{{SAMHD1,RNA}},
            {{DDX3X,RNA}},{{SSH1,RNA}},
            {{DNAJC13,RNA}},{{NDUFS1,RNA}},
            {{AGO1,RNA}},{{FAM168A,RNA}},{{CLPX,RNA}},
            {{ACLY,RNA}},{{SYNJ2,RNA}},{{INPP4B,RNA}},
            {{PRKCA,RNA}},{{PTPLB,RNA}},{{PRKAR2A,RNA}},
            {{CASK,RNA}},{{RUNX1,RNA}},{{PTGER4,RNA}},
            {{EHD4,RNA}},{{PSME3,RNA}},{{WBP1L,RNA}},
            {{ADNP2,RNA}},{{AAGAB,RNA}},{{PDK3,RNA}},
            {{GTF3C3,RNA}},{{HSPA14,RNA}},
            {{TBC1D22B,RNA}}}}},
```
GO:0005829→{{1.06226×10^{-7},0.0000438182,**True**},
   {77,3476,47241,21},
   {{cytosol,cellular_component},
    {{{PADI4,RNA}},{{AGO2,RNA}},{{JAK1,RNA}},
     {{CEP78,RNA}},{{PRKDC,RNA}},{{ARHGEF6,RNA}},
     {{MAPK1,RNA}},{{GSK3B,RNA}},{{AHCYL2,RNA}},
     {{ERAP1,RNA}},{{AGO1,RNA}},{{ACLY,RNA}},
     {{SYNJ2,RNA}},{{INPP4B,RNA}},{{PRKCA,RNA}},
     {{PRKAR2A,RNA}},{{CASK,RNA}},{{PSME3,RNA}},
     {{IARS2,RNA}},{{AAGAB,RNA}},{{HSPA14,RNA}}}}}|>

As described in Chap. 3, the output is an association which has the form:

<|GO:Term$_j$]→{{p–value$_j$,
   multiple hypothesis adjusted p–value$_j$,
   True/False for statistical significance},
   {{number of members in group being tested,
     number of successes for Term$_j$ in population,
     total number of members in population,
     number of members (or more) in current group
     being tested associated to Term$_j$},
   {{GO Term$_j$ description,
     ontology category for Term_j},
    {input IDs associated to Term$_j$]}}}}|>

MathIOmica lets us export the reports as Excel spreadsheets,

In[154]:= EnrichmentReportExport[goAnalysisRNA,
         OutputDirectory -> $UserDocumentsDirectory,
         AppendString-> "GOAnalysisRNA"];

The reports are written into your $UserDocumentsDirectory, which you can evaluate:

In[155]:= $UserDocumentsDirectory
Out[155]= /Users/user/Documents

An Excel spreadsheet is generated for each class separately, named after the class key (and by default a string with the current date is appended as well). Within each spreadsheet, sheets are created for and named after each group in that class containing the enrichment output for that Group.

We can also carry out a KEGG: Kyoto Encyclopedia of Genes and Genomes [9] pathway analysis for all the classes and groups/subgroups. We should note here that

KEGG requires licensing if used for commercial purposes (please check the licenses prior to use).

We will use the **KEGGAnalysis** function in **MathIOmica**, and only report terms for which there are at least 2 members.

In[156]:= keggAnalysisRNA=**KEGGAnalysis**[rnaClusters,
 ReportFilter-> 2];

In[157]:= **Query**[**All**, **All**, **Length**]@keggAnalysisRNA
Out[157]= <|SpikeMax→
 <|G1S1→0,G1S2→0,G1S3→1,G1S4→0,G1S5→0,
 G1S6→0,G1S7→0,G1S8→5,G1S9→0,G1S10→0,
 G1S11→0,G1S12→0,G1S13→0,G1S14→0|>,
 SpikeMin→ <|G1S1→145,G1S2→8|>,
 f1→ <|G1S1→1,G1S2→0,G2S1→0,G2S2→0|>,
 f2→ <|G1S1→0,G2S1→0|>,
 f3→ <|G1S1→0,G1S2→0,G2S1→0,G2S2→0,
 G3S1→0,G3S2→0,G4S1→0,G4S2→0|>,
 f4→ <|G1S1→0,G1S2→0,G2S1→0|>,
 f5→ <|G1S1→0,G1S2→0,G1S3→0,
 G2S1→0,G2S2→0,G2S3→0|>,
 f6→ <|G1S1→0,G1S2→0,G2S1→0,G2S2→0|>,
 f7→ <|G1S1→5,G1S2→1,G2S1→0,G2S2→0|>|>

Many classes have no pathway enrichments. Please note that not all genes are annotated in terms of pathways, and the enrichment analysis is very dependent on the actual set of genes with existing annotations in pathways.

We can view results for any of the groups (and also check out the behavior using the heatmaps generated in the previous section). Let us list the first 10 pathways in the "SpikeMin" class:

In[158]:= **Query**["SpikeMin","G1S1",1;;10,3,1]@
 keggAnalysisRNA
Out[158]= <|path:hsa04142→Lysosome − Homo sapiens (human),
 path:hsa04120→Ubiquitin mediated
 proteolysis − Homo sapiens (human),
 path:hsa04660→T cell receptor signaling
 pathway − Homo sapiens (human),
 path:hsa04662→B cell receptor signaling
 pathway − Homo sapiens (human),
 path:hsa04210→Apoptosis − Homo sapiens (human),
 path:hsa01100→
 Metabolic pathways − Homo sapiens (human),
 path:hsa05169→Epstein−Barr virus
 infection − Homo sapiens (human),
 path:hsa05161→Hepatitis B − Homo sapiens (human),
 path:hsa01200→
 Carbon metabolism − Homo sapiens (human),
 path:hsa04722→Neurotrophin signaling
 pathway − Homo sapiens (human)|>

We can also get how many members are in each of these patways:

11.4 Unevenly Sampled Time Series Classification Examples

```
In[159]:= Query["SpikeMin","G1S1",1;;10,3,2/*Length]@
          keggAnalysisRNA
Out[159]= <|path:hsa04142→81,path:hsa04120→82,
          path:hsa04660→68,path:hsa04662→51,
          path:hsa04210→79,path:hsa01100→423,
          path:hsa05169→100,path:hsa05161→78,
          path:hsa01200→65,path:hsa04722→66|>
```

Let us look at the third pathway:

```
In[160]:= Query["SpikeMin","G1S1",{3}]@keggAnalysisRNA
Out[160]= <|path:hsa04660→
          {{1.8551×10⁻¹⁷,1.83655×10⁻¹⁵,True},
          {1810,<<3>>},{T cell receptor signaling
              pathway - Homo sapiens (human),
          {{{GRAP2,RNA}},{{NCK2,RNA}},<<64>>,{{NFKBIA,RNA}},{{JUN,
  RNA}}}}|>
```

We can extract the members. The full list is in the third list second element, and from all sublists we extract the first part which is the gene symbol:

```
In[161]:= pathwaymembers=
          Query["SpikeMin","G1S1",3,3,2,All,1,1]@
          keggAnalysisRNA;

In[162]:= pathwaymembers//Short
Out[162]//Short= {GRAP2,NCK2,ITK,CD40LG,<<61>>,CD8A,NFKBIA,JUN}
```

We can also obtain a link to a KEGG pathway of interest:

```
In[163]:=pathKey=
          First @
          Keys @ Query["SpikeMin","G1S1",{3}]@
          keggAnalysisRNA
Out[163]= path:hsa04660

In[164]:= KEGGPathwayVisual[pathKey]
Out[164]= <|Pathway→path:hsa04660,Results→
          {http://www.kegg.jp/kegg-bin/show_pathway?map=
              hsa04660}|>
```

And we can highlight the genes in the results in the pathway if we want. To see the image, just follow the link produced in the "Results":

```
In[165]:= keggVisual=KEGGPathwayVisual[pathKey,MemberSet->
          pathwaymembers];

In[166]:= keggVisual//Short[#,10]&
Out[166]//Short= <|Pathway→path:hsa04660,Results→
  {http://www.kegg.jp/kegg-bin/show_pathway?map=hsa04660&multi_query=
      hsa%3A9402+%2380b2ff%2C%23000000%0D%0Ahsa%3A8440+%2380b2ff%2C
      %23000000%0D%0Ahsa%3A3702+%2380b2ff%2C%23000000%0D%0Ah ...
      915+%2380b2ff%2C%23000000%0D%0Ahsa%3A387+%2380b2ff%2C%23000000%0D%0
      Ahsa%3A925+%2380b2ff%2C%23000000%0D%0Ahsa%3A4792+%2380b2ff%2C
      %23000000%0D%0Ahsa%3A3725+%2380b2ff%2C%23000000%0D%0A}|>
```

In[167]:= **SystemOpen** @@ **Query**["Results"] @ keggVisual

The figure can also be downloaded directly (assuming you have the right permissions - not shown here.):

In[168]:= **KEGGPathwayVisual**["path:hsa04668",
ResultsFormat-> "Figure",
MemberSet->pathwaymembers]

Additional information with the different options for the enrichment analysis and visualization is available in **MathIOmica**'s manual.

References

1. Akaike, H.: Information theory and an extension of the maximum likelihood principle. In: 2nd Inter. Symp. on Information Theory, Akademiai Kidao, 1973 (1973)
2. Ashburner, M., Ball, C.A., Blake, J.A., Botstein, D., Butler, H., Cherry, J.M., Davis, A.P., Dolinski, K., Dwight, S.S., Eppig, J.T.: Gene ontology: tool for the unification of biology. Nat. Genet. **25**(1), 25–29 (2000)
3. Bar-Joseph, Z., Gitter, A., Simon, I.: Studying and modelling dynamic biological processes using time-series gene expression data. Nat. Rev. Genet. **13**(8), 552–64 (2012)
4. Box, G.E.P., Pierce, D.A.: Distribution of residual autocorrelations in autoregressive-integrated moving average time series models. J. Am. Stat. Assoc. **65**(332), 1509–1526 (1970). https://doi.org/10.1080/01621459.1970.10481180
5. Bretthorst, G.L.: Generalizing the lomb-scargle periodogram. pp. 241–245. IOP INSTITUTE OF PHYSICS PUBLISHING LTD
6. Brockwell, P.J., Davis, R.A.: Time Series: Theory and Methods. Springer Series in Statistics, 2nd edn. Springer, Berlin, New York (1991)
7. Chatfield, C.: The Analysis of Time Series: an Introduction. CRC press, Boca Raton (2016)
8. Chen, R., Mias, G.I., Li-Pook-Than, J., Jiang, L., Lam, H.Y., Chen, R., Miriami, E., Karczewski, K.J., Hariharan, M., Dewey, F.E., Cheng, Y., Clark, M.J., Im, H., Habegger, L., Balasubramanian, S., O'Huallachain, M., Dudley, J.T., Hillenmeyer, S., Haraksingh, R., Sharon, D., Euskirchen, G., Lacroute, P., Bettinger, K., Boyle, A.P., Kasowski, M., Grubert, F., Seki, S., Garcia, M., Whirl-Carrillo, M., Gallardo, M., Blasco, M.A., Greenberg, P.L., Snyder, P., Klein, T.E., Altman, R.B., Butte, A.J., Ashley, E.A., Gerstein, M., Nadeau, K.C., Tang, H., Snyder, M.: Personal omics profiling reveals dynamic molecular and medical phenotypes. Cell **148**(6), 1293–307 (2012)
9. Kanehisa, M., Goto, S.: Kegg: kyoto encyclopedia of genes and genomes. Nucl. Acids Res. **28**(1), 27–30 (2000)
10. Kirchgässner, G., Wolters, J., Hassler, U.: Introduction to Modern Time Series Analysis. Springer Science & Business Media, Berlin (2012)
11. Ljung, G.M., Box, G.E.P.: On a measure of lack of fit in time series models. Biometrika **65**(2), 297–303 (1978). https://doi.org/10.1093/biomet/65.2.297
12. Lomb, N.: Least-squares frequency analysis of unequally spaced data. Astrophys. Space Sci. **39**(2), 447–462 (1976)
13. Madsen, H.: Time Series Analysis. CRC Press, Boca Raton (2007)
14. Mias, G., Snyder, M.: Personal genomes, quantitative dynamic omics and personalized medicine. Quant. Biol. **1**(1), 71–90 (2013)
15. Mias, G.I., Snyder, M.: Multimodal dynamic profiling of healthy and diseased states for future personalized health care. Clin. Pharmacol. Ther. **93**(1), 29–32 (2013)
16. Mias, G.I., Yusufaly, T., Roushangar, R., Brooks, L.R., Singh, V.V., Christou, C.: Mathiomica: An integrative platform for dynamic omics. Sci. Rep. **6**, 37–237 (2016)

17. Scargle, J.: Studies in astronomical time series analysis. ii-statistical aspects of spectral analysis of unevenly spaced data. Astrophys. J. **263**, 835–853 (1982)
18. Scargle, J.: Studies in astronomical time series analysis. iii-fourier transforms, autocorrelation functions, and cross-correlation functions of unevenly spaced data. Astrophys. J. **343**, 874–887 (1989)
19. The UniProt Consortium: Uniprot: the universal protein knowledgebase. Nucl. Acids Res. **45**(D1), D158–D169 (2017)
20. UniProt, C.: Uniprot: a hub for protein information. Nucl. Acids Res. **43**(Database issue), D204–12 (2015)
21. Van Dongen, H.P., Ruf, T., Olofsen, E., VanHartevelt, J.H., Kruyt, E.W.: Analysis of problematic time series with the lomb-scargle method, a reply to 'emphasizing difficulties in the detection of rhythms with lomb-scargle periodograms'. Biol. Rhythm Res. **32**(3), 347–54 (2001)
22. Wolfram Alpha LLC: Wolfram|Alpha (2017). Accessed Nov 2017
23. Wolfram Research, Inc.: Mathematica, Version 11.2. Champaign, IL (2017)
24. Zhao, W., Agyepong, K., Serpedin, E., Dougherty, E.R.: Detecting periodic genes from irregularly sampled gene expressions: A comparison study. EURASIP J. Bioinform. Syst. Biol. **2008** (2008)

Chapter 12
Epilog: Bioinformatics Development with Mathematica

12.1 Bioinformatics Development With the Wolfram Language

Throughout this book we have seen how we can use the Wolfram Language to analyze data, access information online and visualize results. The functions we have used are specific to the task at hand, and in some ways customized to aid us. We have occasionally written functions for repetitive tasks, and used packages such as MathIOmica where code has been already written to help our analyses. As you develop your own code base in the Wolfram Language [1–3], you will find that often you will reuse notebooks and simply modify them for specific analysis, or you will have your own functions for specific tasks. As your functions mature in complexity you may want to share your work with the community and help others that are analyzing similar work. You may decide that you want to write your own package and we will briefly introduce packages in the next section.

When developing applications, in addition to figuring out how your various functions will fit together, you will also need to be aware of your audience.

12.2 Loading Packages

We have seen how to use certain internal packages, and have also extensively used the **MathIOmica** package. This has allowed us to extend the capabilities of the Wolfram Language, or to use custom solutions that facilitate our workflow. You may also be interested in installing other packages, and many are available online (check for example `"http://packagedata.net"`) You may additionally want to invest time in

---

Electronic supplementary material The online version of this chapter (https://doi.org/10.1007/978-3-319-72377-8_12) contains supplementary material, which is available to authorized users.

writing your own packages, particularly for specialized repetitive tasks, or for sharing with the community. A package consists of code (one or many files depending on the complexity) that can be imported into the Wolfram Language using the shorthand << packageName `. We have done this repeatedly with MathIOmica:

In[1]:= <<MathIOmica`

You can also use the long form of:

In[2]:= **Get**["MathIOmica`"]

Alternatively you can also use **Needs**:

In[3]:= **Needs**["Mathiomica`"]

  (*The following message is printed*)
  Needs::nocont: **Context** Mathiomica` was not created when **Needs** was evaluated.

Needs will actually load the package only if it has not been loaded into your current session already. Once the package is loaded, functions that are defined therein become available for use. Code is stored in files with extension .m.

12.3   The examplePackage

We have included a simplistic example with the distributed files, in a file called examplePackage. The following will open the file in your default editor for .m files:

In[4]:= **SystemOpen**["examplePackage.m"]
Out[4]= examplePackage.m

Our example package has some comments at the start, enclosed within (* and *) so they are ignored - and includes a distribution license. The actual code is:

BeginPackage["examplePackage`",{"ANOVA`","ErrorBarPlots`"}]

exampleWolframLanguageFunction::usage="exampleWolframLanguageFunction is our
    first function"

Begin["`Private`"]
Options[exampleWolframLanguageFunction]={"optional" -> "preset"};
exampleWolframLanguageFunction[input_,**OptionsPattern**[]]:=
  Module[{in=input,someOptions=OptionValue["optional"]},
    Print["Success!"];
    Print["Current OptionValue: "<>someOptions];
    Return[in+1]]

End[]

EndPackage[]

12.3 The examplePackage

The function **BeginPackage**["packageContext`",{Necessary packages}] has to be placed first. A corresponding **EndPackage**[] value is placed a the end. The {Necessary packages} list contains a list of packages that are required for the current package, and will be loaded prior to the execution of the remaining commands. In our example the packages **ANOVA** and ErrorBarPlots are loaded.

As you can see in the code, we have only one function. The function is defined within the **Begin**["`Private`"] and **End**[] space. The code in this space is private and not seen outside. To get our function to be available, its usage is set before the beginning of the private context by:

exampleWolframLanguageFunction::usage="exampleWolframLanguageFunction is our first function"

The file is saved as a text file, and it must be placed in Mathematica's **$Path**. You can check which locations are available by calling the $Path variable:

In[5]:= **$Path**
Out[5]={ a list of paths specific to your system will be output}

If you also place the file in your current directory it will still work:

In[6]:= **SetDirectory**[**NotebookDirectory**[]]
Out[6]= (*your current notebook directory will be output*)

The package is now ready to import:

In[7]:= <<examplePackage`

It is important to remember the accent, `, at the end of the name. Let's check that our imported package works by checking the function defined in the package:

In[8]:= exampleWolframLanguageFunction[3]

    (*the following lines will be printed*)
    Success!
    Current OptionValue: preset
Out[8]= 4

Additionally, the option values are available for our function:

In[9]:= **Options**[exampleWolframLanguageFunction]
Out[9]= {optional→preset}

And the information we created using the usage declaration in the file:

In[10]:= ?exampleWolframLanguageFunction
    exampleWolframLanguageFunction is our first function

12.3.1 Contexts

You may have noticed when we looked up information on a function or variable we have defined that a context information was available:

In[11]:= a=3
Out[11]= 3

In[12]:= ?a
         Global`a
         a=3

Here the context is "Global" and notice how there is a grave accent following the context. Context is essentially a namespace, where symbols loaded can have many origins and different contexts. You can evaluate the current context:

In[13]:= $Context
Out[13]= Global`

We have loaded **MathIOmica** package before, let us look up one of its functions:

In[14]:= Context[**CreateTimeSeries**]
Out[14]= MathIOmica`

You can see that the context for this function is MathIOmica`. In fact, when a package is read in the package adds its context to the **$ContextPath** :

In[15]:= $ContextPath
Out[15]= {ErrorBarPlots`,HypothesisTesting`,ANOVA`,
         examplePackage`,MathIOmica`,
         RLink`,RDataTypeTools`,RLink`DataTypes`Common`,
         RLink`DataTypes`Base`,RLink`,RLink`RCodeHighlighter`,
         WebServices`,Security`,XMLSchema`,DatabaseLink`,
         HierarchicalClustering`,DocumentationSearch`,
         CURLLink`URLResponseTime`,CURLLink`Utilities`,
         CURLInfo`,CURLLink`Cookies`,CURLLink`HTTP`,
         OAuthSigning`,CURLLink`URLFetch`,CURLLink`,
         WolframAlphaClient`,Macros`,
         DocumentationSearch`Skeletonizer`,JLink`,
         GetFEKernelInit`,JSONTools`,CloudObjectLoader`,
         InterpreterLoader`,IntegratedServicesLoader`,
         IconizeLoader`,HTTPHandlingLoader`,AuthenticationLoader`,
         SystemTools`,StreamingLoader`,GeneralUtilitiesLoader`,
         ResourceLocator`,PacletManager`,
         PersistenceLocations`,System`,Global`}

As you can see many contexts are loaded, including our examplePackage` and MathIOmica` contexts. The **$ContextPath** is actually a list of context that can be searched prior to evaluating the current context **$Context** to determine the information for a symbol. So, the order is very important in the case of common short names in functions that can clash between packages and contexts.

Going back to our example above, you can call functions by their long names:

In[16]:= examplePackage`exampleWolframLanguageFunction[3]

         (*the following lines are printed during evaluation*)
         Success!
         Current OptionValue: preset
Out[16]= 4

12.3 The examplePackage

However, this is not necessary and we usually use the short name. The long name becomes useful in case multiple packages use the same short name for their functions, in which case we can call the full context to call the correct functionality.

You can see the packages that are available and loaded by using **$Packages**

In[17]:= $Packages
Out[17]= {QuantityUnits`,CloudObject`,URLUtilities`,MailReceiver`, Iconize`,UUID`,ErrorBarPlots`,HypothesisTesting`,<<35>>, StreamingLoader`,GeneralUtilitiesLoader`, ResourceLocator`,PacletManager`, PersistenceLocations`,System`,Global`}

For additional information consult the ContextsTutorial:

In[18]:= **SystemOpen**["paclet:tutorial/Contexts"]

12.4 Odds and Ends

There is a tremendous amount of functionality that we have not mentioned. In this section we direct to a few more resources for bioinformatics development.

12.4.1 More Information on Packages

If you will be developing packages, you may want to consider the following tutorials:

In[19]:= **SystemOpen**["paclet:tutorial/WolframLanguagePackages"]

In[20]:= **SystemOpen**[
        "paclet:tutorial/SettingUpWolframLanguagePackages"]

For development of packages that are multi-function, and to include documentation as part of the package, in the same way the Wolfram Documentation is presented, you may consider development using Wolfram Workbench:

In[21]:= **SystemOpen**["https://www.wolfram.com/workbench/"]

12.4.2 Dynamic Interfaces and Manipulate

We should point out the extensive availability of tools to create dynamic interfaces for users in the Wolfram Language. These include using **Dynamic** and **Manipulate**, and the best place to look these up is on the Wolfram Demonstrations page:

In[22]:= **SystemOpen**["http://demonstrations.wolfram.com"]

For **Manipulate** please consult:

In[23]:= **SystemOpen**["paclet:tutorial/IntroductionToManipulate"]

For **Dynamic**, additional information is found in the documentation hands-on tutorial:

In[24]:= **SystemOpen**["paclet:tutorial/IntroductionToDynamic"]

As well as a more advanced introduction:

In[25]:= **SystemOpen**[
          "paclet:tutorial/AdvancedDynamicFunctionality"]

12.5 The Wolfram Language Community

Finally, we should make special mention of the Wolfram Language community. A very active set of users are excited to help and participate with coding and trouble shooting. Excellent feedback is always found in both the Wolfram Community at:

In[26]:= **SystemOpen**["http://community.wolfram.com"]

And at StackExchange:

In[27]:= **SystemOpen**["https://mathematica.stackexchange.com"]

Having benefited from the immense expertise available in the communities, you are also encouraged to seek answers and feedback for your projects as you develop new bioinformatics solutions for your own target audience.

Additionally, information related to the Wolfram language can be found in the library which contains links relevant to the Wolfram Language, learning, research and developments.

In[28]:= **SystemOpen**["http://library.wolfram.com"]

Finally, and most importantly, keep coding!

References

1. Wolfram, S.: An Elementary Introduction to the Wolfram Language. Wolfram Media, Champaign (2015)
2. Wolfram Alpha LLC: Wolfram|Alpha (2017). Accessed November 2017
3. Wolfram Research, Inc.: Mathematica, Version 11.2. Wolfram Research, Inc., Champaign (2017)

Index

Symbols
/@, 42
@@, 43
@@@, 43

A
Adenine, 171
AdjacencyMatrix, 308
Amino acids, 227
Analysis of Variance, 213
And, 33
Anonymous functions, 40
ANOVA, 213
Apply, 43
ArrayPlot, 122
Ascending operator, 56
Association, 54
AssociationThread, 55
Assumptions, 88
AtomQ, 13

B
BarabasiAlbertGraphDistribution, 325
Benjamini Hochberg, 221
BenjaminiHochbergFDR, 221
BernoulliDistribution, 72
Betweenness, 313
BetweennessCentrality, 313
BinomialDistribution, 75, 76
Bipartite graph, 321
Blank, 38
BLAST, 182
BLAST API, 182
BLASTp, 248
Block, 36

BMI, USA, 341

C
Cases, 40
Centrality, 313
CentralMoment, 126
Character Codes, 46
Characters, 45
ChemSpider, 268
ChiSquaredDistribution, 94
Classification, 286
Classify, 286
Clear, 14
CloseSQLConnection, 170
Clustering, 283
Clustering coefficients, 314
Complement, 24
ConditionalExpression, 88
Confusion matrix, 289
Contexts, 377
Correlation, 122
Correlations, 120
Covariance, 122
Cycle, 311
CycleGraph, 319
Cytosine, 171

D
Data Sets, 97
DatabaseLink, 167
Datasets, 54
DateListPlot, 332
Degree, 307
Degree centrality, 313
Descending operator, 56

Det, 26
Differences, 332
Differential gene expression, 205
Dimensional reduction, 285
Dimensions, 70
Directory, 48
DiscretePlot, 73
DistibutionFitTest, 127
Distributed, 121
DNA, 171
Dynamic, 379

E
E-utilities, 134
E-utilities API Guidelines, 162
E-utilities protein, 243
Eccentricity, 312
ECitMatch, 162
Edge, 297
EdgeAdd, 302
EdgeDelete, 302
EdgeWeight, 305
EFetch, 152
EGQuery, 160
EInfo, 134
ELink, 154
Empty graphs, 315
EndOfFile, 53
Entities, 106
Entity, 106
EntityList, 106, 171
EntityValue, 106
Entrez Programming Utilities, 134
EPost, 147
ESearch, 137
ESpell, 161
ESummary, 149
Euler, 299
EulerianGraphQ, 300
ExampleData, 97
Expectation, 120
Expectation, 121
ExponentialDistribution, 87
Export, 49
Export, $ExportFormats, 50

F
FASTA, 174
FindClusters, 284
FindGraphIsomorphism, 323
FindShortestPath, 311

First, 70
Flatten, 24
For, 34
FromCharacterCode, 46
FromCharacterCode, 169
Function, 40
Functions, 30

G
Gather, 100
GenBank, 176
Geodesic, 311
GeometricDistribution, 78
Get, 376
Getting Started, 10
GlobalClusteringCoefficient, 314
Golub ALL AML Data Set, 108
Golub combined data, 198
Golub dataset, 193
GolubAssociation, 109
Graph, 297
Graph, 297
Graph center, 313
Graph diameter, 312
Graph radius, 313
GraphData, 315
GraphDiameter, 312
GraphPlot, 303
GraphRadius, 313
Guanine, 171

H
Head, 13
History Server, 145
Hu6800IDtoAnnotation, 109
HypergeometricDistribution, 80
Hypothesis Testing, 115
HypothesisTestData, 128

I
If, 33
Import, 48
Import, $ImportFormats, 49
Integrative Personal Omics, 115
Intersection, 24
IPOP, 115
Iris dataset, 289
Isomorphism, 322

J
JSON, 134

Index

K
KEGG, 274
KEGGAnalysis, 370
KEGGDictionary, 275
Key, 64
KeyMap, 353
Koningsberg, 299
Kurtosis, 126

L
Length, graph, 311
Leukemia combined data, 198
Leukemia data, 193
LinearModelFit, 123
ListPlot, 68
LocalClusteringCoefficient, 314
LocationTest, 120
LongestCommonSubsequence, 181
LowValueTag, 356

M
Machine learning, 283
Manipulate, 379
MannWhitneyTest, 118
Map, 42
MapThread, 43
Marcobal et al., 252
MassDictionary, 274
MassMatcher, 274
MatchQ, 38
MathIOmica, 62
MathIOmica, example data, 114
MathIOmica, installation, 62
MatrixClusters, 122
MatrixDendrogramHeatmap, 224
MatrixForm, 24
Max, 68
Mean, 68
Median, 68
Message, 56
Metabolomics, 251
Min, 68
Missing, 63
Module, 36
Molecular weight search, 246
Mouse metabolomics data, 252

N
Nest, 43
Networks, 297
NormalDistribution, 88

NotebookDirectory, 48

O
OmicsObject, 63
OpenAppend, 52
OpenRead, 53
OpenWrite, 51
Options, 31, 47
OptionsPattern, 47
Or, 33
ORA, 367
Ordering, 119
OutputStream, 52

P
Packages, 375
Path, 311
PathGraph, 311
Patterns, 38
Pavlidis, 111
Pavlidis ANOVA approach, 213
PCA, 257
Periphery, 312
Plot, 57
PoissonDistribution, 80
Postfix, 100
Prefix, 41
Principal Component Analysis, 257
Protein alignment, 246
ProteinData, 237
Pure functions, 40

Q
Quantile, 93
Quantile, 93
QuantileNormalization, 355
Query, 55

R
Random graphs, 324
RandomChoice, 69
RandomGraph, 324
RandomInteger, 35
RandomInteger, 72
RandomReal, 72
RandomVariate, 91
RawJSON, 134
Read, 53
ReadLine, 53
Regular graphs, 318
Remove, 14

ReplaceAll, 28
ReplaceRepeated, 29
ResourceData, 104

S
Sandberg, 111
Sandberg Data, 213
SeedRandom, 36
SeedRandom, 72
Sequence, 41
SequenceAlignment, 179
ServiceConnect, 268
ServiceDisconnect, 274
ServiceExecute, 269
SetDelayed, 27
SetDirectory, 48
Short, 100
Show, 60
Show, 125
Skewness, 126
SortBy, 70
SQLConnections, 167
SQLExecute, 168
Streams, 51
StringContainsQ, 45
StringContainsQ, 110
StringCount, 45
StringJoin, 45
StringLength, 45
StringMatchQ, 45
StringReplace, 45
StringReverse, 47
Strings, 44
StringTake, 45
StudentTDistribution, 95
Switch, 34
Syntax, 12

T
Take, 70
Tally, 70
Thread, 29
Time series, 329
Time series heatmaps, 364
TimeSeries, 329
TimeSeriesModelFit, 342
ToCharacterCode, 46

ToExpression, 17
ToUpperCase, 44
Tr, 26
Trail, 311
Transpose, 59
Trees, 321
TTest, 118

U
UCSC Genome Browser, 163
UniformDistribution, 85
Union, 23
UniProt, 237
UniProt API Query, 239
UniProt identifier mapping, 242
UniProt random entry generation, 242
URLExecute, 134

V
Variables, 13
Variance, 78
Variance, 68, 78
Vertex, 297
VertexAdd, 302
VertexDegree, 307
VertexDelete, 302
VertexEccentricity, 312
VertexInDegree, 307
VertexIndex, 302
VertexList, 302
VertexOutDegree, 307
VertexWeight, 305
Volcano plot, 209

W
Wolfram|Alpha, 10
Walk, 311
WeightedAdjacencyMatrix, 310
While, 35
With, 36
Wolfram Community, 11

Z
ZTest, 117

CPSIA information can be obtained
at www.ICGtesting.com
Printed in the USA
LVHW021811210419
614972LV00001B/76/P